Sustainable Technologies for Energy Efficient Buildings

The text begins by discussing sustainable buildings, energy-efficient technologies, advanced materials, advances in renewable energy for the building sector, green intelligent infrastructure, policies on sustainable infrastructure, and life cycle assessment. It further presents design considerations, challenges, and applications of net zero energy buildings with a global perspective. The book covers renewable energy technologies for energy-efficient buildings.

This book:

- Discusses the importance of developing new materials for energy and heat transfer optimization in sustainable buildings and life cycle assessment of sustainable building materials.
- Investigates the city gas system, sustainable smart cities infrastructure, and data mining techniques in green building for evaluation of energy cost, grades, and adoption.
- Highlights the development and application of net zero energy buildings, energy policies and infrastructure requirements, building performance prediction and optimization, and energy planning and thermal comfort in buildings.
- Presents renewable energy policies, social, economic, and environmental issues associated with sustainable buildings, and emerging trends in smart green building technologies.
- Covers energy-efficient urban infrastructure, earth-air heat exchanger, and retrofitting of existing buildings to achieve energy-efficient buildings.

It is primarily written for senior undergraduates, graduate students, and academic researchers in the fields of energy engineering, environmental science and engineering, materials science, mechanical engineering, and civil engineering.

Smart Innovations and Technological Advancements in Mechanical and Materials Engineering

Ashwani Kumar and *Kishor Gajrani*

The book series editors are inviting edited, reference, and textbook proposals for consideration in the book series. The main objective of this book series is to invite quality content in the field of smart materials applications, cutting-edge manufacturing techniques, innovative designs, and the use of advanced computational tools like AI and ML to solve engineering problems. The series includes recent innovative topics in four different sections which are as follows:

Technological Advancements: This section covers recent developments in nanotechnology, nonlinear dynamics, dynamics of complex systems, MEMS, green and sustainable technologies, vibration control, AI in power station, analog-digital hybrid modulation, advancement in inverter technology, adaptive piezoelectric, pattern recognition, data mining and knowledge discovery, control, robotics, automation and vision, knowledge-based simulation, sensor networks, and big data analytics.

Bioinformatics Techniques: This section covers bioinformatics and biomedical engineering, biomedical imaging, image processing and visualization, biomedical robotics, computer-assisted medical diagnostic systems, biometrics and bio-measurements, innovative technologies in medicine, nanoscience, and advanced computing.

Smart Materials: This section covers bioinspired materials, semiconductor materials, application of phase change materials, biodegradable composites, biomaterials, material synthesis and processing, energy materials, polymers, and biodegradable composites.

Digital Manufacturing: This section covers recent developments in additive manufacturing, 3D printing, nanomanufacturing, manufacturing of bioimplants, micromachining, sustainable manufacturing, green manufacturing, environmentally friendly manufacturing, application of soft computing techniques in manufacturing, and adaptive technologies for sustainable growth.

This book series compiled all aspects of manufacturing, design, and computational intelligence techniques from fundamental principles to current advanced concepts. This book series is an effort to address the changes occurring in mechanical and materials engineering and the application of new techniques in this field. The book series covers a variety of topics from conventional to advanced ones.

Sustainable Technologies for Energy Efficient Buildings

Chandan Swaroop Meena, Ashwani Kumar, Varun Pratap Singh, and *Aritra Ghosh*

Sustainable Technologies for Energy Efficient Buildings

Edited by

Chandan Swaroop Meena, Ashwani Kumar, Varun Pratap Singh, and Aritra Ghosh

CRC Press is an imprint of the
Taylor & Francis Group, an **informa** business

Designed cover image: shutterstock

First edition published 2024
by CRC Press
2385 NW Executive Center Drive, Suite 320, Boca Raton FL 33431

and by CRC Press
4 Park Square, Milton Park, Abingdon, Oxon, OX14 4RN

CRC Press is an imprint of Taylor & Francis Group, LLC

ISBN: 9781032742892 (hbk)
ISBN: 9781032804064 (pbk)
ISBN: 9781003496656 (ebk)

DOI: 10.1201/9781003496656

Typeset in Sabon
by Deanta Global Publishing Services, Chennai, India

Contents

Aim and scope

The aim of the book *Sustainable Technologies for Energy Efficient Buildings* is to highlight the multifaceted aspects of sustainable construction practices, focusing on innovation in the green building sector to foster a sustainable future. The chapters within this book span a wide spectrum, providing a thorough exploration of the fundamental concepts as well as cutting-edge advancements in green building technologies. The primary objective is to equip researchers, practitioners, and policymakers with a profound understanding of the key elements shaping the sustainable built environment.

This introductory chapter sets the stage by providing the details of the foundational principles of sustainable construction and subsequently delving into the latest innovations that drive the sector forward. The chapter concentrates on the pivotal role of materials in sustainable building practices, examining the development of new, eco-friendly materials that contribute to the longevity and environmental compatibility of buildings. Focusing on cutting-edge materials, the chapters explore the integration of smart materials in construction, emphasizing their role in optimizing energy consumption and heat transfer within buildings and the life cycle assessment of materials, providing a holistic understanding of their environmental impact from extraction to disposal.

The subsequent chapters extend the exploration into diverse aspects such as city gas systems, sustainable infrastructure in India and Oman, data mining techniques for energy evaluation, net zero energy buildings, building performance prediction, energy policies, and the application of renewable energy technologies. The book provides a detailed examination of social, economic, and environmental issues associated with sustainable buildings and a focus on emerging trends in smart green building technologies. By critically evaluating these sustainable technologies, the book aims to contribute to a nuanced discourse that informs stakeholders about the complexities, opportunities, and potential pitfalls inherent in the pursuit of energy-efficient buildings for a sustainable future.

This book *Sustainable Technologies for Energy Efficient Buildings* critically addresses the need for transformative approaches in the building sector with sustainable technologies that hold the promise of significantly enhancing energy efficiency within buildings. Each chapter within this book is meticulously designed to scrutinize specific facets of sustainable technologies, ranging from smart building materials to comprehensive life cycle assessments. This diverse collection aims to be an invaluable resource for scholars, professionals, and policymakers engaged in the pursuit of sustainable urban development.

Editor
Dr. Chandan Swaroop Meena
Dr. Ashwani Kumar
Dr. Varun Pratap Singh
Dr. Aritra Ghosh

Preface

Sustainable Technologies for Energy Efficient Buildings highlights the importance of sustainable buildings, energy-efficient building technologies, advanced materials, advances in renewable energy for building sector, green intelligent infrastructure, policies on sustainable infrastructure, and life cycle assessment (LCA). The introductory Chapter 1 serves as a comprehensive primer on the essential principles, concepts, and advancements in energy-efficient building design and technology. Beginning with a global perspective, the chapter highlights fundamental principles, architectural considerations, insulation materials, HVAC systems, lighting technologies, and emphasizing the interconnectedness of these elements for life cycle analysis. In continuation, Chapter 2 is a detailed exploration of the sustainable building materials. It addresses the current need for environmentally conscious construction and explains innovative materials, spanning from bio-based composites to cutting-edge nanotechnologies, reshaping traditional construction methods.

Chapter 3 outlines the key materials that are currently contributing to heat transfer and energy optimization in sustainable buildings. It covers from traditional organic and inorganic materials to the utility of mechanical processes and computational fields that can upgrade existing smart architecture. Chapter 4 explores the concept of life cycle assessment (LCA) in the context of sustainable building materials and provides a detailed examination of the life cycle stages of these materials, assessing their environmental, social, and economic impacts from extraction to disposal.

Moving toward the advanced sustainable infrastructure, Chapter 5 examines the essential components of city gas systems in detail, including their design, installation, and wide range of uses in modern buildings. In continuation, Chapter 6 reviews the initiatives (missions, programs, projects, policies, and schemes) of the Indian Government with the perspective of the sustainable development of urban and rural infrastructure. Ministry of Environment, Forest and Climate Change; Housing and Urban Affairs; Rural Development; Road Transport and Highways; Railways; and other ministries directly or indirectly impacting the sustainability aspect in infrastructure development are considered in this chapter.

A planned smart infrastructure can mitigate the requirement of NZEBs. Chapter 7 highlights an important issue of modern infrastructure, i.e., net zero energy buildings (NZEBs). It focuses on NZEBs' design considerations, challenges, and diverse applications. The chapter also covers the importance of energy efficiency, renewable energy integration, and holistic planning for NZEBs. The core aim of this Chapter 8 is to evaluate, critically analyze, and assess the strategic, operational, and management aspects of smart cities development in Oman. Also, to investigate the current urbanization challenges that are encountered to convert the central cities in the Sultanate of Oman to be smart sustainable cities by domesticating intelligent technologies and implementing sustainable solutions.

Chapter 9 delves into various data mining methodologies, including clustering, regression analysis, and machine learning algorithms, to extract valuable insights. It highlights the significance of data-driven approaches in unraveling complex patterns that influence energy costs and the effectiveness of green building technologies.

Chapter 10 deals with energy policies and infrastructure requirement policies that comprise a range of efforts, including tax incentives, rebates, and strict building rules that facilitate the incorporation of renewable energy sources, energy-efficient technologies, and sustainable materials. In continuation, Chapter 11 explores the multifaceted realm of building performance prediction and optimization (BPPO), encompassing the integration of advanced technologies, data analytics, and simulation tools. The focus is on enhancing energy efficiency, occupant comfort, and overall sustainability in built environments. Chapter 12 explores the intricate relationship between energy planning and thermal comfort in the context of building design and operation.

Chapter 13 examines case studies and current trends to demonstrate the various advantages of incorporating renewable energy sources into building infrastructure. These benefits include a decrease in environmental impact, improved energy efficiency, and economic gains. In continuation, Chapter 14 highlights economic perspectives on sustainable construction that are scrutinized, considering factors such as initial costs, long-term savings, and job creation. Environmental considerations form a pivotal focus, exploring green building materials, energy-efficient technologies, and waste reduction strategies.

Chapter 15 explores that efficient energy management through smart systems like the Internet of Things, renewables, and AI-driven controls is vital for sustainable buildings. Integrating energy-efficient systems, Internet of Things, and renewables optimizes resource consumption, reduces costs, and curtails environmental impact.

Chapter 16 explores renewable energy sources in city planning helps not only to sustainably satisfy increasing energy demands but also to build resilient, environmentally conscious communities that prioritize economic growth alongside social and environmental sustainability. Chapter 17 covers various modeling methods, including numerical, analytical, and data-driven approaches for earth-air heat exchanger (EAHE). Advancements in EAHE technology, exploring the adoption of hybrid technologies to enhance system efficiency, minimize energy consumption and ensure optimal thermal comfort.

Chapter 18 attempts to analyze sustainable wall, floor, and roof assemblies for retrofitting in a single-story residential building in Vijayawada, India. The study analyzes alternative assemblies using four natural, six bio, and five salvaged materials which significantly reduces the embodied and operational energy of buildings, thereby reducing the negative impact on the environment.

The applicability of the book *Sustainable Technologies for Energy Efficient Buildings* covers a wide range of industries such as building construction, design engineering, research and development engineering, civil engineering, researchers in energy sectors and policymakers for renewable energy etc., and it will help to audience conducting research in these industries with renewable energy for sustainable buildings.

Editor
Dr. Chandan Swaroop Meena
Dr. Ashwani Kumar
Dr. Varun Pratap Singh
Dr. Aritra Ghosh

Acknowledgments

We would like to thank all the contributors of this book for their dedication and quality work submission. Our heartfelt gratitude to CRC Press (Taylor & Francis Group) and the editorial team for their support during completion of this book. We are sincerely grateful to reviewers for their suggestions and illuminating views on each book chapter presented in the book *Sustainable Technologies for Energy Efficient Buildings.*

This book is dedicated to all academician, researchers, and energy scientist.

Editors

Dr. Chandan Swaroop Meena is presently working as a Scientist in Building Energy Efficiency Department and is an Assistant Professor at the Academy of Scientific and Innovative Research (AcSIR) at CSIR-Central Building Research Institute, Roorkee (Ministry of Science and Technology, Government of India). He earned his Ph.D. from IIT Roorkee and M.Tech. from NIT Kurukshetra, India, both in Thermal Engineering. He conducts research in renewable energy utilization techniques, geothermal systems, building energy efficiency, two-phase flow, and heat transfer. He is a Life Member of the Indian Society of Heat and Mass Transfer (ISHMT), a Life Member of the Indian Buildings Congress (IBC), and a Life Member of the Institution of Engineers India (IEI). He has published 29 research articles in reputed journals and international conferences. He is associated with international conferences as Invited Speaker/Advisory Board/Review Board member. He has delivered many invited talks in webinar, FDP, and Workshops.

Dr. Ashwani Kumar received Ph.D. (Mechanical Engineering) in the area of Mechanical Vibration and Design. He is currently working as Senior Lecturer, Mechanical Engineering (Gazetted Officer Group B) at Technical Education Department Uttar Pradesh (under Government of Uttar Pradesh) Kanpur, India, since December 2013. He has worked as Assistant Professor in the Department of Mechanical Engineering, Graphic Era University Dehradun India (NIRF Ranking 55) from July 2010 to November 2013. He has more than 13 years of research, academic, and administrative experience including Coordinator for AICTE-Extension of Approval, Nodal officer for PMKVY-TI Scheme (Government of India), Internal Coordinator-CDTP scheme (Government of Uttar Pradesh), Industry Academia Relation Officer, Assistant Centre Superintendent (ACS)-Institute Examination Cell, Zonal Officer to conduct Joint Entrance Examination (JEE-Diploma), Sector Magistrate for Lok Sabha-Vidhan Sabha Election, etc. Being an academician and researcher, he is Series Editor of five book series Advances in Manufacturing, Design and Computational Intelligence Techniques[1]/ Renewable and Sustainable Energy Developments[2]/ Smart

Innovations and Technological Advancements in Mechanical and Materials Engineering[3]/ Solar Thermal Energy Systems: Advancements in Engineering, Ergonomics, and Sustainable Development[4]/ Computational Intelligence and Biomedical Engineering[5] published by CRC Press (Taylor & Francis USA)[1-4] and Wiley Scrivener Publishing[5] USA. He is Guest Editor of special issue titled Sustainable Buildings, Resilient Cities and Infrastructure Systems (Buildings *ISSN: 2075-5309*, I.F. 3.8). He is Editor-in-Chief for *International Journal of Materials, Manufacturing and Sustainable Technologies* (IJMMST, *ISSN: 2583-6625*), and Editor of *International Journal of Energy Resources Applications* (IJERA, *ISSN: 2583-6617*). He is Guest Editor and Editorial Board Member of eight international journals and acts as Review Board Member of 20 prestigious (Indexed in SCI/SCIE/Scopus) international journals with high impact factor, i.e., *Applied Acoustics*, *Measurement*, *JESTEC*, *AJSE*, *SV-JME*, and *LAJSS*. In addition, he has published 100+ research articles in journals, book chapters, and conferences. He has authored/co-authored cum edited 30+ books on Mechanical, Materials, and Renewable Energy Engineering. He has published three patents. He is associated with international conferences as Invited Speaker/Session Chair/Advisory Board/Review Board member/ Program Committee Member. He has delivered many invited talks in webinar, FDP, and Workshops. He has been awarded as Best Teacher for excellence in academic and research. He has successfully guided 15 B.Tech., M.Tech., and Ph.D. theses. He is external doctoral committee member of S.R.M. University, New Delhi. He is currently involved in the research area of AI and ML in mechanical engineering, smart materials and manufacturing techniques, thermal energy storage, building efficiency, renewable energy harvesting, sustainable transportation, and heavy vehicle dynamics. His Orcid is ***0000-0003-4099-935X***, Google Scholar web link is ***https://scholar.google.com/citations?hl=en&user=KOILpEkAAAAJ***, and research gate web link is ***https://www.researchgate.net/profile/Ashwani-Kumar-45***.

Dr. Varun Pratap Singh received Ph.D. (Mechanical Engineering) in the area of Solar Energy. He is currently working as Assistant Professor in the Department of Mechanical Engineering and Head of Solar Energy Centre at the University of Petroleum and Energy Studies, Dehradun, India. He has more than 11 years of research and academic experience in mechanical engineering and renewable energy. He has published three patents, five research articles in SCI journals and conferences and completed five projects as Principal Investigator (PI) and two projects as Co-PI with a total research grant of more than 20 lakh INR for government as well as industry partners. He has delivered many invited talks in webinar, FDP, and Workshops and attended 15+ MOOCs and 15+ FDPs from various national bodies like IITs, AICTE, and international bodies like Delft University of Technology-Netherlands, Tsinghua University-China, National Chiao Tung University-Taiwan, and The International Monetary Fund (IMF).

He has been awarded for outstanding mentoring of students and Perfect faculty award for excellence in academic and research. He has successfully guided 16 B.Tech., M.Tech., and Ph.D. theses. In administration, he is working as Head of Centre of Excellence, COER. As an academician, renewable energy, solar-thermal application, automation, and special propose machining (SPM) are his interest areas of association. His ORCID ID is 0000-0002-2148-5604.

Dr. Aritra Ghosh, Assistant Professor specializing in integrated renewables at the Renewable Energy Research Group within the esteemed University of Exeter in the United Kingdom, has emerged as a distinguished figure in the field of Applied Energy. In 2016, Dr. Ghosh achieved the pinnacle of academic excellence by obtaining his Ph.D. in Applied Energy from the Technological University of Dublin in Ireland laying the foundation for a remarkable career in renewable energy research. Dr. Ghosh has significantly contributed to the scientific community through the publication of more than 150 articles in leading peer-reviewed journals. He is not only a prolific researcher but also a dynamic and engaged academic. His active participation in international conferences, where he presents his groundbreaking work, has solidified his presence on the global stage. Furthermore, he is sought after as an expert speaker, delivering invited talks that disseminate knowledge and insights garnered from his extensive research endeavors. In addition to his role as a researcher and speaker, he contributes significantly to the academic community by serving as a Guest Editor, Board Member, and Reviewer for esteemed international journals. His research focus encompasses material development and energetic analysis tailored for solar applications and energy-efficient building-related technologies. His work extends beyond theoretical frameworks, finding practical applications in diverse fields such as building infrastructure, water resources, and agricultural practices. A key facet of Dr. Ghosh's research involves the integration of nanoparticles in various applications, with a particular emphasis on enhancing the functionality of smart windows, optimizing solar cells, and developing self-cleaning coatings. His exploration of these cutting-edge technologies positions him at the forefront of innovative solutions for sustainable energy utilization. In summary, Dr. Aritra Ghosh's dedication to advancing renewable energy, evident in his academic journey and prolific contributions to Applied Energy, is underscored by his extensive publications, global conference presence, and influential roles in academic editorial boards, solidifying his status as a distinguished scholar and a catalyst for sustainable energy solutions.

Contributors

Adarsh Kumar Aarya
Department of Chemical Engineering
HBTU, Kanpur
Uttar Pradesh, India

Mohammad Ahmadizadeh
Faculty of Engineering
Department of Mechanical Engineering
Persian Gulf University
Bushehr, Iran

Eman Al Naamani
Department of Mechanical Engineering
Global College of Engineering and Technology (GCET)
Muscat, Sultanate of Oman

Naval Kishor Banjara
CSIR—Central Building Research Institute
Roorkee, India

Sumanta Bhattacharya
Department of Textile Technology
MAKAUT Kolkata
West Bengal, India

Bhupendra
Department of Computer Science
School of Computer Science
UPES
Dehradun, India

Ashish Kumar Bilatiya
Dayalbagh Educational Institute
Deemed University
Agra, India

Chandrajita Chakraborty
School of Engineering and Technology
Indira Gandhi National Open University
New Delhi, India

Veena Chaudhary
CSIR—Central Building Research Institute
Roorkee, India

Mira Chitt
Department of Mechanical Engineering
Global College of Engineering and Technology (GCET)
Muscat, Sultanate of Oman

Aritra Ghosh
Faculty of Environment, Science and Economy (ESE), Renewable
Energy, Electric and Electronic Engineering
University of Exeter
Penryn, Cornwall

Milad Heidari
Department of Mechanical Engineering
Global College of Engineering and Technology (GCET)
Muscat, Sultanate of Oman

Preet Kaur
Department of Electronics and Communication Engineering
YMCA University of Science and Technology
Faridabad, India

Morteza Khashehchi
Department of Mechanical Engineering
Global College of Engineering and Technology (GCET)
Muscat, Sultanate of Oman

Aditya Kumar
National Institute of Technology
Tiruchirappalli, India

Ashwani Kumar
Department of Mechanical Engineering
Technical Education Department
Kanpur, India

Chandan Swaroop Meena
CSIR-Central Building Research Institute
Roorkee, India

K. Parameswari
Civil and Architectural Engineering Section
Department of Civil Engineering
University of Technology and Applied Sciences
Muscat, Oman

Jaimin Sureshbhai Parmar
Birla Vishvakarma Mahavidyalaya
V.V. Nagar, Anand, Gujarat, India
and
CSIR – Central Building Research Institute
Roorkee, India

Pooyan Rahmanivahid
Global College of Engineering and Technology (GCET)
Muscat, Sultanate of Oman

A. Lilly Rose
Department of Architecture
School of Planning and Architecture
Vijayawada, India

Bhavneet Kaur Sachdev
Department of Humanities,
Suresh Gyan Vihar University
Jaipur, India

Wasswa Shafik
School of Digital Science Universiti
Brunei Darussalam
Jalan Tungku
Link, Gadong, Bandar Seri Begawan,
Brunei Darussalam
and
Dig Connectivity Research Laboratory (DCRLab)
Kampala, Uganda

Jagruti Piyushkumar Shah
Birla Vishvakarma Mahavidyalaya
Gujarat, India

Bhalchandra Shingan
Department of Chemical & Petroleum, Engineering
University of Petroleum and Energy Studies
Dehradun, India

Bhupender Singh
Department of Mechanical Engineering
YMCA University of Science and Technology
Faridabad, India

Varun Pratap Singh
Department of Mechanical Engineering
School of Advanced Engineering UPES
Dehradun, India

Aravamudhan Swaminathan
Civil and Architectural Engineering Section
Department of Architectural Engineering
University of Technology and Applied Sciences
Salalah, Oman

Sivasakthivel Thangavel
Department of Mechanical Engineering
Global College of Engineering and Technology (GCET)
Muscat, Sultanate of Oman

Shweta Tripathi
School of Engineering and Technology
Indira Gandhi National Open University
New Delhi, India

Ashok Kumar Yadav
Department of Mechanical Engineering
Raj Kumar Goel Institute of Technology
Ghaziabad, India

Jitendra Yadav
Department of Mechanical Engineering
School of Advanced Engineering UPES
Dehradun, India

Chapter 1

Innovation in green building sector for sustainable future

Fundamentals to Advances

Varun Pratap Singh, Ashwani Kumar, Chandan Swaroop Meena, Sivasakthivel Thangavel, and Aritra Ghosh

1.1 INTRODUCTION TO ENERGY EFFICIENCY IN THE BUILT ENVIRONMENT

The opening section of this chapter sets the stage by addressing the imperative for energy efficiency in the built environment on a global scale. It starts out by examining the growing significance of sustainable building techniques and emphasizing the pressing need for solutions in light of climate change and the depletion of energy supplies. The conversation explores the complex relationship between building design, energy use, and environmental effect, highlighting the significant impact that building operations and construction have on the entire energy environment. This part also examines the potential and difficulties associated with attaining energy efficiency, providing background information for the discussion of the essential ideas and elements of energy-efficient building design that follows. This introduction section seeks to interest readers by outlining the larger background and emphasizing the role that adopting sustainable technology will have in determining the future of the built environment [1].

1.1.1 Global imperative for sustainable construction

There is an urgent need to address the significant environmental effect of the building sector globally, which is what is driving the global need for sustainable construction. The industry is responsible for about 40% of carbon dioxide emissions connected to energy and a startling 36% of the world's energy consumption, according to the International Energy Agency (IEA). The need for building materials and energy-intensive building techniques is growing as the globe continues to urbanize quickly, aggravating problems like pollution, deforestation, and resource depletion. The exhaustion of limited resources, growing worries about environmental deterioration, and the realization that conventional building methods are no longer practical in a world facing climate change all serve to emphasize this necessity. As a result, using sustainable building techniques becomes essential rather than

DOI: 10.1201/9781003496656-1

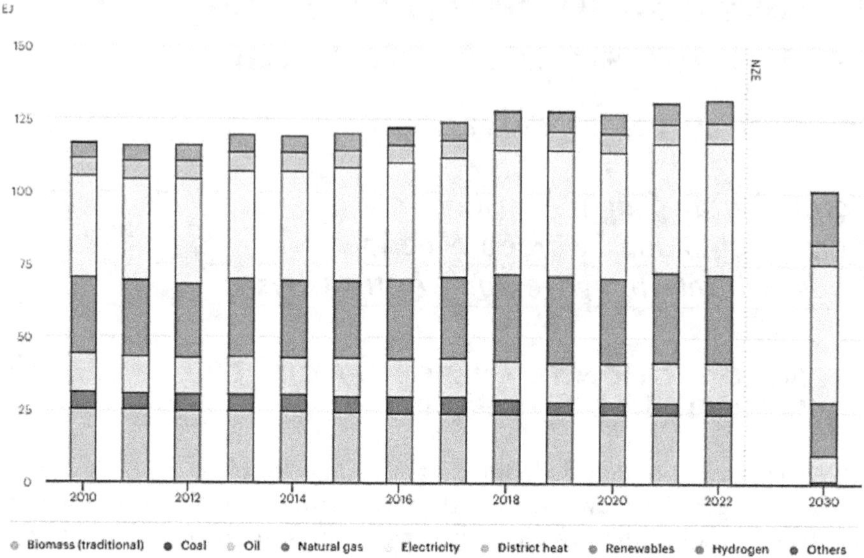

Figure 1.1 Global energy consumption trends in buildings by different fuels in the net zero scenario, for 2010–2030 ["IEA, Energy consumption in buildings by fuel in the Net Zero Scenario, 2010-2030, IEA, Paris. https://www.iea.org/data-and-statistics/charts/energy-consumption-in-buildings-by-fuel-in-the-net-zero-scenario-2010-2030-2, *IEA. License: CC BY 4.0"]*

merely an option. International programs and agreements that highlight the necessity of significant cuts in carbon emissions, like the Paris Agreement and the Sustainable Development Goals (SDGs) of the United Nations, reflect this urgency [2, 3]. The adoption of sustainable building methods is becoming more and more important for governments, corporations, and communities globally in order to reduce their impact on the environment, improve energy efficiency, and develop robust structures that can resist the difficulties posed by a changing climate [4]. The imperative for sustainable manufacturing is not just a moral or ethical one; it is a strategic response to a global crisis that demands immediate and concerted action to ensure the long-term viability of our built environment. Figure 1.1 shows the global energy consumption trends in buildings by different fuels in the Net-Zero Scenario, for 2010–2030.

1.1.2 Linking building design, energy consumption, and environmental impact

Emphasizing the complex links between architectural decisions, energy use, and environmental repercussions, building design, and its energy consumption has an influence on the environment and is a critical point of convergence in the debate on sustainable construction. Building designs have a big impact on how much energy they use and how much of an environmental impact they have during their lifetime. A thorough life cycle evaluation, which takes into account the extraction of raw materials, manufacture, building, occupation, and final demolition, is frequently used to illustrate this link. Principles of sustainable building design are essential for reducing these effects. For example, passive solar architecture minimizes the demand for artificial lighting and heating by maximizing the amount of natural light and heat. Thermal efficiency is increased by energy-efficient insulating materials, which reduces the need for heating, ventilation, and air conditioning (HVAC) systems [5]. Real-time energy consumption optimization and monitoring are made possible by smart building technology. These links may be successfully communicated through visual aids like diagrams that illustrate important design ideas and show the life cycle of a structure [6]. By understanding these relationships, decision-makers in the fields of architecture, engineering, and policymaking may improve energy efficiency, lessen their influence on the environment, and strengthen the built environment's overall resilience and sustainability [7].

An illustration of a building's life cycle, as shown in Figure **1.2**, from raw material extraction through construction, occupation, and final destruction, may be used to show the different phases of energy consumption and environmental effect generation. It is possible to graphically illustrate the functions of sustainable building design concepts, such as energy-efficient insulation, smart HVAC systems, and passive solar design, in reducing energy use and environmental impact [8–12].

1.1.3 Challenges and opportunities in the era of climate change

In the age of climate change, the building sector must contend with the effects of a shifting climate in addition to its substantial contribution to global greenhouse gas emissions. It is a difficult undertaking to retrofit old structures for increased energy efficiency; creative solutions are needed to adapt buildings to changing sustainability standards. The industry also has to deal with changing weather patterns, which calls for durable building designs that can resist harsh occurrences. The Intergovernmental Panel on Climate Change (IPCC) has highlighted the necessity of cutting carbon emissions and switching to more sustainable practices, which exacerbates these difficulties [13]. But these difficulties also present a wealth of opportunity. The increased awareness of climate change has prompted green technology and environmentally friendly building material research and development. Sustainable design innovations, such as circular economy strategies and passive solar concepts, offer chances to improve energy

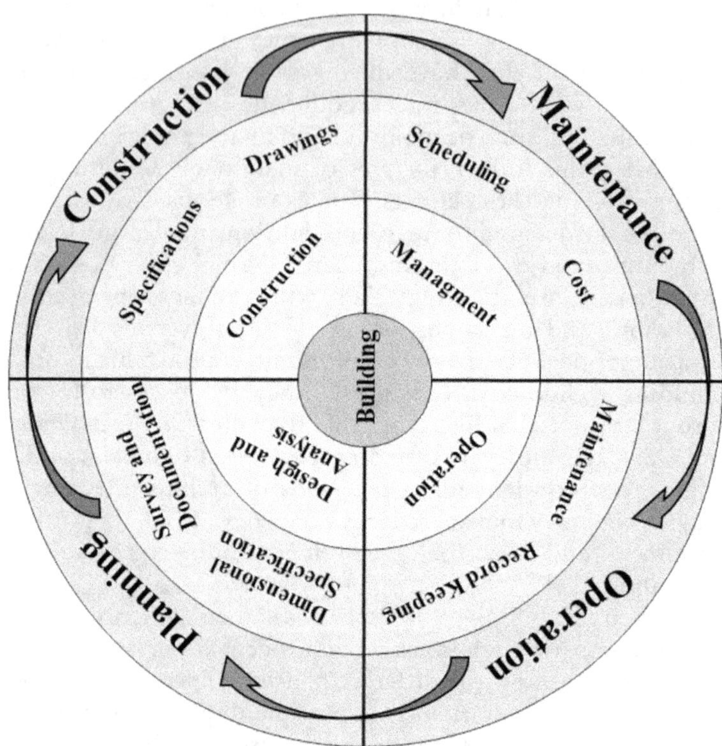

Figure 1.2 A schematic representation of the entire life cycle of a building, showcasing key points of different stages

efficiency and lessen environmental effects. Businesses are compelled to use sustainable building methods as environmentally concerned customers shape consumer preferences, offering benefits to both the economy and the environment. With the development of 3D printing, smart building technology, and the incorporation of renewable energy sources into designs, sustainable construction is expected to have a bright future. These developments will present revolutionary chances to reinterpret the link between the built environment and the ecology [14]. Visual aids such as graphs showing trends in greenhouse gas emissions, diagrams showing the use of circular economy concepts in architecture, and conceptual representations of sustainable building designs for the future help improve comprehension of these issues and potential [12]. This all-encompassing viewpoint highlights that although climate change poses significant obstacles for the building industry, it also provides revolutionary chances that may help create a built environment that is more robust and sustainable [15]. Figure **1.3** highlights resource efficiency and waste reduction while illuminating the fundamentals of a circular economy in the building industry.

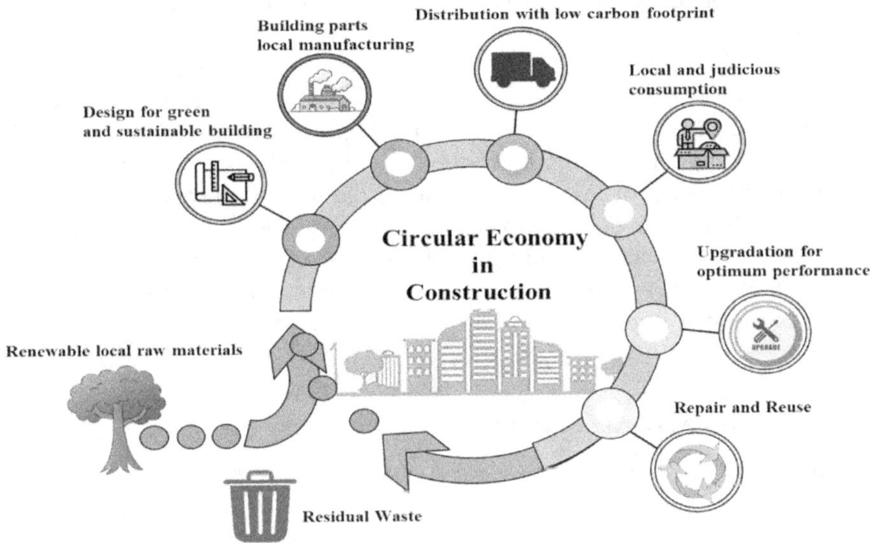

Figure 1.3 Circular economy in construction

1.2 FUNDAMENTALS OF ENERGY-EFFICIENT BUILDING DESIGN

The pursuit of environmentally conscious and sustainable construction techniques is based on the principles of energy-efficient building design. Fundamentally, the goal of energy-efficient building design is to maximize a structure's energy-consumption performance, reducing its impact on the environment and raising overall efficiency [16]. These principles heavily rely on architectural considerations, which include things like building orientation, layout, and the thoughtful positioning of windows to maximize natural light and heat. Reducing the need for mechanical heating and cooling systems requires a thorough integration of passive design concepts, such as the use of thermal mass and adequate insulation [17].

Another essential component of energy-efficient building design is insulation, which serves as a barrier to heat transmission and maintains thermal comfort inside the building envelope. The building process is more environmentally friendly overall and uses less energy because of advancements in insulation technology, which include high-quality, sustainable materials. Energy-efficient design also revolves around optimizing HVAC (heating, ventilation, and air conditioning) systems, which includes choosing energy-efficient appliances, carefully planning ducting, and integrating smart controls for accurate climate control [18]. With the use of daylighting techniques, energy-efficient lighting fixtures, and smart lighting controllers that adjust to occupancy and natural light levels, lighting technologies also become more prominent [19].

The concepts of energy-efficient building design are dynamic and always changing in response to technological breakthroughs and growing awareness of sustainable practices. These foundational principles, which are based on an integrated methodology that takes into account systems, architecture, and materials, are essential to designing structures that not only satisfy the immediate demands of their users but also support the creation of a built environment that is robust, sustainable, and energy-efficient [20]. Diagrams that show HVAC optimizations, insulation materials, and passive design techniques are examples of visual aids that can improve understanding of these principles and act as useful manuals for stakeholders, engineers, and architects who are dedicated to creating a more sustainable built environment [21].

1.2.1 Architectural considerations

Architectural considerations are fundamental to the design and construction of energy-efficient buildings, influencing their overall performance and environmental impact. Several key aspects of architectural design contribute to the energy efficiency and sustainability of structures [22]:

1. **Building orientation:** Proper orientation in relation to the sun and prevailing winds is crucial. Optimal building positioning allows for maximum natural daylighting, reducing the need for artificial lighting, and enables passive solar heating or cooling strategies.
2. **Spatial layout:** Efficient floor plans that promote natural ventilation and circulation contribute to energy savings. Compact designs minimize exposed surface area, enhancing thermal performance and reducing the energy required for heating or cooling.
3. **Window design and placement:** High-performance windows with features like low-emissivity coatings help control heat transfer. Thoughtful placement of windows allows for the effective capture of natural light while minimizing direct sunlight exposure, contributing to both energy efficiency and occupant comfort.
4. **Shading devices:** Incorporating shading elements such as overhangs, awnings, or louvers helps mitigate solar heat gain and enhances comfort. Strategic use of these devices prevents excessive heat build-up in interiors during warmer seasons.
5. **Green building elements:** Integration of green roofs, vertical gardens, and other sustainable landscaping features contributes to insulation, biodiversity, and the mitigation of urban heat island effects. These elements enhance the overall environmental performance of the building.

Visual aids, including diagrams illustrating optimal building orientations, cross-sectional views showcasing natural ventilation strategies, and renderings depicting green building features serve to emphasize the significance of these architectural considerations [23]. By prioritizing these design principles, architects can create buildings that not only reduce energy consumption but also harmonize with their surroundings and promote a more sustainable built environment.

1.2.2 Insulation materials and thermal performance

In the realm of energy-efficient building design, the selection and application of insulation materials stand as paramount considerations for enhancing thermal performance. Insulation plays a crucial role in mitigating heat transfer, ensuring a comfortable indoor environment, and minimizing the reliance on mechanical heating and cooling systems [24]. The thermal performance of a building is contingent upon the insulation's ability to resist the flow of heat, thereby regulating indoor temperatures and reducing energy consumption. Key considerations in this domain include the choice of insulation materials, their R-values, and their compatibility with the specific climate and architectural context. Optimal insulation placement within walls, roofs, and floors further contributes to the overall thermal efficiency of the building envelope [25].

Several critical points underscore the significance of insulation materials and thermal performance in energy-efficient building design [26–30]:

1. **Material selection:** The choice of insulation materials encompasses a spectrum, ranging from traditional options like fiberglass and cellulose to innovative alternatives such as foam boards and reflective foils.
2. **R-values:** Insulation effectiveness is quantified by R-values, representing the material's resistance to heat flow. Higher R-values indicate superior thermal performance.
3. **Climate considerations:** Different climates necessitate tailored insulation strategies. Cold climates may require higher levels of insulation to retain heat, while in warmer climates, the emphasis may shift to minimizing heat gain.
4. **Building envelope integrity:** A holistic approach involves ensuring the integrity of the building envelope, with continuous and properly installed insulation to eliminate thermal bridges and maintain consistent thermal performance.
5. **Environmental impact:** Consideration of insulation materials with minimal environmental impact is pivotal, aligning with sustainable construction practices and reducing the overall carbon footprint of the building.

1.2.3 Optimizing HVAC systems

In the intricate domain of energy-efficient building design, the optimization of heating, ventilation, and air conditioning (HVAC) systems emerges as a critical facet, necessitating a comprehensive strategy to achieve both environmental sustainability and occupant comfort [31, 32]. The intrinsic role of HVAC systems in regulating indoor climate underscores the imperative to streamline their operation for heightened efficiency and reduced energy consumption [33, 34]. A nuanced approach to HVAC optimization entails a thorough examination of various factors:

1. **Appliance efficiency:** Prioritizing HVAC appliances with high seasonal energy efficiency ratio (SEER) ratings ensures optimal performance. Upgrading to newer, energy-efficient models can significantly contribute to reduced energy consumption and operational costs.
2. **Smart controls and automation:** The integration of smart controls adds a layer of sophistication to HVAC systems, allowing for programmable thermostats, occupancy sensors, and adaptive controls. These features enable dynamic adjustments based on real-time occupancy and environmental conditions, minimizing energy wastage.
3. **System zoning strategies:** Implementing zoning strategies involves dividing a building into distinct zones, each with independent climate control. This approach enables targeted heating and cooling in specific areas, optimizing energy use by conditioning only the spaces that require it.
4. **Energy recovery systems:** The incorporation of energy recovery systems, such as heat exchangers, facilitates the transfer of heat between incoming and outgoing air streams. This process enhances energy efficiency by preconditioning fresh air with the thermal energy extracted from the exhaust air.
5. **Regular maintenance protocols:** Consistent and proactive maintenance practices, including the timely replacement of air filters, cleaning of coils, and inspections for potential leaks, are imperative. A well-maintained HVAC system operates at peak efficiency, ensuring optimal performance and extending the system's lifespan.
6. **Occupancy-based controls:** Integrating occupancy-based controls further refines HVAC efficiency by adjusting settings based on real-time occupancy data. This responsive approach prevents unnecessary energy consumption in unoccupied spaces.
7. **Integration of renewable energy:** Exploring the integration of renewable energy sources, such as solar-powered HVAC systems, aligns with broader sustainability goals. Renewable energy integration lessens the reliance on conventional power sources and reduces the overall carbon footprint of HVAC operations.

1.2.4 Advancements in lighting technologies

Advancements in lighting technologies unfold as a dynamic landscape of ongoing innovation and progress in artificial illumination. With a primary focus on enhancing energy efficiency, sustainability, and overall lighting performance, these advancements extend beyond traditional incandescent bulbs. Key players in this evolution include light-emitting diodes (LEDs) and compact fluorescent lamps (CFLs). LEDs, in particular, dominate the scene with their efficiency, extended lifespan, and adaptability, offering directional lighting and compatibility with smart controls for dynamic adjustments based on occupancy and natural light levels [35]. The transition toward energy-efficient lighting solutions, encompassing LEDs and CFLs, aligns seamlessly with sustainability goals, significantly reducing energy consumption and associated greenhouse gas emissions. Complementing this, the integration of smart lighting systems introduces advanced control and automation features, utilizing sensors and connectivity to adjust lighting levels, color temperature, and spectral composition, enhancing user experience and optimizing energy usage [36].

Advancements in daylight harvesting technologies maximize energy savings by harnessing natural daylight, employing sensors to adjust artificial lighting accordingly. The narrative extends to human-centric lighting solutions, acknowledging the impact of lighting on well-being. These technologies mimic natural light patterns, adjusting color temperature and intensity to support circadian rhythms and enhance occupant comfort and productivity [37]. In the era of connected lighting, facilitated by the Internet of Things (IoT), systems can be remotely monitored and controlled, contributing to energy management, predictive maintenance, and data-driven insights for enhanced efficiency and sustainability. Materials science innovations, such as organic light-emitting diodes (OLEDs) and quantum dots, represent a pinnacle in lighting technology, offering improved color rendering, efficiency, and design flexibility [38].

1.2.5 Integration of smart building automation

The integration of smart building automation marks a transformative leap in the realm of modern construction and facility management. At its core, smart building automation involves the seamless incorporation of advanced technologies and interconnected systems to optimize the operational efficiency, sustainability, and occupant experience within a built environment [32]. This intricate web of technologies encompasses a spectrum of components, including sensors, actuators, controllers, and intelligent software, all working in concert to create an environment that adapts intelligently to changing conditions. Smart building automation systems are designed to monitor and control various building functions, such as heating, ventilation, air conditioning (HVAC), lighting, security, and more. The integration extends beyond mere functionality; it involves the gathering and analysis of data in real time, allowing for informed decision-making and predictive maintenance [39].

Key to this integration is the utilization of Internet of Things (IoT) technologies, where devices communicate with each other through interconnected networks, creating a dynamic ecosystem. Smart sensors collect data on occupancy, temperature, and energy usage, enabling the system to adjust settings automatically for energy efficiency and occupant comfort [40]. Furthermore, the integration of artificial intelligence and machine learning enhances the system's ability to adapt and optimize over time, learning from patterns and user preferences. The benefits of smart building automation are manifold, ranging from energy savings and operational cost reductions to enhanced security, improved sustainability, and a more responsive and comfortable environment for occupants. As we navigate the era of smart buildings, the integration of automation becomes a cornerstone in shaping the future of intelligent, adaptive, and sustainable built environments [41].

1.3 KEY COMPONENTS OF ENERGY-EFFICIENT BUILDING SYSTEMS

Energy-efficient building systems comprise key components designed for optimal resource usage and environmental impact reduction. These components include advanced insulation materials, high-efficiency HVAC systems, intelligent lighting solutions like LEDs, and the integration of renewable energy sources such as solar panels. Smart building management systems act as the central hub, coordinating and optimizing the performance of these components through real-time data analytics and sophisticated algorithms [42]. Together, these elements form a cohesive and sustainable approach to building design, promoting energy savings and environmental responsibility [43].

1.3.1 Sustainable materials and construction methods

The pursuit of sustainability in construction has ushered in a critical focus on the selection of materials and construction methods that minimize environmental impact, promote resource efficiency, and enhance the overall lifecycle performance of built structures (Table 1.1). Sustainable materials prioritize eco-friendly sourcing, reduced carbon footprint, and recyclability [44]. Concurrently, sustainable construction methods emphasize energy efficiency, waste reduction, and environmentally conscious practices throughout the building process. The integration of these elements contributes to the creation of structures that not only endure but also harmonize with the natural environment [45].

Table 1.1 Sustainablematerials and construction methods evaluation

	Sustainable Material		*Sustainable Method*	
Criteria	*Bamboo*	*Recycled Steel*	*Passive Design*	*Green Roofs*
Eco-Friendly Sourcing	High	Moderate	N/A	N/A
Recyclability	Yes	Yes	N/A	N/A
Carbon Footprint	Low	Moderate	N/A	N/A
Energy Efficiency	N/A	N/A	High	Moderate
Waste Reduction	Moderate	High	High	Moderate
Durability	High	Moderate	High	High
Cost Efficiency	Moderate	High	Moderate	High

1.3.2 Renewable energy integration

The foundations of renewable energy integration in energy-efficient build ings represent a crucial aspect of sustainable construction, aligning environmental consciousness with efficient energy practices. This integration involves harnessing clean, inexhaustible energy sources to power building operations, reducing reliance on conventional energy grids and mitigating environmental impact [46]. Key principles include comprehensive site assessments to identify renewable energy potential, strategic deployment of microgeneration technologies such as solar panels and wind turbines, and the utilization of solar PV and other renewable sources like green hydrogen systems as a cornerstone for on-site power generation [47, 48]. Wind turbines, geothermal heating and cooling, and energy storage solutions enhance the reliability of renewable sources, while smart building management systems optimize energy use. Regulatory support, life cycle analysis (LCA), and educational initiatives further bolster the integration of renewable energy, contributing to a greener and more resilient infrastructure. These fundamentals provide a strategic blend of site-specific assessments, cutting-edge technologies, and supportive regulatory frameworks, serving as a catalyst for the broader adoption of renewable energy practices in the built environment [49].

1.3.3 Digitalization and data-driven approaches

The advancement of digitalization and data-driven approaches has ushered in a new era in the realm of energy-efficient buildings. Digital technologies are transforming the way buildings are designed, operated, and optimized for energy efficiency. Smart building systems, equipped with sensors, meters, and automation, collect real-time data on various parameters such as occupancy, temperature, and energy consumption. This wealth of data enables sophisticated analytics and machine learning algorithms to identify patterns, optimize energy usage, and predict potential inefficiencies [50]. Building management systems leverage these insights to dynamically adjust lighting, HVAC systems, and other energy-consuming components in real time. The integration of digitalization and data-driven approaches not only enhances operational efficiency but also provides a platform for continuous improvement through performance monitoring and predictive maintenance. In essence, the marriage of technology and data-driven strategies is propelling energy-efficient buildings into a future where precision, adaptability, and sustainability converge for optimal environmental and economic outcomes [51]. Table 1.2 represents a few approaches to digitalization and data-driven strategies in energy-efficient buildings, along with their respective advantages and limitations.

These approaches offer diverse benefits in enhancing energy efficiency, but they also come with certain limitations that need to be carefully addressed for successful implementation and sustained performance improvement in energy-efficient buildings.

Table 1.2 Digitalization and data-driven strategies with their respective advantages and limitations used in energy-efficient buildings

Approach	*Advantages*	*Limitations*
Smart Building Automation Systems [73-77]	Real-time monitoring and control of building systems.	High initial costs for implementation. Complex integration may require specialized expertise.
Energy Management Platforms	Centralized data analytics for optimizing energy usage.	Reliance on accurate and consistent data. Initial setup costs and potential cybersecurity concerns.
Predictive Maintenance Systems	Anticipates equipment failures, reducing downtime.	Dependence on accurate predictive models. Initial setup costs and potential challenges in data quality.
Occupancy Sensors and Analytics	Efficient lighting and HVAC adjustments based on occupancy.	Privacy concerns related to tracking occupancy. Accuracy of sensors may be influenced by factors like furniture arrangement.
Building Energy Modeling [78-80]	Simulates energy performance for design and optimization.	Accuracy depends on data inputs and assumptions. May require skilled professionals for accurate modeling.
Demand Response Systems	Optimizes energy consumption during peak demand periods.	Dependency on external factors such as utility programs. May require investments in equipment compatible with demand response.
Blockchain Technology	Enhances transparency and security in energy transactions.	Complexity of implementation and potential scalability issues. Limited widespread adoption and regulatory challenges.

1.3.4 Life cycle analysis and environmental impact assessment

Life cycle analysis (LCA) and environmental impact assessment (EIA) are comprehensive methodologies employed to evaluate the environmental implications of a product, process, or system throughout its entire life cycle. The goal is to identify and quantify the environmental burdens associated with each life cycle stage, aiding in informed decision-making for sustainability. Both LCA and EIA play crucial roles in advancing sustainability objectives by providing a comprehensive understanding of environmental impacts and guiding the development and implementation of projects with minimal environmental footprints [52]. To create a better understanding of LCA and EIA, detailed steps are discussed here that are used in LCA and EIA [53].

1.3.4.1 Life cycle analysis (LCA) for energy-efficient buildings

In the realm of sustainable construction and environmental responsibility, the life cycle analysis (LCA) process for an energy-efficient building unfolds in a systematic and multifaceted manner. The first phase, goal definition and scope, lays the foundation for the entire assessment. Here, the core objectives of the LCA are meticulously defined, serving as guiding principles throughout the evaluation. This phase involves delineating the precise scope of the analysis, including the establishment of boundaries that encapsulate the areas of interest. Additionally, it necessitates determining the functional unit, often expressed in square meters of building space, providing a standardized metric for evaluation. Crucially, it also involves identifying the distinct life cycle stages, spanning from construction and operation to renovation and end-of-life considerations, ensuring a comprehensive assessment that considers the building's entire life cycle [54].

Following this, the life cycle inventory (LCI) phase demands a rigorous data collection process encompassing all inputs and outputs linked to the energy-efficient building throughout its lifecycle. This entails an exhaustive examination of various aspects, including raw material extraction, manufacturing processes, construction methods, energy consumption during operation, maintenance requirements, and even scenarios involving potential demolition or recycling [81–85]. The subsequent phase, impact assessment, represents a critical juncture where the potential environmental consequences are evaluated, encompassing impact categories such as energy consumption, greenhouse gas emissions, water usage, and other relevant indicators [55]. This assessment also takes into account the energy sources employed during the operational phase, recognizing their substantial influence on the overall environmental footprint. Interpretation, improvement analysis, and reporting phases follow, each contributing to a comprehensive understanding of the building's environmental performance and providing valuable insights into sustainable design and construction practices (Figure 1.4).

1.3.4.2 Environmental impact assessment (EIA) for energy-efficient buildings

The EIA process for the construction or renovation of energy-efficient buildings is a meticulous and multifaceted procedure, designed to ensure that the environmental footprint of such structures is minimized while promoting sustainable practices. The initial phase, screening, involves the crucial determination of whether an EIA is necessary for the proposed project. This decision is predicated on the project's scale and its potential environmental implications [56]. The objective at this stage is to identify if the construction or renovation of the energy-efficient building warrants a detailed

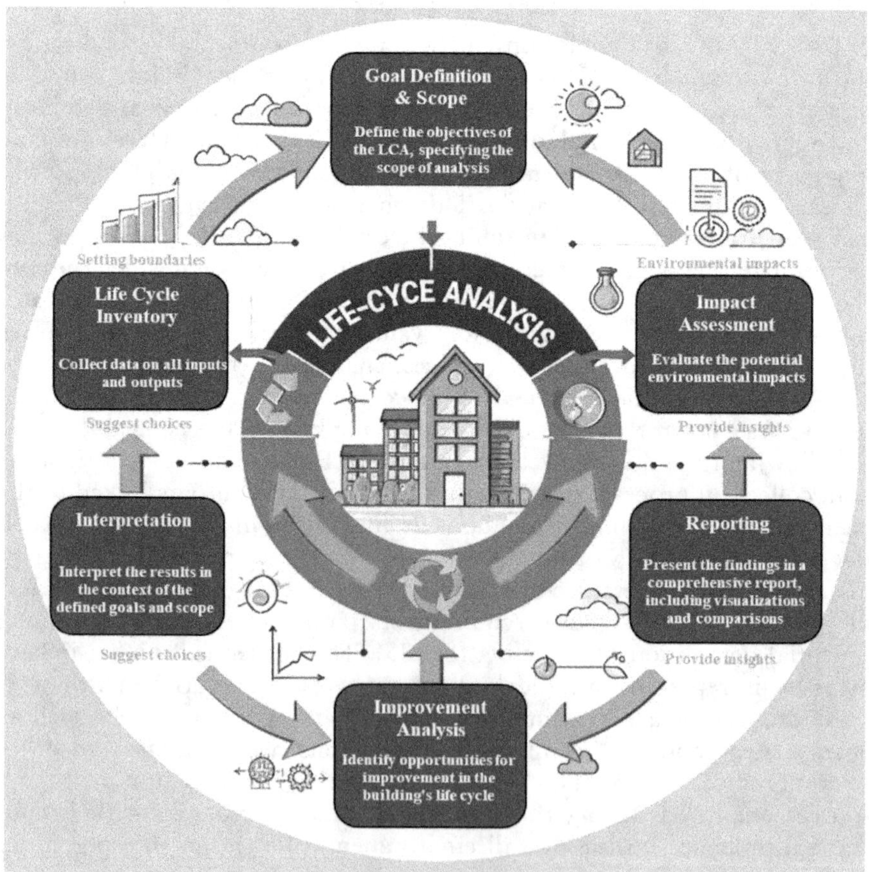

Figure 1.4 Life cycle analysis for energy-efficient buildings

environmental assessment, considering factors like the project's size, location, and the likelihood of significant environmental effects.

Following the screening phase, the scoping process commences. This step is fundamental in delineating the boundaries of the EIA. It involves a thorough consideration of various elements, such as the efficiency of energy use, site selection, and the project's possible impacts on local ecosystems. An essential component of scoping is stakeholder engagement, which aids in pinpointing pertinent concerns and factors that might influence the project. This phase ensures that all relevant environmental aspects, including those that stakeholders find significant, are identified and addressed in the EIA.

The next critical steps involve baseline data collection, impact prediction, mitigation and impact management, and the development of an environmental management plan (EMP). Baseline data collection is about gathering comprehensive information on the existing environmental conditions of the site and its surroundings, including an evaluation of the energy infrastructure, local biodiversity, and any potential environmental restrictions [57]. Impact prediction is focused on forecasting the potential environmental impacts of the project, assessing both direct and indirect effects on various environmental aspects such as air quality, water resources, ecosystems, and energy consumption. The mitigation and impact management phase involves the development of strategies to mitigate the identified environmental impacts. This can include incorporating renewable energy sources, optimizing energy-efficient designs, and sustainable construction practices [58]. The EMP is a strategic plan that outlines how the environmental impacts will be managed and mitigated throughout the construction, operation, and potential decommissioning phases of the building.

The latter stages of the EIA process, including public consultation, decision-making, monitoring and auditing, and reporting, are equally crucial. Public consultation ensures that the views and concerns of the public and stakeholders are incorporated into the decision-making process, enhancing the project's transparency and acceptability. Decision-making is based on the comprehensive assessment, stakeholder feedback, and proposed mitigation measures, leading to informed decisions about the feasibility and execution of the project. Monitoring and auditing are necessary for verifying that the environmental management measures are effectively implemented, ensuring compliance with environmental standards and the project's sustainability goals [59]. Lastly, reporting involves documenting and disseminating the EIA findings, processes, results, and decisions. This step is vital for providing transparency and accountability regarding the environmental considerations and impacts associated with the energy-efficient building (Figure 1.5).

These steps collectively contribute to the holistic assessment of an energy-efficient building, considering its entire life cycle and environmental impact. The integration of LCA and EIA ensures that sustainability considerations are thoroughly addressed from the conceptualization to the operational phases of the building [60].

1.4 CASE STUDIES AND BEST PRACTICES

Case studies and best practices in energy-efficient building projects offer concise and insightful real-world examples [61–64]. These studies highlight successful strategies, technologies, and sustainable practices implemented in construction projects. They provide measurable outcomes, address challenges, and showcase the impact on energy consumption and environmental sustainability [65]. By sharing lessons learned and promoting replicability, these case studies contribute to knowledge sharing, inspire innovation, support decision-making, and benchmark the success of energy-efficient

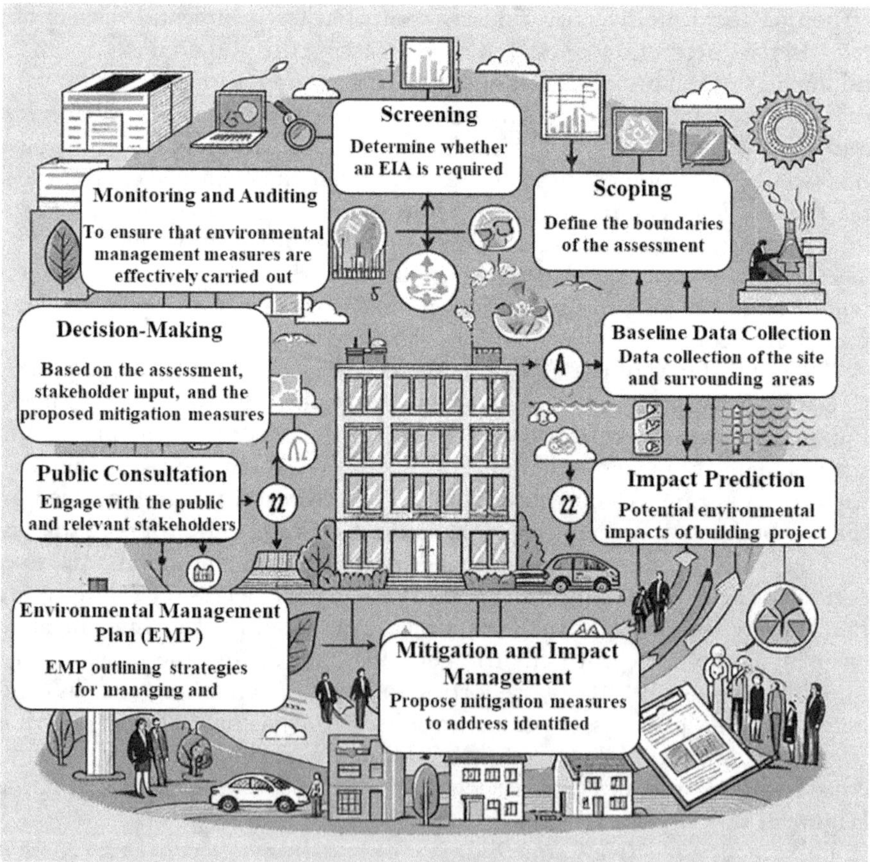

Figure 1.5 Environmental impact assessment (EIA) for energy-efficient buildings

initiatives within the construction industry. Overall, they play a crucial role in advancing sustainable practices for a greener built environment.

1.4.1 Real-world examples of successful energy-efficient building projects

Real-world examples of successful energy-efficient building projects serve as inspiring models for sustainable construction. These projects showcase innovative approaches, technologies, and design strategies that significantly reduce energy consumption and environmental impact. By highlighting measurable outcomes, such as energy savings and certifications (LEED, BREEAM), these examples provide tangible evidence of the benefits of prioritizing energy efficiency in construction. Lessons learned from challenges faced during these projects contribute valuable insights, promoting knowledge sharing and encouraging the adoption of best practices across the industry [64]. These real-world examples play a pivotal role in driving the shift toward more sustainable and environmentally conscious building practices. A real-world example of successful energy-efficient building projects with comparative analysis for their energy efficiency is shown in Table 1.3.

1.4.2 Lessons learned: Challenges and solutions

In the pursuit of energy-efficient buildings, a myriad of challenges has surfaced, offering invaluable lessons that have shaped the trajectory of sustainable construction [86, 87]. Technological integration posed a hurdle, requiring a collaborative approach among architects, engineers, and technology experts to ensure seamless compatibility. Overcoming resistance to behavioral change proved crucial, prompting the implementation of education programs to enlighten occupants about the benefits of energy-efficient practices. The upfront costs associated with such projects emerged as a significant barrier, alleviated by leveraging financial incentives and subsidies to promote adoption. Navigating the complex landscape of evolving energy codes and regulations demanded a continuous monitoring approach, ensuring compliance. The key takeaways emphasize the necessity of a holistic and adaptable approach, with education emerging as a catalyst for change [91–93].

1.4.3 Economic viability and return on investment

The economic viability and return on investment (ROI) for energy-efficient buildings stand as pivotal considerations in their widespread adoption. These structures exhibit economic viability through significant energy savings, resource efficiency, and improved occupant productivity. Despite potentially higher initial costs, the long-term gains in reduced utility bills

Table 1.3 Comparative analysis table of energy-efficient buildings projects around the world

Criteria	*Joset Pavilion*	*HEB at Mueller*	*Exploratorium at Pier 15*	*BIQ Building*	*Solar and Wind Powered Dynamic Tower*	*SunCarrier Omega Solar-Powered Building*	*The ZCB Building in Hong Kong*
Certification	Certified Living Building	Not specified	LEED Platinum	Not specified	Not specified	Not specified	Zero-Carbon Building (ZCB) Certification
Energy Efficiency	Strategic land management	Solar panels and sensors	Natural lighting and water bay	Algae bio-reactors for heat and biogas	Wind turbines and solar energy	Solar energy	Photovoltaic solar panels, energy-efficient designs
Environmental Design	Heritage Live Oak principles	Not specified	800-foot-long roof	Façade algae bio-reactors	Rotating floors with horizontal wind turbines	Off-grid operations with solar energy	Photovoltaic solar panels, energy-efficient designs
Overall Environmental Impact	Promotes healthy watersheds	Significant energy savings	Largest net zero energy	Produces heat and biogas from algae	Generates energy through wind and solar	Runs entirely off the grid with solar energy	Achieves zero-carbon attributes, energy-efficient fixtures

and operational expenses contribute to a positive ROI [32]. Government incentives and tax benefits further sweeten the financial attractiveness of energy-efficient projects. Moreover, these buildings often command higher market values due to their sustainable features, providing additional financial benefits to owners and investors. Strategic considerations include life cycle cost analysis, leveraging incentive programs, and recognizing the economic value of improved occupant well-being, collectively reinforcing the case for investments in energy-efficient building projects [66].

1.5 ADVANCES IN ENERGY-EFFICIENT TECHNOLOGIES

The pursuit of sustainable and energy-efficient buildings has witnessed remarkable strides in technological innovation. These advances span various aspects of construction, from materials to HVAC systems, lighting, and beyond [67].

1.5.1 Cutting-edge materials for sustainable construction

In the realm of sustainable construction, cutting-edge materials have emerged as game-changers. Innovations in insulation materials, such as aerogels and phase-change materials, enhance thermal performance, reducing the need for excessive heating or cooling. Smart materials with self-healing properties contribute to the longevity of structures, minimizing maintenance requirements. Additionally, recycled and bio-based materials, including reclaimed wood, bamboo, and recycled steel, are gaining prominence, promoting environmentally friendly construction practices [68].

1.5.2 Next-generation HVAC and renewable energy systems

The evolution of HVAC systems has been pivotal in optimizing energy consumption. Advanced HVAC technologies, such as variable refrigerant flow (VRF) systems and geothermal heat pumps, provide efficient climate control while minimizing energy usage [69]. Integration with renewable energy sources, such as solar and wind, further reduces the carbon footprint of buildings. Smart HVAC systems, equipped with sensors and automation, dynamically adjust settings based on occupancy and environmental conditions, ensuring optimal efficiency [94–99].

1.5.3 Innovations in lighting and automation

In the realm of lighting, innovations center around energy-efficient technologies and automation. Light-emitting diodes (LEDs) have become the cornerstone of energy-efficient lighting, offering longevity and adaptability. Daylight harvesting systems, equipped with sensors, adjust artificial lighting based on natural light availability, maximizing energy savings. Furthermore, connected lighting systems, part of the Internet of Things (IoT), enable remote monitoring and control, contributing to energy management, predictive maintenance, and data-driven insights [70].

1.5.4 Future trends and emerging technologies

Looking ahead, future trends in energy-efficient building technologies are poised to reshape the landscape. Building-integrated photovoltaics (BIPV) holds the potential to seamlessly integrate solar panels into architectural elements, turning buildings into energy generators. Artificial intelligence (AI) applications in building management systems enhance efficiency by learning and adapting to occupant behavior and environmental conditions. Additionally, advancements in energy storage technologies, such as advanced batteries, contribute to balancing energy supply and demand, fostering resilience in the face of intermittent renewable sources [71].

These advances collectively underscore the transformative impact of technology on the quest for sustainable and energy-efficient buildings [72]. As the construction industry continues to embrace these innovations, the vision of environmentally conscious and high-performing structures is becoming increasingly attainable. Smart city infrastructure will promote intelligent transportation systems which can revolutionize the way people commute with various entities of smart cities. Energy-efficient advanced vehicles will be the central theme of intelligent transportation system which offers a novel approach to connected transportation, traffic, and mobility management solutions [88–91].

1.6 CONCLUSION AND ROADMAP FOR FURTHER EXPLORATION

In conclusion, the exploration of energy-efficient building technologies reveals a dynamic landscape characterized by continuous innovation, integration of cutting-edge materials, and a paradigm shift toward sustainable design practices. The synthesis of key concepts underscores the multifaceted approach required for achieving energy efficiency, encompassing advancements in materials science, HVAC systems, lighting technologies, and beyond. The implications for the future of sustainable building design are profound, as the identified technologies not only enhance energy performance but also contribute to reduced environmental impact and improved occupant well-being.

Looking ahead, the roadmap for further exploration is guided by a commitment to addressing the evolving challenges and opportunities in the field. The intricate interplay between technological advancements, regulatory frameworks, and societal awareness forms the foundation for in-depth exploration in subsequent chapters. As we navigate this roadmap, an emphasis will be placed on emerging technologies, such as artificial intelligence and building-integrated photovoltaics, shaping the future trajectory of energy-efficient buildings.

Moreover, the roadmap extends to the economic viability of these technologies, considering return on investment, life cycle analyses, and the role of government incentives in fostering widespread adoption. The synthesis of these multifaceted elements sets the stage for a comprehensive understanding of energy-efficient building technologies, laying the groundwork for architects, engineers, policymakers, and stakeholders to collaboratively propel the industry toward a sustainable and resilient future.

In subsequent chapters, the exploration will delve into specific facets, offering a deeper analysis of materials, systems, and design principles. By leveraging the knowledge synthesized in this comprehensive overview, the ensuing chapters will provide a nuanced examination, paving the way for actionable insights and transformative strategies. As we embark on this journey, the roadmap serves as a guide, illuminating the path toward a future where energy-efficient buildings not only meet the needs of the present but also ensure a sustainable legacy for generations to come.

REFERENCES

1. P. Wilkinson, K. R. Smith, S. Beevers, C. Tonne, and T. Oreszczyn, "Energy, energy efficiency, and the built environment," *Lancet*, vol. 370, no. 9593, pp. 1175–1187, 2007.
2. C. Du Plessis, "A strategic framework for sustainable construction in developing countries," *Constr. Manag. Econ.*, vol. 25, no. 1, pp. 67–76, 2007.
3. J. D. Sachs, "From millennium development goals to sustainable development goals," *Lancet*, vol. 379, no. 9832, pp. 2206–2211, 2012.
4. T. Kober, H.-W. Schiffer, M. Densing, and E. Panos, "Global energy perspectives to 2060 – WEC's World Energy Scenarios 2019," *Energy Strateg. Rev.*, vol. 31, p. 100523, 2020. doi: 10.1016/j.esr.2020.100523.
5. T. Hong, Y. Chen, X. Luo, N. Luo, and S. H. Lee, "Ten questions on urban building energy modeling," *Build. Environ.*, vol. 168, p. 106508, 2020. doi: 10.1016/j.buildenv.2019.106508.
6. V. P. Singh *et al.*, "Recent developments and advancements in solar air heaters: A detailed review," *Sustain.*, vol. 14, no. 19, pp. 1–57, Sep. 2022. doi: 10.3390/ su141912149.
7. G. Foster, "Circular economy strategies for adaptive reuse of cultural heritage buildings to reduce environmental impacts," *Resour. Conserv. Recycl.*, vol. 152, p. 104507, 2020. doi: 10.1016/j.resconrec.2019.104507.
8. V. P. Singh *et al.*, "Heat transfer and friction factor correlations development for double pass solar air heater artificially roughened with perforated multi-V

ribs," *Case Stud. Therm. Eng.*, vol. 39, no. September, p. 102461, 2022. doi: 10.1016/j.csite.2022.102461.

9. V. P. Singh, S. Jain, and A. Kumar, "Establishment of correlations for the Thermo-Hydraulic parameters due to perforation in a multi-V rib roughened single pass solar air heater," *Exp. Heat Transf.*, vol. 35, no. 5, pp. 1–20, 2022. doi: 10.1080/08916152.2022.2064940.
10. V. P. Singh, S. Jain, and J. M. L. Gupta, "Analysis of the effect of perforation in multi-v rib artificial roughened single pass solar air heater: - Part A," *Exp. Heat Transf.*, pp. 1–20, Oct. 2021. doi: 10.1080/08916152.2021.1988761.
11. V. P. Singh, S. Jain, A. Karn, A. Kumar, and G. Dwivedi, "Mathematical modeling of efficiency evaluation of double pass parallel flow solar air heater," *Sustainability*, vol. 14, no. 17, pp. 1–22, 2022. doi: 10.3390/su141710535.
12. V. P. Singh, S. Jain, and J. M. L. Gupta, "Performance assessment of double-pass parallel flow solar air heater with perforated multi-V ribs roughness – Part B," *Exp. Heat Transf.*, pp. 1–18, 2022. doi: 10.1080/08916152.2021.2019147.
13. J. A. Fonseca, I. Nevat, and G. W. Peters, "Quantifying the uncertain effects of climate change on building energy consumption across the United States," *Appl. Energy*, vol. 277, p. 115556, 2020.
14. K. Verichev, M. Zamorano, and M. Carpio, "Effects of climate change on variations in climatic zones and heating energy consumption of residential buildings in the southern Chile," *Energy Build.*, vol. 215, pp. 109–124, 2020.
15. P. R. Shukla *et al.*, "IPCC, 2019: Climate Change and Land: An IPCC special report on climate change, desertification, land degradation, sustainable land management, food security, and greenhouse gas fluxes in terrestrial ecosystems," , IPCC report. 2019. https://www.ipcc.ch/srccl/.
16. V. P. Singh, S. Jain, and J. M. L. Gupta, "Analysis of the effect of variation in open area ratio in perforated multi-V rib roughened single pass solar air heater- Part A," *Energy Sources, Part A Recover. Util. Environ. Eff.*, vol. 44, no. Jan, pp. 1–21, 2022. doi: 10.1080/15567036.2022.2029976.
17. M. Sajjad *et al.*, "Towards efficient building designing: Heating and cooling load prediction via multi-output model," *Sensors*, vol. 20, no. 22, pp. 1–19, 2020. doi: 10.3390/s20226419.
18. M. Saini, A. Sharma, V. P. Singh, G. Dwivedi, and S. Jain, "Solar thermal receivers – A review," *Adv. Mater. Manuf. Energy Eng. Vol. II*, vol. II, pp. 310–325, 2022. doi: 10.1007/978-981-16-8341-1.
19. E. M. Erebor, E. O. Ibem, I. C. Ezema, and A. B. Sholanke, "Energy efficiency design strategies in office buildings: A literature review," *IOP Conf. Ser. Earth Environ. Sci.*, vol. 665, no. 1, 2021. doi: 10.1088/1755-1315/665/1/012025.
20. V. P. Singh, A. Karn, G. Dwivedi, T. Alam, and A. Kumar, "Experimental assessment of variation in open area ratio on thermohydraulic performance of parallel flow solar air heater," *Arab. J. Sci. Eng.*, vol. 41, no. 12, pp. 1–17, 2022. doi: 10.1007/s13369-022-07525-7.
21. S. O. Eromobor, D. K. Das, and F. Emuze, "Influence of building and indoor environmental parameters on designing energy-efficient buildings," *Int. J. Build. Pathol. Adapt.*, vol. 39, no. 3, pp. 507–524, 2021.
22. P. F. de A. F. Tavares and A. M. de O. G. Martins, "Energy efficient building design using sensitivity analysis—A case study," *Energy Build.*, vol. 39, no. 1, pp. 23–31, 2007. doi: 10.1016/j.enbuild.2006.04.017.

23. N. Dutt, A. Binjola, A. J. Hedau, A. Kumar, V. P. Singh, and C. S. Meena, "Comparison of CFD results of smooth air duct with experimental and available equations in literature," *Int. J. Energy Resour. Appl.*, vol. 1, no. 1, pp. 40–47, 2022. doi: 10.56896/IJERA.2022.1.1.006.
24. R. K. Shubham Srivastava, D. Verma, S. Thusoo, A. Kumar, and V. P. Singh, "Nanomanufacturing for energy conversion and storage devices," in *Nanomanufacturing and Nanomaterials Design: Principles and Applications*, 1st ed., S. Singh, S. K. Behura, A. Kumar, K. Verma, Eds. Boca Raton, CRC Press, 2022, pp. 165–174 doi: 10.1201/9781003220602.
25. G. Heravi and M. Qaemi, "Energy performance of buildings: The evaluation of design and construction measures concerning building energy efficiency in Iran," *Energy Build.*, vol. 75, pp. 456–464, 2014.
26. V. A. Dakwale, R. V Ralegaonkar, and S. Mandavgane, "Improving environmental performance of building through increased energy efficiency: A review," *Sustain. Cities Soc.*, vol. 1, no. 4, pp. 211–218, 2011.
27. Y. Gori, A Kumar, A. K. Singh Gangwar, A. Kumar, C. S. Meena, V. P. Singh, N. Dutt, and A. Prasad, "Biomedical study of femur bone fracture and healing," in *Advanced Materials for Biomedical Applications*, 1st ed., A. Kumar, Y. Gori, A. Kumar, C. S. Meena, N. Dutt, Eds. Boca Raton, CRC Press, 2022, pp. 212–235.
28. A. Datta, A. Kumar, A. Kumar, A. Kumar, and V. P. Singh, "Advanced materials in biological implants and surgical tools," in *Advanced Materials for Biomedical Applications*, 1st ed., N. D. Ashwani Kumar, Yatika Gori, Avinash Kumar, Chandan Swaroop Meena, Ed., Boca Raton, CRC Press, 2022, pp. 283–298.
29. R. Kumar *et al.*, "Experimental and RSM-based process-parameters optimisation for turning operation of EN36B steel," *Materials*, vol. 16, no. 1, pp. 1–18, 2023. doi: 10.3390/ma16010339.
30. V. P. Singh *et al.*, "Nanomanufacturing and design of high-performance piezoelectric nanogenerator for energy harvesting," in *Nanomanufacturing and Nanomaterials Design: Principles and Applications*, 1st ed., S. Singh, S. K. Behura, A. Kumar, K. Verma, Eds. Boca Raton, CRC Press, 2022, pp. 241–272, doi: 10.1201/9781003220602.
31. M. M. Haghighi and A. L. Sangiovanni-Vincentelli, "Modeling and optimal control algorithm design for HVAC systems in energy efficient buildings," Masters Rep., 2011.
32. A. K. Sharma, M. Nayal, S. Jain, and V. P. Singh, "Chapter 07 optimization techniques of solar thermal and hybrid energy systems," in *Highly Efficient Thermal Renewable Energy Systems Design, Optimization and Applications*, 1st ed., R. W. V. Verma, S. Thangavel, N. Dutt, and A. Kumar, Eds. CRC Press, 2024, pp. 1–14.
33. V. Vakiloroaya, Q. P. Ha, and B. Samali, "Energy-efficient HVAC systems: Simulation–empirical modelling and gradient optimization," *Autom. Constr.*, vol. 31, pp. 176–185, 2013. doi: 10.1016/j.autcon.2012.12.006.
34. V. P. Singh and G. Dwivedi, "Technical analysis of a large – Scale solar updraft tower power," *Energies*, vol. 16, no. 1, pp. 1–29, 2023. doi: 10.3390/en16010494.
35. H. J. Han, Y. I. Jeon, S. H. Lim, W. W. Kim, and K. Chen, "New develop ments in illumination, heating and cooling technologies for energy-efficient buildings," *Energy*, vol. 35, no. 6, pp. 2647–2653, 2010.

36. F. G. Montoya, A. Peña-García, A. Juaidi, and F. Manzano-Agugliaro, "Indoor lighting techniques: An overview of evolution and new trends for energy saving," *Energy Build.*, vol. 140, pp. 50–60, 2017. doi: 10.1016/j.enbuild.2017.01.028.
37. V. R. M. Lo Verso, G. Mihaylov, A. Pellegrino, and F. Pellerey, "Estimation of the daylight amount and the energy demand for lighting for the early design stages: Definition of a set of mathematical models," *Energy Build.*, vol. 155, pp. 151–165, 2017. doi: 10.1016/j.enbuild.2017.09.014.
38. R. Pode, "Organic light emitting diode devices: An energy efficient solid state lighting for applications," *Renew. Sustain. Energy Rev.*, vol. 133, p. 110043, 2020. doi: 10.1016/j.rser.2020.110043.
39. V. Marinakis, H. Doukas, C. Karakosta, and J. Psarras, "An integrated system for buildings' energy-efficient automation: Application in the tertiary sector," *Appl. Energy*, vol. 101, pp. 6–14, 2013. doi: 10.1016/j.apenergy.2012.05.032.
40. C. K. Metallidou, K. E. Psannis, and E. A. Egyptiadou, "Energy efficiency in smart buildings: IoT approaches," *IEEE Access*, vol. 8, pp. 63679–63699, 2020. doi: 10.1109/ACCESS.2020.2984461.
41. J. Haase, G. Zucker, and M. Alahmad, "Energy efficient building automation: A survey paper on approaches and technologies for optimized building operation," in *IECON 2014 - 40th Annual Conference of the IEEE Industrial Electronics Society*, 2014, pp. 5350–5356. doi: 10.1109/IECON.2014.7049317.
42. N. Dutt, A. Jageshwar, H. Ashwani, K. Mukesh, K. Awasthi, and V. Pratap, "Thermo Hydraulic performance of solar air heater having discrete D Shaped ribs as artificial roughness," *Environ. Sci. Pollut. Res.*, 2023. doi: 10.1007/s11356-023-28247-9.
43. R. Jia *et al.*, "Design automation for smart building systems," *Proc. IEEE*, vol. 106, no. 9, pp. 1680–1699, 2018. doi: 10.1109/JPROC.2018.2856932.
44. L. F. Cabeza, C. Barreneche, L. Miró, M. Martínez, A. I. Fernández, and D. Urge-Vorsatz, "Affordable construction towards sustainable buildings: Review on embodied energy in building materials," *Curr. Opin. Environ. Sustain.*, vol. 5, no. 2, pp. 229–236, 2013. doi: 10.1016/j.cosust.2013.05.005.
45. M. A. Kamal, "Sustainable building materials and materials for energy efficiency," *Key Eng. Mater.*, vol. 650, pp. 235–257, 2015.
46. A. A. Hassan and K. El-Rayes, "Optimizing the integration of renewable energy in existing buildings," *Energy Build.*, vol. 238, p. 110851, 2021. doi: 10.1016/j.enbuild.2021.110851.
47. M. Nayal, A. K. Sharma, S. Jain, and V. P. Singh, "Chapter 11 green hydrogen production methods, designs and applications," in *Highly Efficient Thermal Renewable Energy Systems Design, Optimization and Applications*, 1st ed., R. W. V. Verma, S. Thangavel, N. Dutt, and A. Kumar, Eds. CRC Press, 2024, pp. 178–193.
48. A. Kumar, V. P. Singh, C. S. Meena, and N. Dutt, "Preface thermal energy systems," in *Thermal Energy Systems: Design, Computational Techniques, and Applications*, Technical Education Department, Kanpur, CRC Press, 2023, pp. x–xiii.
49. A. Chel and G. Kaushik, "Renewable energy technologies for sustainable development of energy efficient building," *Alexandria Eng. J.*, vol. 57, no. 2, pp. 655–669, 2018. doi: 10.1016/j.aej.2017.02.027.

50. M. Nayal, A. K. Sharma, S. Jain, and V. P. Singh, "Chapter 06 design and modelling of solar, geothermal and hybrid energy systems.pdf," in *Highly Efficient Thermal Renewable Energy Systems Design, Optimization and Applications*, 1st ed., R. W. V. Verma, S. Thangavel, N. Dutt, and A. Kumar, Eds. CRC Press, 2024, pp. 1–15.
51. N. Batra, A. Singh, P. Singh, H. Dutta, V. Sarangan, and M. Srivastava, "Data driven energy efficiency in buildings," 2014 [Online]. Available: http://arxiv.org/abs/1404.7227.
52. F. Asdrubali and G. Grazieschi, "Life cycle assessment of energy efficient buildings," *Energy Reports*, vol. 6, pp. 270–285, 2020. doi: 10.1016/j.egyr.2020.11.144.
53. M. Hu, "Building impact assessment–A combined life cycle assessment and multi-criteria decision analysis framework," *Resour. Conserv. Recycl.*, vol. 150, p. 104410, 2019. doi: 10.1016/j.resconrec.2019.104410.
54. S. Bawankar *et al.*, "Environmental impact assessment of lithium ion battery employing cradle to grave," *Sustain. Energy Technol. Assessments*, vol. 60, no. May, 2023. doi: 10.1016/j.seta.2023.103530.
55. A. Kundu, A. Kumar, V. P. Singh, C. S. Meena, and N. Dutt, "Chapter 1: Introduction to thermal energy resources and their smart applications," in *Thermal Energy Systems: Design, Computational Techniques, and Applications*, 1st ed., A. Kumar, V. P. Singh, C. S. Meena, N. Dutt, Eds. Boca Raton, CRC Press, 2023, pp. 1–15.
56. G. Singh, S. Rathore, A. Karn, and V. P. Singh, "Recent progress in solar dryers using phase changing material: A review," *Int. J. Energy Resour. Appl.*, vol. 2, no. 1, pp. 57–77, 2023. doi: 10.56896/IJERA.2023.2.1.005.
57. A. Saxena, A. Prajapati, G. Pant, C. Meena, A. Kumar, and V. Singh, "Water consumption optimization of hybrid heat pump water heating system," in *Lecture Notes in Mechanical Engineering- Recent Advances in Mechanical Engineering Select Proceedings of FLAME 2022*, A. Shukla, B. Sharma, A. Arabkoohsar, and P. Kumar, Eds. Springer Nature Singapore, 2023, pp. 721–732.
58. A. Kundu, A. Kumar, N. Dutt, V. P. Singh, and C. S. Meena, "Chapter 7: Modelling and simulation of thermal energy system for design optimization," in *Thermal Energy Systems: Design, Computational Techniques, and Applications*, A. Kumar, N. Dutt, V. Singh, and C. Meena, Eds. CRC Press, 2023, pp. 103–137.
59. A. Kumar, V. P. Singh, C. S. Meena, and N. Dutt, *Thermal Energy Systems: Design, Computational Techniques, and Applications*, 1st ed. CRC Press, 2023.
60. M. Žigart, R. Kovačič Lukman, M. Premrov, and V. Žegarac Leskovar, "Environmental impact assessment of building envelope components for low-rise buildings," *Energy*, vol. 163, pp. 501–512, 2018. doi: 10.1016/j.energy.2018.08.149.
61. Z. Zhou, S. Zhang, C. Wang, J. Zuo, Q. He, and R. Rameezdeen, "Achieving energy efficient buildings via retrofitting of existing buildings: A case study," *J. Clean. Prod.*, vol. 112, pp. 3605–3615, 2016. doi: 10.1016/j.jclepro.2015.09.046.
62. R. Jank, "Annex 51: Case studies and guidelines for energy efficient communities," *Energy Build.*, vol. 154, pp. 529–537, 2017. doi: 10.1016/j.enbuild.2017.08.074.

63. V. Martinaitis, E. K. Zavadskas, V. Motuzienė, and T. Vilutienė, "Importance of occupancy information when simulating energy demand of energy efficient house: A case study," *Energy Build.*, vol. 101, pp. 64–75, 2015. doi: 10.1016/j.enbuild.2015.04.031.
64. J. Carlander and P. Thollander, "Drivers for implementation of energy-efficient technologies in building construction projects – Results from a Swedish case study," *Resour. Environ. Sustain.*, vol. 10, p. 100078, 2022. doi: 10.1016/j.resenv.2022.100078.
65. G. Pant, C. S. Meena, A. Saxena, A. Kumar, V. P. Singh, and N. Dutt, "Study the temperature variation in alternate coils of insulated condenser cum storage tank: Experimental study," in *Lecture Notes in Mechanical Engineering*. Springer, Singapore, 2023, pp. 627–638. doi: 10.1007/978-981-99-2382-3_52.
66. W. Kuckshinrichs, T. Kronenberg, and P. Hansen, "The social return on investment in the energy efficiency of buildings in Germany," *Energy Policy*, vol. 38, no. 8, pp. 4317–4329, 2010.
67. A. Kumar, V. P. Singh, C. S. Meena, and N. Dutt, "Aim and scope of thermal energy systems: Design, computational techniques, and applications," in *Thermal Energy Systems: Design, Computational Techniques, and Applications*, A. Kumar, V. P. Singh, C. S. Meena, and N. Dutt, Eds. CRC Press, 2022, pp. 9–10.
68. C. J. Kibert and R. Srinivasan, "Sustainable construction: The cutting edge and emerging challenges," in *Analytics for Building-Scale Sustainable Ecosystems*, Yongxin Tao, Yi Jiang, Eds., Begell House, 2016.
69. C. T. Veje *et al.*, "NeGeV: Next generation energy efficient ventilation system using phase change materials," *Energy Informatics*, vol. 2, no. 1, p. 2, 2019. doi: 10.1186/s42162-019-0067-1.
70. M. Frascarolo, S. Martorelli, and V. Vitale, "An innovative lighting system for residential application that optimizes visual comfort and conserves energy for different user needs," *Energy Build.*, vol. 83, pp. 217–224, 2014. doi: 10.1016/j.enbuild.2014.03.072.
71. V. J. L. Gan, I. M. C. Lo, J. Ma, K. T. Tse, J. C. P. Cheng, and C. M. Chan, "Simulation optimisation towards energy efficient green buildings: Current status and future trends," *J. Clean. Prod.*, vol. 254, p. 120012, 2020. doi: 10.1016/j.jclepro.2020.120012.
72. A. Allouhi, Y. El Fouih, T. Kousksou, A. Jamil, Y. Zeraouli, and Y. Mourad, "Energy consumption and efficiency in buildings: Current status and future trends," *J. Clean. Prod.*, vol. 109, pp. 118–130, 2015. doi: 10.1016/j.jclepro.2015.05.139.
73. A. K. Dewangan, S. Q. Moinuddin, M. Cheepu, S. K. Sajjan, and A. Kumar, "Thermal energy storage: Opportunities, challenges and future scope," in *Thermal Energy Systems: Design, Computational Techniques and Applications*, A. Kumar, V. P. Singh, C. S. Meena, and N. Dutt, Eds. Boca Raton, FL: CRC Press, 2023, pp. 17–28, Chapter 2. doi: 10.1201/9781003395768-2.
74. C. S. Meena, A. Kumar, S. Jain, A. U. Rehman, and S. Mishra, "Innovation in green building sector for sustainable future," *Energies*, vol. 15, p. 6631, 2022. doi: 10.3390/en15186631.
75. V.Verma, C. S. Meena, S.Thangavel, A. Kumar, T.Choudhary, and G. Dwivedi, "Ground and solar assisted heat pump systems for space heating and cooling applications in the northern region of India – A study on energy and CO2

saving potential," *Sustainable Energy Technologies and Assessments*, vol. 59, p. 103405, 2023. ISSN 2213-1388. doi: 10.1016/j.seta.2023.103405.
76. P. Kr. Kushwaha, N. Kr. Sharma, A. Kumar, and C. S. Meena, "Recent advancements in augmentation of solar water heaters using nanocomposites with PCM: Past, present, and future," *Buildings*, vol. 13, p. 79, 2023. doi: 10.3390/buildings13010079.
77. C. S. Meena, A. N. Prajapati, A. Kumar, and M. Kumar, "Utilization of solar energy for water heating application to improve building energy efficiency: An experimental study," *Buildings*, vol. 12, p. 2166, 2022. doi: 10.3390/buildings12122166.
78. C. S. Meena, A. Kumar, S. Roy, A. Cannavale, and A. Ghosh, "Review on boiling heat transfer enhancement techniques," *Energies*, vol. 15, p. 5759, 2022. doi: 10.3390/en15155759.
79. Y. K. Prajapati, P. K. Gupta, and A. Kumar, "Free surface undulation and air entrainment in a rectangular tank," *International Journal of Applied Engineering Research*, vol. 6, no. 4, pp. 409–419, 2011.
80. N. Dutt, A. Hedau, A. Kumar, M. K. Awasthi, S. Hedau, and C. S. Meena, "Thermo-hydraulic performance investigation of solar air heater duct having staggered D-shaped ribs: Numerical approach," *Heat Transfer*, pp. 1–31, 2024. doi: 10.1002/htj.22998.
81. V. K. Sharma, A. Kumar, M. Gupta, V. Kumar, D. K.Sharma, and S. Sharma, *Additive Manufacturing in Industry 4.0: Methods, Techniques, Modeling, and Nano Aspects.* Taylor & Francis (CRC Press), 2022. doi: 10.1201/9781003360001.
82. A. Kumar, A. Kumar, and A. Kumar, *Laser Based Technologies for Sustainable Manufacturing.* Taylor & Francis (CRC Press), 2023. ISBN: 9781003402398. doi: 10.1201/9781003402398.
83. R. Kumar, A. Prasad, and A. Kumar, *Sustainable Smart Manufacturing Processes in Industry 4.0.* Taylor & Francis (CRC Press). ISBN: 9781003436072, 2023. doi: 10.1201/9781003436072.
84. G. Chakraborty, V. Pandey, A. Prasad, and A. Kumar, "Introduction to sustainable manufacturing for industries 4.0," in *Sustainable Smart Manufacturing Processes in Industry 4.0*, R. Kumar, A. Prasad, and A. Kumar, Eds. Boca Raton, FL: CRC Press, 2023; Chapter 01; pp. 01–17. doi: 10.1201/9781003436072-1.
85. A. Pathak, A. Kumar, A. Kumar, A. Kumar, and F. L. King M, "Application of laser technology in the mechanical and machine manufacturing industry," in *Laser-based Technologies for Sustainable Manufacturing*, A. Kumar, A. Kumar, and A. Kumar, Eds. Boca Raton, FL: CRC Press, 2023; Chapter 06; pp. 107–155, doi: 10.1201/9781003402398-6.
86. M. K. Awasthi, N. Dutt, A. Kumar, and A. Hedau, "Introduction to mathematical and computational methods," in *Computational Fluid Flow and Heat Transfer: Advances, Design, Control and Applications*, M. K. Awasthi, A. Kumar, N. Dutt, and S. Singh, Eds. Boca Raton, FL: CRC Press, 2024; Chapter 01. ISBN 9781032603186.
87. M. K. Awasthi, A. Kumar, N. Dutt, and S. Singh, Eds, *Computational Fluid Flow and Heat Transfer: Advances, Design, Control and Applications.* Boca Raton, FL: CRC Press, 2024. ISBN 9781032603186.
88. V. P. Singh, A. Kumar, C. S. Meena, and G. Dwivedi, "Introduction to energy efficient vehicles for sustainable transportation: Transitions and challenges,"

In *Energy Efficient Vehicles: Technologies and Challenges*, V. P. Singh, A. Kumar, C. S. Meena, and G. Dwivedi, Eds. Boca Raton, FL: CRC Press, 2024; Chapter 01. ISBN 9781032548111.
89. R. K. Upadhyay, V. P. Singh, and A. Kumar, "Sustainable transportation: Policy, planning and implementation," in *Energy Efficient Vehicles: Technologies and Challenges*, V. P. Singh, A. Kumar, C. S. Meena, and G. Dwivedi, Eds. Boca Raton, FL: CRC Press, 2024; Chapter 05. ISBN 9781032548111.
90. V. P. Singh and A. Kumar, "Techno-economic and future aspects of the HEV-EV-FCV: Decarbonisation, digitalization and sustainability," in *Energy Efficient Vehicles: Technologies and Challenges*, V. P. Singh, A. Kumar, C. S. Meena, and G. Dwivedi, Eds. Boca Raton, FL: CRC Press, 2024; Chapter 11. ISBN 9781032548111.
91. V. P. Singh, A. Kumar, C. S. Meena, and G. Dwivedi, Eds. *Energy Efficient Vehicles: Technologies and Challenges*. Boca Raton, FL: CRC Press, 2024. ISBN 9781032548111.
92. A. Kumar, Y. Singla, and T. Namboodri, "Globalization and international issues in sustainable manufacturing," in *Sustainability in Smart Manufacturing: Trends, Scope, and Challenges*, S. Shah, H. Nautiyal, G. Gugliani, A. Kumar, T. Namboodri, and Y. K. Singla, Eds. Boca Raton, FL: CRC Press, 2024; Chapter 01. ISBN 9781032 740713.
93. S. Shah, H. Nautiyal, G. Gugliani, A. Kumar, T. Namboodri, and Y. K. Singla, Eds. *Sustainability in Smart Manufacturing: Trends, Scope, and Challenges*. Boca Raton, FL: CRC Press, 2024. ISBN 9781032 740713.
94. V. Verma, S. Thangavel, N. Dutt, A. Kumar, and R. Weerasinghe, "Recent development of thermal energy storage: Solar, geothermal and hydrogen energy," in *Highly Efficient Thermal Renewable Energy Systems: Design, Design, Optimization and Applications*, V. Verma, S. Thangavel, N. Dutt, A. Kumar, and R. Weerasinghe, Eds. Boca Raton, FL: CRC Press, 2024; Chapter 01. ISBN 9781032595641.
95. M. Ahmadizadeh, M. Heidari, S. Thangavel, E. A. Naamani, M. Khashehchi, V. Verma, and A. Kumar, "Technological advancements in sustainable and renewable solar energy systems," in *Highly Efficient Thermal Renewable Energy Systems: Design, Design, Optimization and Applications*, V. Verma, S. Thangavel, N. Dutt, A. Kumar, and R. Weerasinghe, Eds. Boca Raton, FL: CRC Press, 2024; Chapter 02. ISBN 9781032595641.
96. M. Chitt, S. Thangavel, V. Verma, and A. Kumar, "Green hydrogen productions: Methods, designs and smart applications," in *Highly Efficient Thermal Renewable Energy Systems: Design, Design, Optimization and Applications*, V. Verma, S. Thangavel, N. Dutt, A. Kumar, and R. Weerasinghe, Eds. Boca Raton, FL: CRC Press, 2024; Chapter 16. ISBN 9781032595641.
97. M. Khashehchi, S. Thangavel, P. Rahmanivahid, M. Heidari, T. Moazzeni, V. Verma, and A. Kumar, "Solar desalination techniques: Challenges and opportunities," in *Highly Efficient Thermal Renewable Energy Systems: Design, Design, Optimization and Applications*, V. Verma, S. Thangavel, N. Dutt, A. Kumar, and R. Weerasinghe, Eds. Boca Raton, FL: CRC Press, 2024; Chapter 19. ISBN 9781032595641.
98. V. Verma, S. Thangavel, N. Dutt, A. Kumar, and R. Weerasinghe, Eds, *Highly Efficient Thermal Renewable Energy Systems: Design, Design,*

Optimization and Applications. Boca Raton, FL: CRC Press, 2024. ISBN 9781032595641.

99. Z. Razaviyn, M. Heidari, V. Verma, S. Thangavel, A. Kumar, K. K. Saxena, and G. Dwivedi, "Numerical simulation of a marine energy convertor based on vortex induced vibrations," *Proceedings of the Institution of Mechanical Engineers, Part E: Journal of Process Mechanical Engineering*, vol. 283, pp. 441–4512024.

Chapter 2

Development of new materials for sustainable buildings

Mohammad Ahmadizadeh, Milad Heidari, Sivasakthivel Thangavel, Morteza Khashehchi, Pooyan Rahmanivahid, Varun Pratap Singh, and Ashwani Kumar

2.1 INTRODUCTION

The evolution of sustainable building materials represents a significant stride in the construction industry, marking a departure from conventional practices toward more ecologically conscious solutions. This paradigm shift is propelled by the urgent need to mitigate environmental impact, reduce resource depletion, and address climate change challenges. The development of new materials tailored for sustainable buildings epitomizes a fusion of scientific innovation, engineering precision, and environmental awareness. It spans a wide spectrum of material categories, from bio-based composites to recycled aggregates and cutting-edge nanotechnologies. The quest for these materials is driven not only by their capacity to enhance structural integrity and performance but also by their ability to substantially decrease carbon footprints and optimize resource utilization. Through ongoing research and interdisciplinary collaboration, these materials not only promise to redefine construction methodologies but also stand as a testament to the harmonious integration of human ingenuity with ecological preservation.

In recent years, the variability of political, economic, and social issues has led people to study the issue of sustainability. The study of sustainability in today's issues has led to the creation of knowledge of sustainable development in various aspects such as social, economic, and environmental, which the United Nations has also approved a commission in this regard in 2018 [1]. When we measure anything that can be changed, we are required to choose a quantity so that we and others can understand it and make its measurements concrete for the general public. We measure sustainability and sustainable development by determining the quantity or extent of development of a thing, its durability, and the tools needed to maintain it. The development of production and economy is one of the infrastructures required for sustainable development and solving environmental problems.

Sustainable methods consider a more general approach and sometimes go beyond the environmental study to the point where it is necessary to examine the buying and selling market in addition to the construction industry

 DOI: 10.1201/9781003496656-2

[2]. Buildings are considered one of the main causes of climate change by allocating 25% of the world's carbon dioxide. Due to this amount of impact, buildings can have a high capacity to improve and reduce carbon production, which can also change and improve energy consumption.

2.2 SUSTAINABLE BUILDINGS

Sustainable buildings stand as beacons of innovation and responsibility in the modern architectural landscape. They encapsulate a holistic construction approach, integrating design, materials, and systems that prioritize environmental stewardship, occupant health, and resource efficiency. At their core, these structures minimize negative impacts on the environment while enhancing the well-being of inhabitants. The ethos of sustainability pervades every stage, from conception to operation, fostering a symbiotic relationship between human habitats and the natural world. These buildings leverage cutting-edge technologies, renewable energy sources, passive design strategies, and eco-friendly materials, ushering in an era where architecture harmonizes seamlessly with nature. As a testament to conscious design and responsible construction, sustainable buildings serve as catalysts, inspiring a paradigm shift toward a built environment that not only sustains but also regenerates the ecosystems it inhabits.

Sustainable buildings exemplify the fusion of innovation and environmental consciousness, redefining the very fabric of construction practices. This chapter delves into the intricate realm of sustainable architecture, exploring the myriad strategies employed to engineer structures that balance functionality, durability, and environmental integrity. These buildings embrace passive design principles, optimizing orientation, natural ventilation, and daylighting to reduce energy consumption. Moreover, they employ green building materials, sourced responsibly and often incorporating recycled or rapidly renewable elements. Integrated systems for water conservation, waste management, and renewable energy generation further augment their sustainability quotient. By amalgamating architectural ingenuity with a commitment to ecological responsibility, sustainable buildings not only mitigate environmental impact but also serve as exemplars of resilience, adaptability, and human-centric design.

Renovating buildings and turning them into green and sustainable buildings can be an important factor in reducing harmful gases such as greenhouse gases and reducing other environmental and surface pollution [3]. One of the factors that can promote sustainable buildings is upgrading building materials and increasing their resistance against water and weather, which can be promising for reducing environmental pollution and strengthening the infrastructure of sustainable buildings [4]. Sustainable practices in architecture can reduce energy consumption to a great extent and emit less greenhouse gases into the air [5].

In the definition of sustainable or green buildings, it can be said as follows: buildings that are built to improve the quality of life of people and the environment. These buildings are designed and implemented to preserve natural resources and reduce damage to human habitations, from excessive wastage of water, energy, and materials and reducing damage to human health. Sustainable buildings have technologies to reduce energy loss and increase productivity, which is also effective in the real estate market. This effect causes developers to turn toward sustainable buildings to be able to cope with the demands made by the people to provide buildings with high technology. After extensive research and examination of the affected elements in the field of sustainable development, including green and sustainable buildings, they reached this conclusion. The need to expand this and reach sustainable buildings has a direct impact on the development of cities and a sustainable future [6–10].

2.2.1 Sustainable building costs

The economics of sustainable building represent a complex interplay between initial investment, long-term operational savings, and the broader socio-environmental impact. While the upfront costs of sustainable construction often appear higher than conventional methods, a deeper analysis reveals a multifaceted financial landscape. Sustainable buildings prioritize energy efficiency, resource conservation, and reduced environmental footprint, influencing operational expenses significantly over their lifecycle. Though initial construction expenses may outweigh traditional counterparts, long-term benefits emerge through lower utility bills, reduced maintenance, and enhanced occupant health and productivity. Moreover, government incentives, rebates, and increasingly competitive pricing for sustainable materials and technologies are gradually narrowing the cost gap. Comprehensive life cycle assessments encompassing construction, operations, maintenance, and end-of-life considerations underscore the holistic financial picture, illustrating the compelling economic case for sustainable buildings beyond their apparent upfront expenses.

Cost efficiency in sustainable buildings is a multifaceted paradigm that transcends the conventional perspective of immediate financial outlay. It encompasses a strategic shift toward investments that prioritize resilience, resource optimization, and environmental responsibility. Sustainable construction fosters a redefinition of value, where expenditures align with enhanced performance, reduced environmental impact, and improved quality of life for occupants. Initiatives like green financing, which consider environmental benefits alongside financial returns, are reshaping the economic landscape of sustainable building projects. Furthermore, advances in technology and increasing market demand are steadily driving down costs associated with sustainable materials and technologies. As the industry continues to innovate and refine best practices, the economic argument for sustainable buildings becomes increasingly compelling, positioning them not merely as costlier alternatives but as prudent, forward-thinking investments that yield enduring economic, environmental, and social dividends.

In addition, it should be noted that the project of green and sustainable buildings still faces problems such as the high cost of implementation and construction [11]. The real estate developers are looking to place the green building industry among the current construction industries so that they can eventually replace this industry with the previous one. In this sense, researchers believe that while the price of suitable mechanical systems is high, sustainable building materials are still efficient [12, 13].

2.3 CATEGORIES OF SUSTAINABLE BUILDING MATERIALS

The architecture and construction industries stand at the precipice of transformation, compelled by an urgent mandate to reevaluate material choices and construction practices in response to pressing environmental concerns. At the heart of this evolution lies a spectrum of categories encapsulating sustainable building materials, a multifaceted array that converges nature-inspired innovation, engineering ingenuity, and resource-conscious design. These categories form the pillars upon which a sustainable built environment is being sculpted, each offering distinct characteristics, advantages, and contributions to the overarching goal of mitigating environmental impact while fostering resilient and efficient structures. From renewable and recycled materials, embodying the ethos of resource efficiency and circularity, to innovative composites that seamlessly blend natural elements with cutting-edge technology, these categories exemplify a paradigm shift in material science. Furthermore, the spectrum encompasses energy-efficient and insulating materials that redefine thermal comfort and energy conservation within built spaces. This comprehensive exploration delves into these categories, unraveling their significance, properties, and transformative potential in shaping a built environment that harmonizes with nature, setting the stage for a discourse that redefines construction methodologies while championing environmental stewardship.

2.3.1 Renewable and recycled materials

Sustainable building materials encompass a diverse array of categories, with renewable and recycled materials standing as cornerstones of environmentally conscious construction. Renewable materials derive from replenishable sources, showcasing the regenerative capacity of nature. Bamboo, for instance, known for its rapid growth and durability, serves as an exemplary renewable material for various structural applications. Additionally, recycled materials champion the concept of circularity by repurposing waste streams into valuable construction resources. Recycled steel, reclaimed wood, and glass cullet exemplify this category, embodying the principles of resource efficiency and waste reduction. These materials not only mitigate the environmental burden associated with extraction and manufacturing but also possess properties that often rival or surpass those of traditional counterparts, fostering a sustainable approach without compromising structural integrity or functionality.

2.3.2 Innovative composites

Another pivotal category within sustainable building materials comprises innovative composites that blend natural elements with advanced engineering. Bio-based composites, fashioned from renewable resources like agricultural byproducts or natural fibers, offer versatility and strength while significantly reducing embodied energy and carbon footprint. Engineered timber, such as cross-laminated timber (CLT), emerges as a sustainable alternative to conventional building materials, exhibiting exceptional strength-to-weight ratios and sequestering carbon within its structure. Advanced polymers, derived from renewable feedstocks or designed for biodegradability, showcase immense potential in diverse applications, encapsulating durability and eco-friendliness. These composites exemplify the fusion of nature-inspired solutions with cutting-edge technology, redefining construction materials by prioritizing sustainability without compromising performance.

2.3.3 Energy-efficient and insulating materials

Energy efficiency stands as a fundamental criterion in sustainable building materials, driving innovation in insulating systems and energy-efficient components. High-performance insulation materials, such as expanded cork, aerogels, or recycled denim, play a pivotal role in reducing heat transfer and minimizing energy consumption for heating and cooling. Phase-change materials, capable of storing and releasing heat, contribute to passive temperature regulation within buildings, optimizing comfort while curbing reliance on mechanical heating and cooling systems. Smart glazing technologies, encompassing electrochromic or thermochromic glass, dynamically control light transmission and heat gain, enhancing daylighting and reducing energy demands for artificial lighting and climate control. These materials not only elevate indoor comfort but also significantly contribute to energy conservation and operational cost savings, underlining their essential role in sustainable building design for optimal thermal performance and reduced environmental impact

2.4 GEOGRAPHICAL AREAS

Sustainable building design doesn't exist in a vacuum; it's deeply intertwined with the geographical context in which structures are erected. Geographic areas play a pivotal role in shaping the parameters and possibilities of sustainable construction. From arid regions necessitating passive cooling techniques to coastal areas mandating resilience against rising sea levels, each locale demands a tailored approach to sustainable building. Understanding local climates, available resources, cultural practices, and environmental vulnerabilities is paramount. Strategies such as utilizing indigenous materials, harnessing renewable energy sources specific to the region, and implementing designs that respond to local climate patterns are essential. Geographic considerations not only dictate the structural elements of sustainable buildings but also inform the socio-environmental impacts, aligning construction practices with the unique needs and challenges of specific landscapes.

The establishment of green building councils in some countries indicates that they will grow the sustainable building movement and bring it to the stage of global implementation. These councils include the United States Green Building Council (USGBC), the Hong Kong Green Building Council (HKGBC), the Green Building Council of India (IGBC), and the Saudi Green Building Council (SGBC). With this, we will see the ranking and construction of standard sets for the implementation of sustainable buildings. With policies and rankings, councils can bring the optimal use of energy, water, material efficiency, and waste reduction and increase material efficiency closer to their best state [14, 15]. In addition to the encouragement of the World Council, the European Union has also signed an agreement on the economic sustainability of Europe in the field of implementing green and sustainable buildings [16].

Europe plans to reach a neutral and ideal climate in another 27 years, that is, by 2050. Also, the government of Cyprus has approved the new governance system based on the implementation and adherence to the European Green Agreement mentioned above and the implementation of the National Energy Plan (NECP) in the Cyprus Energy Union for 2020. This means that Cyprus undertakes to increase the use of renewable energy by more than 20% by 2030 by implementing sustainable buildings, which reduces the final energy consumption by more than 32% [17].

Saudi Arabia also has a good perspective in the direction of sustainable buildings, which can be considered aligned with FSCP programs. Saudi Arabia intends to create buildings in 17 of its big cities that have sufficient parameters in terms of compatibility with the environment, according to the standards of international organizations and councils of sustainable buildings. Saudi Arabia's growing economic capacity allows it to create a better balance in its urban development. Among the cities in which Saudi Arabia plans to implement the sustainable building program, Taif can be considered the most diverse and populated city. Because the city of Taif has a good relative height and attracts many tourists in terms of entertainment. Despite the environmental problems, Saudi cities have a good capacity to upgrade green buildings [18].

2.5 ADVANCEMENTS AND TECHNOLOGIES

The pursuit of new materials for sustainable buildings represents an intricate tapestry woven from the threads of scientific ingenuity, engineering acumen, and an unwavering commitment to environmental preservation. From nanotechnology and 3D printing revolutionizing material design and manufacturing to biodegradable polymers and carbon-negative materials reshaping the ecological footprint of structures, these technological leaps collectively represent a symphony of possibilities, transforming the construction paradigm toward a future where sustainability and resilience harmoniously converge. This comprehensive journey navigates through the complexities and nuances of these advancements, illuminating their transformative potential and their pivotal role in shaping a built environment that not only endures but thrives in symbiosis with our planet.

Advancements and technologies in new materials for sustainable buildings epitomize a pioneering frontier in the construction industry, propelled by a convergence of scientific breakthroughs, engineering prowess, and a steadfast commitment to environmental stewardship. This realm is characterized by a spectrum of transformative innovations, encompassing an array of groundbreaking materials and technologies. Nanotechnology, for instance, offers unprecedented potential in enhancing material properties at a molecular scale, fostering superlative durability, self-cleaning surfaces, and enhanced structural strength. Additionally, 3D printing techniques revolutionize manufacturing processes, enabling intricate designs and customized building components while minimizing material waste. Emerging advancements in smart materials, responsive to environmental stimuli such as temperature or light, pave the way for structures that dynamically adapt to changing conditions, optimizing energy efficiency and occupant comfort. These cutting-edge innovations collectively represent a tapestry of possibilities, reshaping the very fabric of construction materials and methodologies toward a more sustainable and resilient built environment.

Material science breakthroughs serve as catalysts propelling sustainable building materials toward unprecedented heights. Biodegradable polymers, derived from renewable sources, offer an eco-friendly alternative to conventional plastics, exhibiting impressive strength and biocompatibility while minimizing environmental impact. Furthermore, carbon-negative materials, designed to sequester more carbon dioxide during their lifecycle than emitted during production, showcase immense promise in mitigating greenhouse gas emissions associated with construction. Moreover, advancements in bio-inspired materials draw inspiration from nature's design principles, mimicking the resilience and efficiency found in natural systems. These innovations harness biomimicry to create materials capable of self-repair, adaptive thermal regulation, and enhanced structural performance, showcasing a fusion of cutting-edge science with ecological consciousness.

Sustainable technologies integrated into building materials mark a pivotal shift in construction paradigms, fostering a dynamic relationship between functionality, efficiency, and environmental responsibility. Photovoltaic-integrated materials, such as solar roof tiles or translucent solar panels, exemplify the seamless integration of renewable energy generation within building components, offsetting energy demands and reducing reliance on traditional power sources. Additionally, advancements in aerogel-based insulation systems offer ultra-efficient thermal barriers, significantly reducing heat transfer and enhancing building envelope performance. Furthermore, the convergence of sensor technologies and smart materials facilitates the development of responsive building systems capable of real-time monitoring, optimizing energy consumption, and enhancing occupant comfort. These integrated sustainable technologies underscore a holistic approach to material innovation, bridging the gap between technological advancement and environmental sustainability within the built environment.

2.6 CASES INVESTIGATED IN THE FIELD OF SUSTAINABLE BUILDING

One of the effective factors in the implementation of green and sustainable buildings is the level of acceptance by the real estate industry. Examining this class can make the economic explanations, implementation restrictions, and detailed opinions of the building consumers, which have a significant and final role in the implementation of green buildings, clearer for us. In addition to working on creating green and sustainable buildings, we can create an understanding of the importance of its use in the minds of people and consumers [19].

After studying various aspects of development such as economic, social, and climate which was done with the help of TBL theory, it can be suggested that sustainable design and the creation of sustainable buildings can ultimately reduce operating costs by increasing the number of people's assets and also by improving the welfare and comfort of the public, making the final part acceptable to a good extent [20]. The rate of stabilization of buildings in the 12 months of the year is only 1%. Considering this and increasing environmental awareness, we need to speed up the conversion of buildings and their renovation so that we can reduce the production of pollutants on the ground faster [21–23].

One of the ways to reduce environmental pollutants in high areas and sustainable cities is green roof technology, which can be called a new and innovative solution [24]. Green roofs can create security and stability against climate change [25]. Green roofs offer a range of environmental benefits. They can reduce noise pollution and help cities adapt to climate change by mitigating the urban heat island effect [26–28]. Additionally, green roof technology helps reduce carbon dioxide, lower air and indoor temperatures in buildings, improve water recycling, and promote sustainable development [29–32].

One approach to reducing energy consumption in buildings and mitigating environmental problems is the concept of green buildings. These structures are designed to minimize energy loss, reduce flood risk, and prevent pollution. Green building principles can be applied to both new construction and the renovation of existing buildings, transforming them into more sustainable spaces [33–36].

Installing solar and photovoltaic panels on the roofs of buildings creates an insulating cover against temperature and also provides a suitable platform for the growth of smart technologies in buildings. Some governments support these projects through loans and projects that support the design and implementation of solar panels and photovoltaic systems. Some countries, like Cyprus, have made improvement projects mandatory for themselves [37, 38]. Some countries, like Iran, lend 70% of the total cost of the project on the roofs of buildings to people who are eager to install solar panels, and these loans have a very low interest and repayment of several years. This will help the people as an asset to earn income from selling their excess electricity to the government after a few years and after the return on investment. In some regions of the world, including Europe, more than a third of buildings are over 50 years old, and most of these statistics are inefficient in terms of energy and sustainability.

The promotion of sustainability standards in a building is one of the most important parts of design, which can be called passive building design. Passive means the use of natural and renewable or so-called green resources to reduce energy loss and pollution, maintaining the appropriate temperature range in the building. This increase in sustainability leads to the reduction of two-thirds of current energy in tropical countries such as Saudi Arabia [39]. Recently, the principles of passive temperature adjustment in hot and humid areas have been investigated. Among these things, we can refer to insulations, natural ventilation systems, the glass industry, and creating shade in areas that must be prevented from absorbing heat [40, 41].

2.7 BENEFIT OF NEW MATERIALS FOR SUSTAINABLE BUILDINGS

New materials designed for sustainable buildings herald a paradigm shift in the construction landscape, offering a myriad of profound benefits that extend far beyond mere structural integrity. These materials serve as catalysts in fostering environmentally conscious and resilient built environments while significantly mitigating the ecological footprint of construction practices (Table 2.1).

Table 2.1 Benefit of new materials in sustainable buildings

Benefit	*Key Points*
Resource Conservation	New sustainable materials, sourced from renewable or recycled resources, reduce the strain on natural resources by minimizing extraction and depletion. This includes materials like bamboo, cork, or recycled steel, which require fewer raw materials for production.
Reduced Environmental Footprint	These materials often possess lower embodied energy and carbon footprint, contributing to mitigating greenhouse gas emissions during their manufacturing processes. Bio-based polymers or low-energy production techniques exemplify this aspect, reducing the environmental burden associated with construction.
Enhanced Insulation	New materials designed for optimal insulation, such as aerogels or phase-change materials, significantly reduce heat transfer, enhancing a building's thermal performance. This translates to reduced reliance on mechanical heating and cooling systems, leading to lower energy consumption and operational costs.
Renewable Energy Integration	Some sustainable materials, like photovoltaic-integrated surfaces or translucent solar panels, facilitate renewable energy generation within building components. This integration offsets energy demands and promotes self-sufficiency.
Durability and Longevity	Innovative materials often exhibit enhanced durability, resilience, and longevity. Engineered timber, for instance, offers exceptional strength-to-weight ratios, reducing maintenance needs and extending a building's lifespan.
Adaptability and Innovation	Self-healing materials or adaptive technologies integrated into building materials showcase innovation, contributing to structures that can repair themselves or adapt to changing environmental conditions, ensuring long-term functionality.
Improved Indoor Air Quality	Low-emission materials and finishes minimize indoor pollutants, fostering better indoor air quality. VOC-free paints and formaldehyde-free composites contribute to healthier indoor environments.
Occupant Comfort	Materials regulating indoor humidity levels or enhancing natural lighting contribute to occupant comfort, productivity, and well-being. Bio-based materials often create a more pleasant and conducive indoor climate.

2.8 OPTIMIZING BUILDINGS WITH INSULATING MATERIALS

Insulation stands as a cornerstone of building performance, regulating indoor temperatures, reducing energy consumption, and mitigating heat transfer through walls, roofs, and floors. The selection and application of insulating materials, ranging from traditional fiberglass and foam boards to innovative aerogels and phase-change materials, play a pivotal role in minimizing thermal bridging and heat loss, thereby enhancing a structure's overall energy performance. This optimization not only translates into reduced reliance on mechanical heating and cooling systems but also curtails operational costs and carbon emissions, fostering a more sustainable and cost-effective building operation over its lifecycle. Moreover, beyond energy conservation, effective insulation augments occupant comfort by maintaining consistent indoor temperatures, reducing drafts, and mitigating moisture-related issues, underscoring the multifaceted benefits of strategically integrating insulating materials into building design and construction.

Furthermore, the optimization of buildings with insulating materials extends beyond mere thermal regulation to encompass a holistic approach that integrates environmental consciousness and structural efficiency. Contemporary advancements in insulating technologies prioritize materials that not only exhibit exceptional thermal performance but also align with sustainability goals. Bio-based insulation materials sourced from renewable resources, recycled content insulation, and eco-friendly aerogels exemplify this trend, embodying a commitment to reducing embodied energy and environmental impact. This optimization journey involves meticulous considerations of insulation R-values, installation techniques, and compatibility with various construction systems, ensuring that insulating materials seamlessly integrate into building envelopes to maximize efficiency. As buildings evolve to meet higher energy performance standards, the strategic application of insulating materials emerges as an indispensable element in the pursuit of sustainable, resilient, and comfortable built environments that harmonize with ecological responsibility.

The last covering layer of buildings is an insulation against heat, noise, and moisture that separates the internal and external environment of the building from each other and reduces the heat transfer rate. This insulating layer can be composed of several layers, each with a different material and properties [42]. Cement with lower density can have a lower heat transfer rate. Autoclave aeration can reduce the temperature load that enters the building by increasing the porosity of AAC in the material [43]. Sustainable buildings that are managed by mechanical, electronic, and electromechanical devices with the help of computers are known as intelligent buildings or Sustainable Automated Building Systems (SABS) [44]. Concrete Ultra-high-performance concrete (UHPC) has been considered in the construction of sustainable houses in recent years due to its high strength and specific mechanical properties. However, this type of concrete is expensive and it causes pollution, which has led to increased restrictions around its use in construction [45]. The use of mixed concretes has led to various optimizations in the cement industry and green buildings, the presence of these fine grains in concrete reduces the carbon produced by cement. Eight percent of carbon emissions in the world are related to cement industries [46].

2.9 ENVIRONMENTAL IMPACT OF NEW MATERIALS ON SUSTAINABLE BUILDINGS

The environmental impact of new materials on sustainable buildings embodies a profound departure from traditional construction practices, heralding a transformative era characterized by conscientious resource stewardship and reduced ecological footprints. These materials, often sourced from renewable or recycled origins, mitigate resource depletion, significantly lowering the strain on ecosystems. Embedded within their adoption is a testament to reduced embodied energy and diminished carbon footprints during production, illuminating a path toward curbing greenhouse gas emissions and lessening environmental strain [47–52]. Moreover, these materials foster improved indoor air quality by minimizing pollutants, advocating for healthier living environments. Additionally, their emphasis on recyclability and biodegradability orchestrates a shift toward circularity, effectively reducing waste generation and landfill contributions which promote sustainable manufacturing [53–58]. In the grand tapestry of sustainable construction, these materials stand as pivotal agents, catalyzing a harmonious synergy between human habitation and the natural world, culminating in a built environment that champions environmental responsibility and resilience. Table 2.2 outlines the environmental impact of new materials on sustainable buildings.

Table 2.2 Environmental impact of new materials on sustainable buildings [47–55]

Impact	*Key Points*
Recycled Content	Utilizing recycled content, such as reclaimed wood or recycled steel, minimizes waste and reduces the need for additional raw materials, decreasing environmental impact.
Renewable Sources	Materials sourced from renewable resources, like bamboo or cork, reduce reliance on finite resources, preserving ecosystems and biodiversity.
Embodied Energy	New sustainable materials often have lower embodied energy, requiring less energy for extraction, manufacturing, and transportation, contributing to reduced environmental impact.
Carbon Footprint	Life cycle assessments (LCAs) of these materials frequently reveal lower carbon footprints compared to conventional materials, lessening greenhouse gas emissions.
Reduced Pollutants	Sustainable materials often involve fewer harmful substances or emissions during production, enhancing indoor air quality and reducing overall environmental pollution.
Waste Reduction	Emphasis on recyclability and biodegradability minimizes waste generation, aligning with principles of a circular economy and reducing landfill contributions.
Ecosystem Integrity	Materials from sustainable sources contribute to preserving ecosystems by reducing deforestation, habitat destruction, and ecosystem degradation.
Climate Resilience	Building with sustainable materials contributes to climate resilience, as some materials, like engineered timber, sequester carbon and mitigate climate change impacts.

2.10 FUTURE TRENDS AND OUTLOOK

The future of sustainable materials converges at the intersection of technological innovation, biomimicry, circular economy principles, and a commitment to minimizing environmental impact. The envisioned advancements hold the potential to revolutionize construction practices, yielding structures that not only endure but harmonize with nature, paving the way for a more sustainable and resilient built environment.

The future of sustainable materials is poised to witness a convergence with nanotechnology, ushering in an era of smart, responsive materials that dynamically adapt to environmental stimuli. Nanomaterials offer exceptional structural properties, allowing for increased strength, durability, and flexibility while utilizing minimal resources. These materials enable the development of smart coatings capable of self-cleaning, self-healing, or adapting to temperature variations. Further advancements may lead to the integration of sensors within building materials, creating structures that can actively monitor and respond to changing conditions, optimizing energy use, and enhancing occupant comfort. The potential of nanotechnology extends to enhancing the performance of renewable materials as well, unlocking their full potential in construction by augmenting their strength and durability [59–64].

The exploration of biological systems continues to inspire innovative materials by mimicking nature's designs and processes. Biomimicry in material science involves replicating structures and functionalities found in nature, leading to materials with remarkable resilience, adaptability, and sustainability. Examples include materials inspired by spider silk for exceptional strength or biomimetic membranes for water purification. This trend is likely to expand further, unlocking new avenues for materials that emulate nature's efficiency, resourcefulness, and regenerative capabilities. The integration of biomimetic principles into construction materials may lead to structures that possess self-regulating capabilities, enhanced energy efficiency, and reduced environmental impact.

Future sustainable materials are increasingly expected to embody carbon-negative attributes, actively sequestering more carbon dioxide than their production emits. Innovations in materials, such as carbon-negative concrete or engineered timber that sequesters carbon, showcase the immense potential in combating climate change by turning buildings into carbon sinks. Furthermore, a deeper embrace of circular economy practices within material production and construction is anticipated. This approach prioritizes closed-loop systems where materials are reused, recycled, or repurposed, minimizing waste and resource depletion. Materials designed with end-of-life considerations, ensuring their easy disassembly and reutilization, would transform construction into a regenerative and waste-free industry [65–67].

The future trajectory of sustainable materials and sustainable manufacturing emphasizes advancements in recycling technologies, enabling the conversion of diverse waste streams into high-quality construction materials. Innovative recycling processes, such as chemical or biological recycling, hold promise in transforming waste plastics or construction debris into durable and functional building components. Additionally, the evolution of bio-based materials, derived from renewable feedstocks like algae or fungi, presents an avenue for sustainable construction. These materials possess unique properties and can be engineered to meet specific structural requirements, offering a renewable alternative to traditional construction materials.

2.11 CONCLUSION

The exploration of sustainable building materials represents a remarkable fusion of human ingenuity and environmental responsibility, mirroring our urgent global need to combat climate change. These innovations, ranging from bio-based composites to cutting-edge nanotechnologies, signify a profound departure from conventional practices toward ecologically conscious solutions.

Beyond mere structural enhancement, sustainable materials signify a commitment to reducing carbon footprints and optimizing resource utilization. This journey, fueled by interdisciplinary collaboration, embodies the harmonious integration of human intellect with ecological preservation. In the realm of sustainable buildings, innovation weaves together design, materials, and systems to prioritize environmental stewardship while enhancing occupant health. These structures leverage passive design strategies, renewable energy sources, and eco-friendly materials, acknowledging the influence of local climates on tailored construction approaches. Global initiatives standardizing sustainable building practices signal a momentum toward energy and material efficiency, with considerations for waste reduction. While initial costs might seem higher, the long-term operational savings, coupled with government incentives and decreasing sustainable material costs, strengthen the financial argument for sustainable buildings. Varied categories of sustainable materials symbolize a shift toward nature-inspired design principles, resilience, and heightened environmental responsibility. Technological advancements, including nanotechnology and smart materials, promise a future of seamlessly converged sustainability and resilience, reshaping construction methodologies.

In summary, the exploration of sustainable building materials charts a dynamic trajectory, converging innovation, environmental consciousness, and societal acceptance. This journey toward sustainable construction forecasts a future where structures thrive harmoniously with nature, promising a resilient and sustainable built environment for generations to come.

REFERENCES

1. Garren SJ (2018) Sustainability definitions, historical context, and frameworks. In *The Palgrave Handbook of Sustainability*. Berlin/Heidelberg: Springer.
2. Doyle TM (2015) *Environment and Politics*. Abingdon: Routledge.
3. Zhang GJ (2018) Facile "Spot-Heating" synthesis of carbon dots/carbon nitride for solar hydrogen evolution synchronously with contaminant decomposition. *Advanced Functional Materials* 28:1706462. [CrossRef]
4. Yang R, Yu R, Shui Z, Gao X, Han J, Lin G, Qian D, Liu Z, He Y (2020) Environmental and economical friendly ultra-high performance concrete incorporating appropriate quarry-stone powders. *Journal of Cleaner Production* 260. https://doi.org/10.1016/j.jclepro.2020.121112.
5. Nocentini K, Biwole P, Achard P (2018) Silica aerogel blankets as superinsulating material for developing energy efficient buildings. In *Sustainability through Energy-Efficient Buildings, by Amritanshu Shukla and Atul Sharma*, A Shukla, A Sharma, Editors, 151–164. Boca Raton: CRC Press.
6. Bontempi E et al (2021) Sustainable materials and their contribution to the Sustainable Development Goals (SDGs): A critical review based on an Italian example. *Molecules* 26:5.
7. Mahmoud A et al (2017) Energy and economic evaluation of green roofs for residential buildings in hot-humid climates. *Buildings* 7(4):30.
8. Mahmoud AS et al (2017) Energy and economic evaluation of green roofs for residential buildings in hot-humid climates. *Buildings* 7(2):30.
9. Omer MAB, Noguchi T (2020) A conceptual framework for understanding the contribution of building materials in the achievement of Sustainable Development Goals (SDGs). *Sustainable Cities and Society* 52(10186):9.
10. Sharma AK, Nigrawal A, Baredar P (2020) Sustainable development by constructing green buildings in India: A review. *Mater Today* 46:5329–5332.
11. Heinzle SL, Yip ABY, Xing MLY (2013) The influence of green building certification schemes on real estate investor behaviour: Evidence from Singapore. *Urban Studies* 50:1970–1987. [CrossRef]
12. Chegut A, Eichholtz P, Kok N (2019) The price of innovation: An analysis of the marginal cost of green buildings. *Journal of Environmental Economics and Management* 98:102248.
13. Nagrale SS (2020) Cost comparison between normal building and green building considering its construction and maintenance phase. *International Journal of Scientific Research and Engineering Development* 3:77–80.
14. Illankoon IMCS, Tam VWY, Le KN, Shen L (2017) Key credit criteria among international green building rating tools. *Journal of Cleaner Production* 164:209–220.
15. Shan M, Hwang B-G (2018) Green Building rating systems: Global reviews of practices and research efforts. *Sustainable Cities and Society* 39:172–180.
16. Tutak M, Brodny J, Bindzár P (2021) Assessing the level of energy and climate sustainability in the European union countries in the context of the European Green Deal strategy and agenda 2030. *Energies* 14:1767. [CrossRef]
17. Cyprus Action Plan (2020) Cyprus' integrated national energy and climate plan. Available online: https://energy.ec.europa.eu/ system/files/2020-01/cy_final_necp_main_en_0.pdf.

18. Katafygiotou M, Serghides D (2014) Analysis of structural elements and energy consumption of school building stock in Cyprus: Energy simulations and upgrade scenarios of a typical school. *Energy and Buildings* 72:8–16.
19. Fuerst FK (2014) Determinants of green building adoption. *Environment and Planning B: Planning and Design* 41:551–570. [CrossRef]
20. Goh CS, Chong H-Y, Jack L, Faris AFM (2019) Revisiting triple bottom line within the context of sustainable construction: A systematic review. *Journal of Cleaner Production* 252:119884. [CrossRef]
21. European Union (2023) Official website of EU. Available online: https://energy.ec.europa.eu/topics/energy-efficiency/energyefficient-buildings/energy-performance-buildings-directive_en (accessed on 12 October 2023).
22. Skordoulis M, Ntanos S, Kyriakopoulos GL, Arabatzis G, Galatsidas S, Chalikias M (2020) Environmental innovation, open innovation dynamics and competitive advantage of medium and large-sized firms. *Journal of Open Innovation: Technology, Market, and Complexity* 6:195.
23. Khan MK, Teng J-Z, Khan MI, Khan MO (2019) Impact of globalization, economic factors and energy consumption on CO2 emissions in Pakistan. *Science of the Total Environment* 688:424–436.
24. Susca T (2019) Green roofs to reduce building energy use? A review on key structural factors of green roofs and their effects on urban climate. *Building and Environment* 162(10627):3.
25. Dauda I, Alibaba HZ (2020) Green roof benefits, opportunities and challenges. *International Journal of Civil and Structural Engineering Research* 7:106–112.
26. Andric I, Kamal A, Al-Ghamdi SG (2020) Efficiency of green roofs and green walls as climate change mitigation measures in extremely hot and dry climate: Case study of Qatar. *Energy Reports* 6:2476–2489.
27. Maghrabi A, Alyamani A, Addas A (2021) Exploring pattern of Green Spaces (GSs) and their impact on climatic change mitigation and adaptation strategies: Evidence from a Saudi Arabian City. *Forests* 12(5):629.
28. Shushunova N, Feoktistova O, Shushunova T (2021) Efficiency of reducing noise pollution by using the greening system of buildings. IOP Conf Ser 1079(4):042001.
29. Ibrahim VAR (2018) Roof planting as a tool for sustainable development in residential buildings in Egypt. *Aerospace Research Communications Publication* 2(4):544.
30. Pradhan S, Al-Ghamdi SG, Mackey HR (2019) Greywater recycling in buildings using living walls and green roofs: A review of the applicability and challenges. *Science of the Total Environment* 652:330–344.
31. Shafique M, Kim R, Rafiq M (2018) Green roof benefits, opportunities and challenges—A review. *Renewable and Sustainable Energy Reviews* 90:757–773.
32. Wahba S, Kamil B, Nassar K, Abdelsalam A (2019) Green envelop impact on reducing air temperature and enhancing outdoor thermal comfort in arid climates. *Civil Engineering Journal* 5:5.
33. Asdrubali F, Evangelisti L, Guattari C (2019) Green roof for zero energy buildings: A pilot project. IOP Conf Ser 609(7):072011.
34. Ayata T, Erdemir D, Ozkan OT (2017) An investigation for predicting the effect of green roof utilization on temperature decreasing over the

roof surface with gene expression programming. *Energy and Buildings* 139:254–262.

35. Ebadati M, Ehyaei MA (2018) Reduction of energy consumption in residential buildings with green roofs in three different climates of Iran. *Advances in Building Energy Research* 14:1.
36. Wahba SM, Kamel BA, Nassar KM, Abdelsalam AS (2018) Effectiveness of green roofs and green walls on energy consumption and indoor comfort in arid climates. *Civil Engineering Journal* 4(10):2284.
37. Katafygiotou M, Serghides D (2014) Analysis of structural elements and energy consumption of school building stock in Cyprus: Energy simulations and upgrade scenarios of a typical school. *Energy and Buildings* 72:8–16. [CrossRef]
38. Serghides D, Dimitriou S, Katafygiotou M (2016) Towards European targets by monitoring the energy profile of the Cyprus housing stock. *Energy and Buildings* 132:130–140. [CrossRef]
39. Khoukhi M, Shaaban AK, Khatib OA, Cameselle C (2020) Passive environmental design of an eco-house in the hot-humid climate of the middle east: A qualitative approach. *Cogent Engineering* 7(1): 1837410.
40. Shafique M, Xue X, Luo X (2020) An overview of carbon sequestration of green roofs in urban areas. *Urban For Urban Green* 47(12651):5.
41. Xiao Z, Ge H, Lacasse MA, Wang L, Zmeureanu R (2023) Naturebased solutions for carbon neutral climate resilient buildings and communities: A review of technical evidence, design guidelines, and policies. *Buildings* 13:1389. https://doi.org/10.3390/buildings13061389.
42. Aye L, Jayalath A (2018) Passive and low energy buildings. In *Sustainability through EnergyEfficient Buildings, by Amritanshu Shukla and Atul Sharma*, A Shukla, A Sharma, Editors, 73–88. FL: Taylor & Francis.
43. Teng L, Addai-Nimoh A, Khayat KH (2023) Effect of lightweight sand and shrinkage reducing admixture on structural build-up and mechanical performance of UHPC. *Journal of Building Engineering* 68. https://doi.org/10.1016/j.jobe.2023.106144.
44. Chaturvedi AK, Jain S, Gupta D, Singh M (2018) Advances in energy-efficient buildings for new and old buildings. In *Sustainability through Energy-Efficient Buildings*, A Shukla, A Sharma, Editors, 235–258. FL: Taylor & Francis.
45. Hamada HM, Shi J, Abed F, Al Jawahery MS, Majdi A, Yousif ST (2023) Recycling solid waste to produce eco-friendly ultra-high performance concrete: A review of durability, microstructure and environment characteristics. *Science of The Total Environment* 876:162804. https://doi.org/10.1016/j.scitotenv.2023.162804.
46. Wang X, Yu R, Song Q, Shui Z, Liu Z, Wu S, Hou D (2019) Optimized design of ultra-high performance concrete (UHPC) with a high wet packing density. *Cement and Concrete Research* 126. https://doi.org/10.1016/j.cemconres.2019.105921.
47. Meena CS, Kumar A, Jain S, Rehman AU, Mishra S (2022) Innovation in green building sector for sustainable future. *Energies* 15:6631. https://doi.org/10.3390/en15186631.
48. Kumar A, Singla Y, Namboodri T (2024) Globalization and international issues in sustainable manufacturing. In *Sustainability in Smart Manufacturing:*

Trends, Scope, and Challenges, S Shah, H Nautiyal, G Gugliani, A Kumar, T Namboodri, YK Singla, Editors. Boca Raton, FL: CRC Press; Chapter 01. ISBN 9781032 740713.
49. Shah S, Nautiyal H, Gugliani G, Kumar A, Namboodri T, Singla YK, Editors (2024) *Sustainability in Smart Manufacturing: Trends, Scope, and Challenges*. Boca Raton, FL: CRC Press. ISBN 9781032 740713.
50. Singh VP, Dwivedi, A, Karn, A, Kumar A, Singh, S, Srivastava, S, Srivastava, K (2022) Nanomanufacturing and design of high-performance piezoelectric nanogenerator for energy harvesting. In *Nanomanufacturing and Nanomaterials Design: Principles and Applications*, S Singh, SK Behura, A Kumar, K Verma, Editors, 241–272, Boca Raton, FL: CRC Press; Chapter 15. https://doi.org/10.1201/9781003220602-15.
51. SrivastavaS,VermaD,ThusooS,KumarA,SinghVP(2022)Nanomanufacturing for energy conversion and storage devices. In *Nanomanufacturing and Nanomaterials Design: Principles and Applications*, S Singh, SK Behura, A Kumar, K Verma, Editors, 165–174. Boca Raton, FL: CRC Press; Chapter 10. https://doi.org/10.1201/9781003220602-10.
52. Bansal B, Bhardwaj HK, Sharma VK, Kumar A (2022) Static and dynamic behavior analysis of Al-6063 alloy using modified Hopkinson bar. In *Additive Manufacturing in Industry 4.0: Methods, Techniques, Modeling and Nano Aspects*, VK Sharma, A Kumar, M Gupta, V Kumar, DK Sharma, SK Sharma, Editors, 107–124. Boca Raton, FL: CRC Press; Chapter 6. https://doi.org/10.1201/9781003360001-6.
53. Sharma VK, Kumar V, Joshi RS, Kumar A (2022) Effect of REOs on tribological behavior of aluminum hybrid composites using ANN. In *Additive Manufacturing in Industry 4.0: Methods, Techniques, Modeling and Nano Aspects*, VK Sharma, A Kumar, M Gupta, V Kumar, DK Sharma, SK Sharma, Editors, 153–168. Boca Raton, FL: CRC Press; Chapter 9. https://doi.org/10.1201/9781003360001-9.
54. Sharma VK, Kumar A, Gupta M, Kumar V, Sharma DK, Sharma S (2022) *Additive Manufacturing in Industry 4.0: Methods, Techniques, Modeling, and Nano Aspects*. Taylor & Francis (CRC Press). https://doi.org/10.1201/9781003360001.
55. Kumar A, Kumar A, Kumar A (2023) *Laser Based Technologies for Sustainable Manufacturing*. Taylor & Francis (CRC Press). ISBN: 9781003402398. https://doi.org/10.1201/9781003402398.
56. Kumar R, Prasad A, Kumar A (2023) *Sustainable Smart Manufacturing Processes in Industry 4.0*. Boca Raton, FL: CRC Press. ISBN: 9781003436072. https://doi.org/10.1201/9781003436072.
57. Chakraborty G, Pandey V, Prasad A, Kumar A (2023) Introduction to sustainable manufacturing for industries 4.0. In *Sustainable Smart Manufacturing Processes in Industry 4.0*, R Kumar, A Prasad, A Kumar, Editors, 1–17. Boca Raton, FL: CRC Press; Chapter 01. https://doi.org/10.1201/9781003436072-1.
58. Pathak A, Kumar A, Kumar A, Kumar A, King M FL (2023) Application of laser technology in the mechanical and machine manufacturing industry. In *Laser-based Technologies for Sustainable Manufacturing*, A Kumar, A Kumar, A Kumar, Editors, 107–155. Boca Raton, FL: CRC Press; Chapter 06. https://doi.org/10.1201/9781003402398-6.

59. Verma V, Thangavel S, Dutt N, Kumar A, Weerasinghe R, Editors (2024) *Highly Efficient Thermal Renewable Energy Systems: Design, Design, Optimization and Applications*. Boca Raton, FL: CRC Press. ISBN 9781032595641.
60. Awasthi MK, Kumar A, Dutt N, Singh S, Editors (2024) *Computational Fluid Flow and Heat Transfer: Advances, Design, Control and Applications*. Boca Raton, FL: CRC Press. ISBN 9781032603186.
61. Verma V, Meena CS, Thangavel S, Kumar A, Choudhary T, Dwivedi G (2023) Ground and solar assisted heat pump systems for space heating and cooling applications in the northern region of India – A study on energy and CO2 saving potential. *Sustainable Energy Technologies and Assessments* 59:103405. ISSN 2213-1388. https://doi.org/10.1016/j.seta.2023.103405.
62. Meena CS, Prajapati AN, Kumar A, Kumar M (2022) Utilization of solar energy for water heating application to improve building energy efficiency: An experimental study. *Buildings* 12:2166. https://doi.org/10.3390/buildings12122166.
63. Dewangan AK, Moinuddin SQ, Cheepu M, Sajjan SK, Kumar A (2023) Thermal energy storage: Opportunities, challenges and future scope. In *Thermal Energy Systems: Design, Computational Techniques and Applications*, A Kumar, VP Singh, CS Meena, N Dutt, Editors, 17–28. Boca Raton, FL: CRC Press; Chapter 2. http://dx.doi.org/10.1201/9781003395768-2.
64. Verma V, Thangavel S, Dutt N, Kumar A, Weerasinghe R (2024) Recent development of thermal energy storage: Solar, geothermal and hydrogen energy. In *Highly Efficient Thermal Renewable Energy Systems: Design, Design, Optimization and Applications*, V Verma, S Thangavel, N Dutt, A Kumar, R Weerasinghe, Editors. Boca Raton, FL: CRC Press; Chapter 01. ISBN 9781032595641.
65. Ahmadizadeh M, Heidari M, Thangavel S, Naamani EA, Khashehchi M, Verma V, Kumar A (2024) Technological advancements in sustainable and renewable solar energy systems. In *Highly Efficient Thermal Renewable Energy Systems: Design, Design, Optimization and Applications*, V Verma, S Thangavel, N Dutt, A Kumar, R Weerasinghe, Editors. Boca Raton, FL: CRC Press; Chapter 02. ISBN 9781032595641.
66. Chitt M, Thangavel S, Verma V, Kumar A (2024) Green hydrogen productions: Methods, designs and smart applications. In *Highly Efficient Thermal Renewable Energy Systems: Design, Design, Optimization and Applications*, V. Verma, S. Thangavel, N. Dutt, A. Kumar, R. Weerasinghe, Editors. Boca Raton, FL: CRC Press. Chapter 16. ISBN 9781032595641.
67. Khashehchi M, Thangavel S, Rahmanivahid P, Heidari M, Moazzeni T, Verma V, Kumar A (2024) Solar desalination techniques: Challenges and opportunities. In *Highly Efficient Thermal Renewable Energy Systems: Design, Design, Optimization and Applications*, V Verma, S Thangavel, N Dutt, A Kumar, R Weerasinghe, Editors. Boca Raton, FL: CRC Press; Chapter 19. ISBN 9781032595641.

Chapter 3

Smart building materials for energy and heat transfer optimization

Shweta Tripathi, Chandrajita Chakraborty, Bhupender Singh, Preet Kaur, and Ashok Kumar Yadav

3.1 INTRODUCTION

Issues over the release of greenhouse gases (GHG) and, as a result, a global surge of temperature, leading to global warming, have arisen due to a tremendous surge in the consumption of energy across the globe. It has been suggested that a substantial amount of energy utilization and harmful emissions of gases originating from fossil fuels come from the construction industry [1–44]. One of the primary causes of this surge in energy consumption and emissions globally is the enhanced need for heating and cooling [5]. Up to 80% of the European Union's total greenhouse gas emissions are attributable to energy-associated releases [6], whereas, on the other hand, the construction industry uses around 40% of the final energy usage in total of the European Union [7].

To help with the worldwide difficulties of minimizing energy use in buildings and greenhouse gas emissions, a variety of relevant technologies, such as passive engineering and active operational solutions, have been proposed and evolved. The energy-saving technologies could be categorized into a number of areas, such as the usage of insulated materials in external walls and planning of the geometry of the outer layout of the building, all of these coming under passive designing [8, 9]. On the other hand, active engineering for energy optimization includes the usage of elements like heating, ventilation, and air conditioning (HVAC) systems, domestic hot water heating systems, and other systems that utilize renewable energy alternatives [10]. Such systems bring about a balanced approach to sustain the energy use and demands of buildings while also combating the on-site distribution of power supply, thus achieving savings.

Despite the availability of numerous definitions across a myriad of resources, smart buildings can be characterized as utilizing intelligent sources of technology to enhance the comfort of residents as well as efficiently manage energy use and demand [11]. These structures can align their functions as a response to adverse events in their surroundings. Hence, they are responsive and can perfectly balance the need for control by human beings with the usage of technology, thus balancing automation.

DOI: 10.1201/9781003496656-3

By adjusting to the demands of the occupants, the internal structure should likewise reflect the adaptable nature of the building.

This chapter discusses the multifaceted materials that are used and can be used for the construction of intelligent smart buildings leading to energy efficiency as well as optimized circulation of heat with them. Study and analysis of these materials serve as the bedrock for constructing eco-conscious structures that curtail energy consumption, thus mitigating the environmental impact of buildings and lowering operational costs. By utilizing innovative supplies that improve insulation, control temperature, and optimize the utilization of energy, these intelligent buildings create cozy interior spaces that increase worker productivity. Further, by assuring adherence to strict energy efficiency rules, such research promotes scientific developments in materials science and has promise for larger applications beyond the building industry, creating a foundation of sustainability for the future.

3.1.1 The origin

Following the beginning of the industrial era, there has been a great deal of progress in the construction materials industry, including breakthroughs in steel, glass manufacturing, and responsive materials [12–13]. Originally, materials were selected based on their physical characteristics and purpose; however, with constant scrutiny and experimentation, unique qualities have been attained throughout time. Building supplies have undergone shifts throughout time, moving from serving only the requirements of architecture to enhancing their practical effectiveness and providing novel stylistic potential. Contemporary building innovations are progressively taking the place of more established ones in the industry. These technologies have several benefits, including superior workmanship, longer lifespan, the capacity to create distinctive layouts, and faster delivery times [14].

3.2 SMART MATERIALS

Temporal behaviors and reactivity with shifting contexts are characteristics of novel materials and technology that have revolutionized architecture. Examples include electroluminescent supplies, shape-retaining alloys, displays of suspended particles, and photochromic and thermochromic materials, which change color by both light and temperature intensity. Architects now have more options to design buildings that adapt to their environment and the demands of their users, thanks to these materials and technology.

It is customary for people to employ the phrases "intelligent," "functional," and "smart" identically. For the first two words, this makes sense, although a little challenging. However, the final one possibly implies a level of cognition that is not present in any system other than biological ones. These materials can be compartmentalized into three classes as depicted in

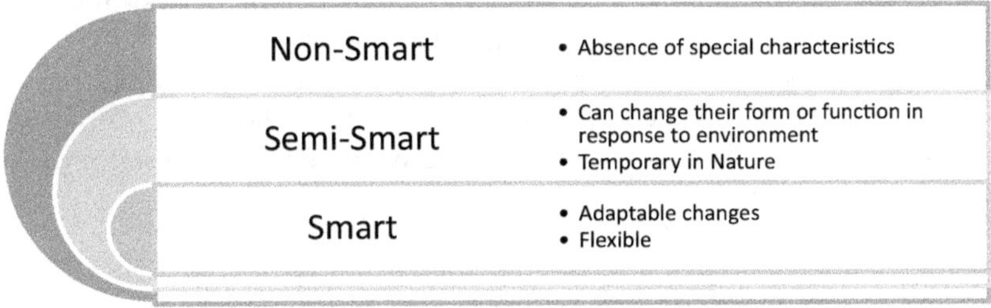

Figure 3.1 Classification of smart materials on the basis of changes [15]

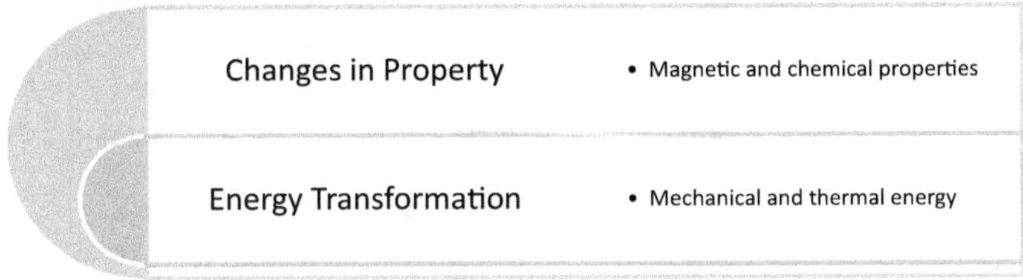

Figure 3.2 Classification of smart materials on the basis of properties and energy transformation [16]

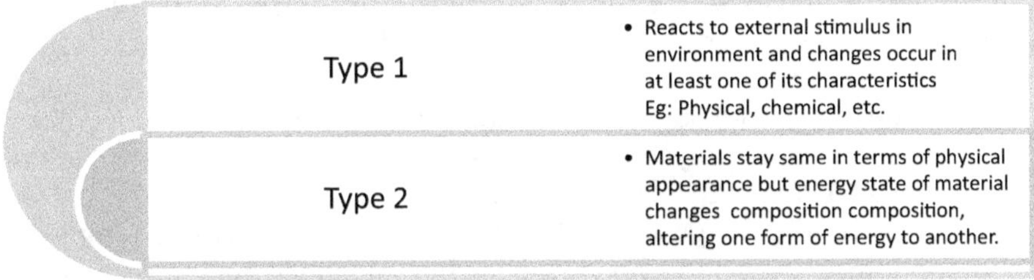

Figure 3.3 Types of smart materials [16]

Figure 3.1 [15]. According to other classification systems, these materials can be classified as shown in Figures 3.2 and 3.3.

3.2.1 Characteristics of smart materials

Smart materials can be characterized on the basis of certain traits [17, 18, 19]. These have to respond immediately to external stimuli or responses in real-time, whenever situations arise. Their transiency allows these materials to change their properties frequently, thus navigating multiple environmental states. Harboring intelligence within, these materials make their responses internal rather than relying on external factors. This intrinsic intelligence ensures a level of autonomy, enhancing their functionality.

Furthermore, the generation of discrete yet predictable responses when exposed to specific triggers sets smart construction materials different from others. Also, they react in a localized manner to the activating event, showcasing a focused and immediate response that amplifies their utility in targeted applications. These properties collectively define the essence of smart materials.

3.3 TRADITIONAL ENERGY-EFFICIENT MATERIALS

There are two primary types of traditional energy-efficient materials that work to provide insulation: (1) organic or frothy materials and (2) inorganic fiber-like materials. Fiberglass is an especially widely used and affordable insulating substance. It is composed of a pultruded combination of interlaced silicon, glass, sand, and other minerals that are subjected to heat until a molten state is obtained, to be spun into fibers using equipment. Fiberglass has been demonstrated to be resistant to the high temperature of flames, having a low coefficient of thermal conductivity. Fiberglass sheets have been used for home insulation in places like the eaves and domes [20, 21]. However, due to potential microscopic fragments, precautionary treatments may be required for optimal usage even if it is not harmful.

Thermally moldable, organic polystyrene is employed in the construction of enclosures of architectural paneling insulation and concrete. It is also made via the polymerization process in which a novel sigma bond is produced due to the breakage of pi-bonds between carbon atoms. As shown in Figure 3.4, ejected polystyrene and expandable polystyrene are the two most popular forms of insulation [22–24].

A permeable organic substance called polyurethane foam can be employed to improve insulation [25–27] in spaces like pipelines and tiny cracks. It may be produced by copolymerizing polyol and di-isocyanate. A significant amount of air and just a small percentage of polymer make up this foam.

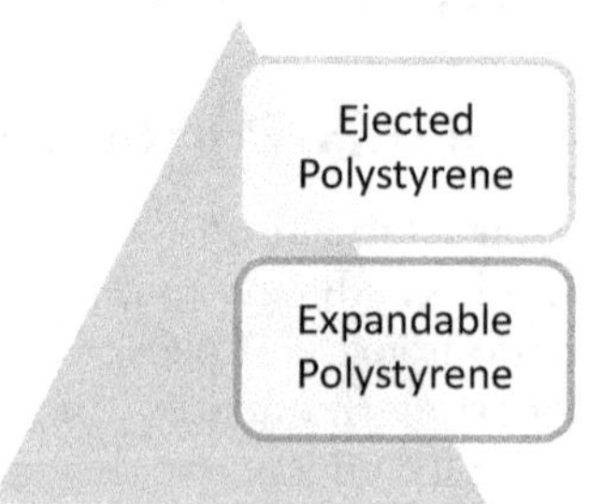

Figure 3.4 Types of organic polystyrene for thermoregulation

For better insulating efficacy, gases that have decreased conductivity can also be used instead. In all these cases, non-chlorofluorocarbon sources of foams are used to avoid detrimental effects on the ozone.

To provide fire retardancy, organic substrates like cellulose can be compacted tightly through existing residences' roof and wall crevices or loosely placed into incomplete roof spaces. An additional 10–15% of the components can be chemical additives like boric acid and ammonium sulfate [28–30].

As depicted in Figure 3.5, rock wool is from minerals that are naturally occurring like basalt or diabase, and slag wool, which is derived from blast slag waste, has more mass than fiberglass. These can be easily inserted and sculpted to the required configuration. Since mineral-based wool has an elevated point of melting and is non-flammable, it is frequently regarded as resilient to fire material that can withstand extreme temperatures. Wearing protective gear when handling is typically necessary to prevent exposure to the skin, consumption, or injury [31–33].

Aerogels are artificially created, highly permeable, not very dense rigid substances that are frequently produced by evaporating off the fluid that makes up a gel and setting behind a solid but porous component that is nearly fully occupied by air. These are possible options to increase sustainability and energy conservation because of their many desirable attributes for use in buildings, such as their significant thermal resistance, chemical composition, resilience, toughness, and low conductivity [34, 35]. As shown in Figure 3.6, organic aerogels are less weaker and more durable than the inorganic ones. However, they bear safety issues due to potential blazing near fire sources, hence are not safe.

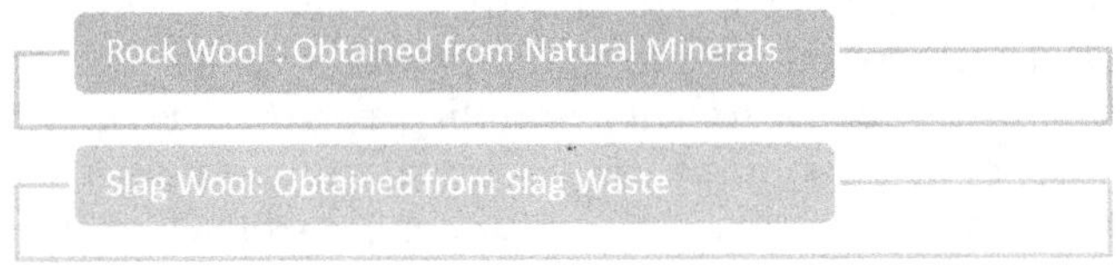

Figure 3.5 Types of resilient wool

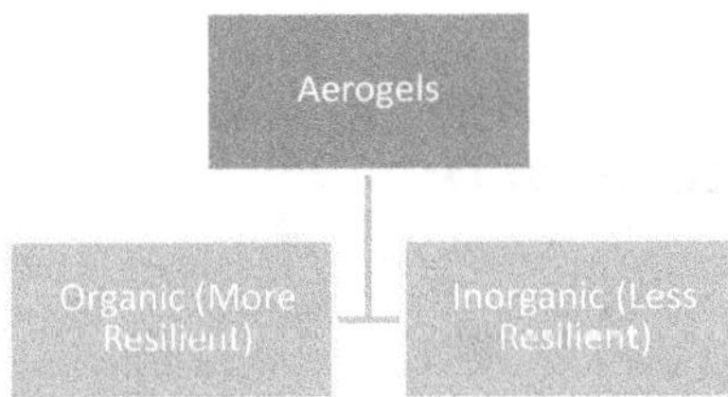

Figure 3.6 Types of organic aerogels

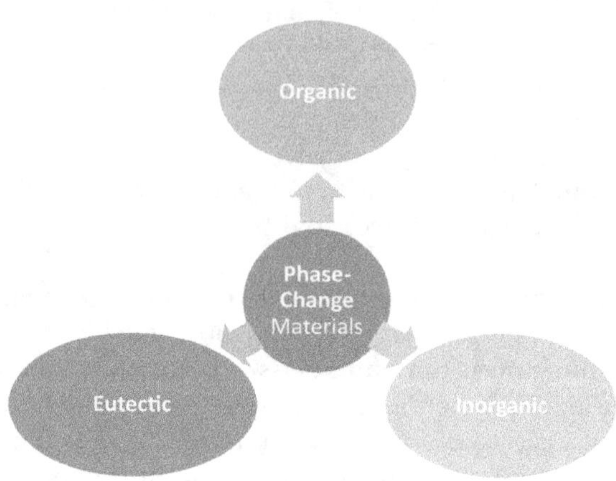

Figure 3.7 Types of PCMs

3.4 PHASE-CHANGE MATERIALS

Inert thermal storage systems use an element that changes phase between two states—solid-liquid or liquid-gas—to either capture or discharge thermal energy. The single-phase shifts used for creating viable thermal operations are solid-liquid ones. Phase-change materials' (PCMs) propensity to take in and release thermal energy is modulated by the materials' quantity, inert heat of fusion, and mass. PCMs have great potential for effective thermal retention in everyday situations because of their considerable energy retention rate as well as their ability to let go or collect heat without triggering an ambient temperature change [36–38]. As depicted in Figure 3.7, three categories are often used to classify PCMs: inorganic, organic, and eutectic PCMs. Compared to inorganic PCMs, organic PCMs offer several benefits. Their lack of toxicity, stability, corrosiveness, and ability to melt and re-solidify without a phase disintegration renders them excellent choices for heat retention in architectural components. The eutectic PCMs are an amalgamation of any two PCMs of decreased melting points [39]. This is more controllable than the other two options as the ratio of its component can allow for a change in properties in accordance with the needs.

3.5 ELECTROCHROMIC MATERIALS AND NANOTECHNOLOGY

Electrochromic substances have been researched and made accessible for purchase for adjusting optical characteristics under various circumstances [40–42]. A clear electrolyte is surrounded by twin oxide coatings in the conventional form of this device. The addition of nanotechnology in these materials has been the latest development in the field [43–47]. Cost-effective smart materials were employed in buildings sector for heat transfer enhancement having greater durability and lesser carbon footprints [64–68]. Smart materials consisting of specialized scaffolds were utilized to improve the dissemination of ions as well as changeover rapidity, and stability, and permit chromatic shifting. Future developments should see an increase in the popularity of such materials as their costs continue to decline and they promote sustainability [69–73].

3.6 MECHANICAL ENGINEERING FOR THERMAL INSULATION

The thermal systems that are dependent on mechanical engineering may be divided into three categories that utilize various control systems: shiftable insulation based on thermal convection [48–50], incorporation of vacuum for insulation [51–53], and a heat switch based on the principles of mechanical contact [54–56], as shown in Figure 3.8. One may reversibly adjust the thermal circulation rate in an enclosure by pressurizing and emptying the gas molecules inside using a specialized pump lassed with a sealed interior. Since thermal conduction across either liquid or solid matter often exceeds that of gaseous ones, a different conventional method of manipulating the transmission of heat is mechanically governing the interaction that exists between two moving components. An actuator that allows for motion and two smooth and rigid surfaces is needed for a standard contact-based heat switch system [57–58].

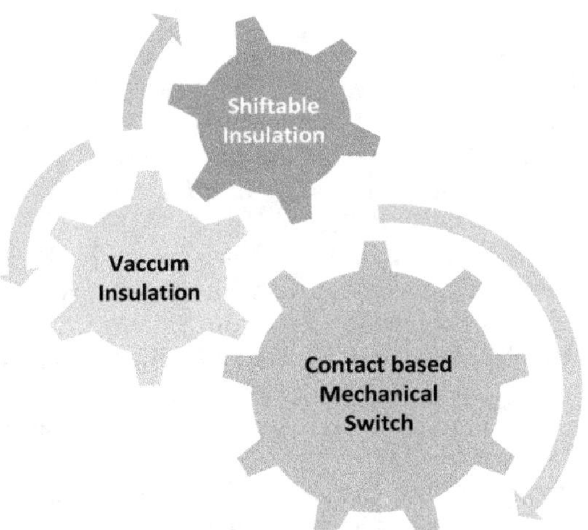

Figure 3.8 Mechanical engineering in heat transfer and thermoregulation

3.7 THE INTERPLAY OF ALGORITHMS AND MACHINE LEARNING

Smart buildings are more comfortable, effective, adaptable, and protected when they can adjust to fluctuating environments thanks to the use of machine learning and artificial intelligence approaches [59–80]. These computations use what they've learned from both previous and present experiences to maximize energy utilization. A significant technical advancement, deep learning imitates the human brain's tiered perception of settings and things, identifying trends in buildings, lighting, and temperatures of spaces [81–98]. By detecting trends and forecasting inclinations, predictive analytics technologies can turn unstructured data into insightful knowledge. These instruments can monitor important energy-saving indicators, offer commentary on extended patterns, draw attention to anomalies, and reveal obscure information, thereby affecting energy management.

3.8 CONCLUSION AND FUTURE SCOPE

The current trend of research for thermoregulating materials is majorly inclined toward homogenous substances. The need is to shift and emphasize more for the heterogenous ones for better and balanced screening. Moreover, with a surge in awareness regarding the tremendous potential of artificial intelligence and other computational channels, diverse energy-optimizing opportunities continue to lurk on the horizon. Additionally, the incorporation of greener non-conventional renewable energy sources is bound to play a pivotal role in the field of thermoregulation to make buildings more intelligent and adaptable.

REFERENCES

1. Crawford, R. H. (2022). Greenhouse gas emissions of global construction industries. In *IOP conference series: Materials science and engineering* (Vol. 1218, No. 1, p. 012047). IOP Publishing.
2. Li, Y. L., Han, M. Y., Liu, S. Y., & Chen, G. Q. (2019). Energy consumption and greenhouse gas emissions by buildings: A multi-scale perspective. *Building and Environment*, *151*, 240–250.
3. Zhong, X., Hu, M., Deetman, S., Steubing, B., Lin, H. X., Hernandez, G. A., ... Behrens, P. (2021). Global greenhouse gas emissions from residential and commercial building materials and mitigation strategies to 2060. *Nature Communications*, *12*(1), 6126.
4. Zhao, S., Song, Q., Duan, H., Wen, Z., & Wang, C. (2019). Uncovering the lifecycle GHG emissions and its reduction opportunities from the urban buildings: A case study of Macau. *Resources, Conservation and Recycling*, *147*, 214–226.

5. Ürge-Vorsatz, D., Cabeza, L. F., Serrano, S., Barreneche, C., & Petrichenko, K. (2015). Heating and cooling energy trends and drivers in buildings. *Renewable and Sustainable Energy Reviews*, *41*, 85–98.
6. Balat, M. (2010). Greenhouse gas emissions and reduction strategies of the European Union. *Energy Sources, Part B: Economics, Planning, and Policy*, *5*(2), 165–177.
7. Pérez-Lombard, L., Ortiz, J., & Pout, C. (2008). A review on buildings energy consumption information. *Energy and Buildings*, *40*(3), 394–398.
8. Liu, C., Sharples, S., & Mohammadpourkarbasi, H. (2023). A review of building energy retrofit measures, passive design strategies and building regulation for the low carbon development of existing dwellings in the hot summer–cold winter region of China. *Energies*, *16*(10), 4115.
9. Lee, J. (2019). Multi-objective optimization case study with active and passive design in building engineering. *Structural and Multidisciplinary Optimization*, *59*(2), 507–519.
10. Bradshaw, V. (2010). *The building environment: Active and passive control systems*. John Wiley & Sons.
11. Kumar, A., Sharma, S., Goyal, N., Singh, A., Cheng, X., & Singh, P. (2021). Secure and energy-efficient smart building architecture with emerging technology IoT. *Computer Communications*, *176*, 207–217.
12. Ahmed, A., Yadav, A. K., & Singh, A. (2023). Optimization of cavitation-assisted biodiesel production and fuel properties from neochloris oleoabundans microalgae oil using genetic algorithm and response surface methodology. *Journal of Proceedings of the Institution of Mechanical Engineers, Part E*. doi.org/10.1177/09544089231159832
13. Zembala, D. M. (1984). *Machines in the glasshouse: The transformation of work in the glass industry, 1820–1915*. The George Washington University.
14. Moe, K. (2008). *Integrated design in contemporary architecture*. Princeton Architectural Press.
15. Gomaa, E. G., Badran, E. E., Mahmoud, M. A., & Saleh, A. M. (2016). The use of smart materials in raising the efficiency performance of buildings. *International Journal of Application or Innovation in Engineering & Management*, *5*, 1–18.
16. Addington, M., & Schodek, D. (2005). Smart materials and technologies. *Architecture and Urbanism*, *5*(3), 8–13.
17. Anand, A., Singh, R., Yadav, A. K., Dewangan, A., Singh, B. & Saha, A. (2024). Experimental investigations on the electrochemical machining of D3 die steel material and multi-objective parameters optimization, *Journal of Surface Review and Letters*, DOI:10.1142/S0218625X24500744
18. Khan, I. A., Singh, S. K., & Yadav, A. K. (2020). Efficient production of biodiesel from Cannabis sativa oil using intensified transesterication (hydrodynamic cavitation) Method. *Energy Sources, Part A*, *42*(20), 2461–2470.
19. Yadav, A. K., Khan, T. A., Khan, T. A., & Kumar, S. (2021). Methyl ester of Gmelina arborea oil as a substitute for petroleum diesel: an experimental study on its performance and emissions in a diesel engine. *Energy Sources, Part A*, *43* (11), 1307–1314
20. Dewangan, A. & Yadav, A. K. (2017). Wax deposition during production of Waxy Crude Oil and its remediation. *Petroleum Science and Technology*, *35*(18), 1831–1838.

21. Alhuwayil, W. K., Mujeebu, M. A., & Algarny, A. M. M. (2019). Impact of external shading strategy on energy performance of multi-story hotel building in hot-humid climate. *Energy, 169*, 1166–1174.
22. Khoukhi, M. (2018). The combined effect of heat and moisture transfer dependent thermal conductivity of polystyrene insulation material: Impact on building energy performance. *Energy and Buildings, 169*, 228–235.
23. Reynoso, L. E., Romero, Á. B. C., Viegas, G. M., & San Juan, G. A. (2021). Characterization of an alternative thermal insulation material using recycled expanded polystyrene. *Construction and Building Materials, 301*, 124058.
24. Ahmed, A., Yadav, A. K., & Singh, A. (2023). Optimization of biogas yield from anaerobic co-digestion of dual waste for environmental sustainability: ANN, RSM, and GA, approach. *International Journal of Oil, Gas and Coal Technology, 33*(1), 75.
25. Somarathna, H. M. C. C., Raman, S. N., Mohotti, D., Mutalib, A. A., & Badri, K. H. (2018). The use of polyurethane for structural and infrastructural engineering applications: A state-of-the-art review. *Construction and Building Materials, 190*, 995–1014.
26. Yadav, A. K., Khan, M. E., & Pal, A. (2018). Ultrasonic assisted production of biodiesel from karabi oil using heterogeneous catalyst. *Biofuels, 9*, 101–112.
27. Yadav, A. K., Khan, M. E., & Pal, A. (2017). Kaner biodiesel production through hybrid reactor and its performance testing on a CI engine at different compression ratios. *Egyptian Journal of Petroleum, 26*(2), 525–532.
28. Dewangan, A., Mallick, A., & Yadav, A. K. (2020). Effect of metal oxide nanoparticles and engine parameters on the performance of a diesel engine: A review. *Materials Today: Proceedings, 21*, 1722–1727.
29. Jiang, S., Zhang, M., Li, M., Zhu, J., Ge, A., Liu, L., & Yu, J. (2021). Cellulose-based composite thermal-insulating foams toward eco-friendly, flexible and flame-retardant. *Carbohydrate Polymers, 273*, 118544.
30. Ahmed, A., Yadav, A. K., & Singh, A. (2023). Enhancing waste cooking oil biodiesel yield and characteristics through machine learning, response surface methodology, and genetic algorithm for optimal utilization in CI Engines. *International Journal of Green Energy.* https://doi.org/10.1080/15435075.2023.2253870
31. Muhieldeen, M. W., Lye, L. C., Kassim, M. S. S., Yen, T. W., & Teng, K. H. (2021). Effect of rockwool insulation on room temperature distribution. *Journal of Advanced Research in Experimental Fluid Mechanics and Heat Transfer, 3*(1), 9–15.
32. Ahmed, A., Yadav, A. K., & Singh, A. (2023). Application of machine learning and genetic algorithm for prediction and optimization of biodiesel yield from waste cooking oil. *Korean Journal of Chemical Engineering.* DOI: 10.1007/s11814-023-1489-9
33. Kumar, D., Alam, M., Zou, P. X., Sanjayan, J. G., & Memon, R. A. (2020). Comparative analysis of building insulation material properties and performance. *Renewable and Sustainable Energy Reviews, 131*, 110038.
34. Yadav, A. K., Pal, A., & Dubey, A. M. (2018). Experimental studies on utilization of Prunus armeniaca L. (Wild Apricot) biodiesel as an alternative fuel for CI engine, *Waste and Biomass Valorization, 9*(10), 1961–1969..

35. Dewangan, A., Mallick, A., Yadav, A. K., Ahmad, A., & Alqahtani, D. (2023). Combined effect of operating parameters and nano particles on performance of a diesel engine: Response surface methodology coupled genetic algorithm approach. *ACS Omega*, *8*(27). DOI:10.1021/acsomega.3c02782
36. Yadav, A. K., Vinay, Singh, B. (2018). Optimization of biodiesel production from annona squamosa seeds oil using response surface methodology and its characterization. *Energy Sources, Part A*, *40*(9), 1051–1059.
37. Dewangan, A., Mallick, A., Yadav, A. K., Islam , S., & Agbulut, U. (2023). Production of oxy-hydrogen gas and the impact of its usability on CI Engine combustion, performance, and emission behaviors. Energy, *278*, 127937.
38. Romdhane, S. B., Amamou, A., Khalifa, R. B., Said, N. M., Younsi, Z., & Jemni, A. (2020). A review on thermal energy storage using phase change materials in passive building applications. *Journal of Building Engineering*, *32*, 101563.
39. Ahmed, A., Yadav, A. K., Singh, A., & Agbulut, U. (2023). A hybrid RSM-GA-PSO approach on optimization of process intensification of linseed biodiesel synthesis using an ultrasonication reactor: Enhancing fuel properties and engine characteristics with ternary fuel blends. *Energy*, *288*(1), 129077.
40. Rai, V., Singh, R. S., Blackwood, D. J., & Zhili, D. (2020). A review on recent advances in electrochromic devices: A material approach. *Advanced Engineering Materials*, *22*(8), 2000082.
41. Pooyodying, P., Ok, J. W., Son, Y. H., & Sung, Y. M. (2021). Electrical and optical properties of electrochromic device with WO3: Mo film prepared by RF magnetron Co-Sputtering. *Optical Materials*, *112*, 110766.
42. Ahmed, A., Yadav, A. K., & Singh, A. (2024). A comprehensive machine learning-coupled response surface methodology approach for predictive modeling and optimization of biogas potential in anaerobic co-digestion of organic waste. *Biomass and Bioenergy*, *180*, 106995.
43. Kumar, R., Pathak, D. K., & Chaudhary, A. (2021). Current status of some electrochromic materials and devices: A brief review. *Journal of Physics D: Applied Physics*, *54*(50), 503002.
44. Wu, W., Wang, M., Ma, J., Cao, Y., & Deng, Y. (2018). Electrochromic metal oxides: Recent progress and prospect. *Advanced Electronic Materials*, *4*(8), 1800185.
45. Yadav, A. K., Dewangan, A., & Mallick, A. (2018). Effect of N-butanol and diethyl ether additives on performance and emission characteristics of a diesel engine fuelled with diesel-pongamia biodiesel blends. *Journal of Energy Engineering*, *144*(6), 04018062.
46. Ma, D., & Wang, J. (2017). Inorganic electrochromic materials based on tungsten oxide and nickel oxide nanostructures. *Science China Chemistry*, *60*(1), 54–62.
47. Ahmed, A., Yadav, A. K., & Singh, A. (2023). Multi-response optimization to improve the performance and emissions characteristics of A VCR engine fueled with microalgae spirulina (L.): A response surface methodology approach coupled with genetic algorithm. *Environment, Development and Sustainability*, https://doi.org/10.1007/s10668-023-04016-z.
48. Cui, H., & Overend, M. (2019). A review of heat transfer characteristics of switchable insulation technologies for thermally adaptive building envelopes. *Energy and Buildings*, *199*, 427–444.

49. Ahmed, A., Yadav, A. K., & Singh, A. (2023). Process optimization of spirulina microalgae biodiesel synthesis using RSM coupled GA technique: a performance study of a biogas-powered dual-fuel engine. *International Journal of Environmental Science and Technology*. Doi: 10.1007/s13762-023-04948-z 20.
50. Ghosh, A., & Norton, B. (2018). Advances in switchable and highly insulating autonomous (self-powered) glazing systems for adaptive low energy buildings. *Renewable Energy*, *126*, 1003–1031.
51. Yadav, A. K., Khan, M. E., & Pal, A. (2018). Performance, emission and combustion characteristics of an indica diesel engine operated with yellow oleander (Thevetia Peruviana) oil biodiesel produced through hydrodynamic cavitation method. *International Journal of Ambient Energy*, *39*(4), 365–371.
52. Zhou, J., Peng, Y., Xu, J., Wu, Y., Huang, Z., Xiao, X., & Cui, Y. (2022). Vacuum insulation arrays as damage-resilient thermal superinsulation materials for energy saving. *Joule*, *6*(10), 2358–2371.
53. Boafo, F. E., Kim, J. H., Ahn, J. G., Kim, S. M., & Kim, J. T. (2023). Vacuum insulation panel: Evaluation of declared thermal conductivity value and implications for building energy. *Energies*, *16*(15), 5841.
54. Yadav, A. K., Khan, M. E., & Pal, A. (2017). Biodiesel production from oleander (Thevetia Peruviana) oil and its performance testing on a diesel engine, *Korean Journal of Chemical Engineering, 34*(2), 340–345.
55. Ahmed, A., Yadav, A. K., & Singh A. (2023). Enhancement of biogas yield from dual organic waste using hybrid statistical approach and its effects on ternary fuel blend (Biodiesel/n-butanol /Diesel) Powered Diesel Engine. *Environmental Progress & Sustainable Energy*, doi.org/10.1002/ep.14163.
56. Dewangan, A., Yadav, A. K., & Mallick, A. (2018). Current scenario of biodiesel development in India: Prospects and challenges, *Energy Sources, Part A*, *40*(20), 2494–2501.
57. Shin, S., & So, H. (2021). Time-dependent motion of 3D-printed soft thermal actuators for switch application in electric circuits. *Additive Manufacturing*, *39*, 101893.
58. Khan, T. A., Khan, T.A., & Yadav, A. K. (2022). A hydrodynamic cavitation-assisted system for optimization of biodiesel production from green microalgae oil using a genetic algorithm and response surface methodology approach. *Environmental Science and Pollution Research*. doi.org/10.1007/s11356-022-20474-w2022.
59. Farzaneh, H., Malehmirchegini, L., Bejan, A., Afolabi, T., Mulumba, A., & Daka, P. P. (2021). Artificial intelligence evolution in smart buildings for energy efficiency. *Applied Sciences*, *11*(2), 763.
60. Mocanu, E., Nguyen, P. H., Gibescu, M., & Kling, W. L. (2016). Deep learning for estimating building energy consumption. *Sustainable Energy, Grids and Networks*, *6*, 91–99.
61. Yadav, A. K., Khan, M. E., & Pal, A. (2023). Performance and emission characteristics of a stationary diesel engine fuelled by Schleichera Oleosa oil Methyl Ester (SOME) produced through hydrodynamic cavitation process. *Egyptian Journal of Petroleum*, *27*, 89–93.
62. Ahmed, A., Yadav, A. K., & Singh, A. (2023). Modelling and optimisation of VCR diesel engine parameters fuelled with microalgae methyl-ester: A RSM-coupled Taguchi approach. *International Journal of Ambient Energy*, *44*(1), 2498–2506.

63. Jung, S., Jeoung, J., & Hong, T. (2022). Occupant-centered real-time control of indoor temperature using deep learning algorithms. *Building and Environment*, *208*, 108633.
64. Meena, C. S., Kumar, A., Jain, S., Rehman, A. U., & Mishra, S. (2022). Innovation in green building sector for sustainable future. *Energies*, *15*, 6631. https://doi.org/10.3390/en15186631
65. Kumar, R., Prasad, A., & Kumar, A. (2023). *Sustainable smart manufacturing processes in industry 4.0*. Taylor & Francis (CRC Press). ISBN: 9781003436072. https://doi.org/10.1201/9781003436072
66. Kumar, A., Singla, Y., & Namboodri, T. (2024). Globalization and international issues in sustainable manufacturing. In S. Shah, H. Nautiyal, G. Gugliani, A. Kumar, T. Namboodri, & Y. K. Singla (Eds.), *Sustainability in smart manufacturing: Trends, scope, and challenges*. Boca Raton, FL: CRC Press. ISBN 9781032 740713.
67. Chakraborty, G., Pandey, V., Prasad, A., & Kumar, A. (2023). Introduction to sustainable manufacturing for industries 4.0. In R. Kumar, A. Prasad, & A. Kumar (Eds.), *Sustainable smart manufacturing processes in industry 4.0* (pp. 1–17). Boca Raton, FL: CRC Press. https://doi.org/10.1201/9781003436072-1.
68. Shah, S., Nautiyal, H., Gugliani, G., Kumar, A., Namboodri, T., & Singla, Y. K. (Eds.). (2024). *Sustainability in smart manufacturing: Trends, scope, and challenges*. Boca Raton, FL: CRC Press. ISBN 9781032740713.
69. Yadav, A. K. & Vinay, Singh, B. (2018). Optimization of biodiesel production from annona squamosa seeds oil using response surface methodology and its characterization.Energy Sources, Part A, 40(9), 1051–1059.
70. Sharma, V. K., Kumar, A., Gupta, M., Kumar, V., Sharma, D. K., & Sharma, S. (2022). *Additive manufacturing in industry 4.0: Methods, techniques, modeling, and nano aspects*. Taylor & Francis (CRC Press). https://doi.org/10.1201/9781003360001
71. Kumar, A., Kumar, A., & Kumar, A. (2023). *Laser based technologies for sustainable manufacturing*. Taylor & Francis (CRC Press). ISBN: 9781003402398. https://doi.org/10.1201/9781003402398.
72. Pathak, A., Kumar, A., Kumar, A., Kumar, A., & F. L. King M. (2023). Application of laser technology in the mechanical and machine manufacturing industry. In A. Kumar, A. Kumar, & A. Kumar (Eds.), *Laser-based technologies for sustainable manufacturing* (pp. 107–155). Boca Raton, FL: CRC Press. https://doi.org/10.1201/9781003402398-6
73. Verma, V., Thangavel, S., Dutt, N., Kumar, A., & Weerasinghe, R. (Eds.). (2024). *Highly efficient thermal renewable energy systems: Design, design, optimization and applications*. Boca Raton, FL: CRC Press. ISBN 9781032595641.
74. Singh, S., Singh, B., Kumar, S., & Yadav, A. K. (2019).Temperature-dependent dynamic hysteresis scaling of ferroelectric hysteresis parameters of lead free [(Ba 0.825 + xCa0.175-x) (Ti1- x Snx)O3]ceramics. *Ferroelectrics*, *551*, 133–142
75. Dewangan, A., Yadav, A. K., & Mallick, A. (2020). Optimization of biodiesel production and engine performance from underutilized simarouba oil in compression ignition Engine. *International Journal of Oil Gas and Coal Technology*, *25*(3), 357.

76. Ahmed, A., Yadav, A. K., & Singh, A. (2023). An environmental impact assessment and optimization study of biodiesel production from microalgae. *International Journal of Global Warming (IJGW)*, *31*(3), 294–313.
77. Yadav, A. K., Khan, M. E., & Pal, A., (2016). A comparative study on ultrasonic cavitation and mechanical stirring method towards efficient production of biodiesel from non-edible oils and performance testing on a C.I. *International Journal of Environment and Waste Management*, *18*, 349–367.
78. Vinay, Singh B. & Yadav, A. K. (2020). Optimization of performance and emission characteristics of CI engine fuelled with Mahua oil methyl ester-diesel blend using response surface methodology. *International Journal of Ambient Energy*, *44*(6), 674–85.
79. Yadav, A. K., Pal, A., Ghosh, U., & Gupta, S. K. (2019) Comparative study of biodiesel production methods from yellow oleander oil and its performance analysis on an agricultural diesel engine. *International Journal of Ambient Energy*, *40*(2), 152–157.
80. Khan, I. A., Singh, S. K., Yadav, A. K., & Sharma, D. (2019). Enhancement in the performance of a diesel engine fueled with pongamia methyl ester and n-Butanol as oxygenated additive. *International Journal of Ambient Energy*, *40*(8), 842–846.
81. Ahmed, A., Yadav, A. K., & Singh, A. (2023). Biodiesel yield optimization from a third-generation feedstock (Microalgae Spirulina) using a hybrid statistical approach. *International Journal of Ambient Energy*, *44*(1), 1202–1213.
82. Ahmed, A. & Yadav, A. K. (2023). Biodiesel production from Mahua oil: Characterization, optimization, and modeling with a hybrid statistical approach. *International Journal of Ambient Energy*, *44*, 2618–2627.
83. Khan, T.A, Khan, M. E., & Yadav, A. K. (2023). Experimental studies on utilization of neochloris oleoabundans microalgae biodiesel as an alternative fuel for diesel engine. *International Journal of Ambient Energy*, *44*(1), 115–123.
84. Yadav, A. K & Ahmed, A. (2024). Parametric analysis of wastewater electrolysis for green hydrogen production: A combined RSM, genetic algorithm, and particle swarm optimization approach. *International Journal of Hydrogen Energy*, *59*, 51–62.
85. Singh, D., Yadav, A. K, Kumar, A., & Samsher. (2023) Energy matrices and life cycle conversion analysis of N-identical hybrid double slope solar distiller unit using Al2O3 nanoparticle. *Journal of Water and Environmental Nanotechnology*, *8*(3), 267–284.
86. Yadav, A. K., Dewangan, A., & Mallick, A. (2018). Synthesis and stability study of biodiesel from Kachnar seed oil. *Journal of Energy Engineering*, *144*(5), 04018053.
87. Yadav, A.K., Khan, O., & Khan, M. E. (2018). Utilization of high FFA landfill waste (leachates) as a feedstock for sustainable biodiesel production: Its characterization and engine performance evaluation. *Environmental Science and Pollution Research*, *25*(32), 32312–32320.
88. Kumar, N., Yadav, A. K., Dewangan, A., & Kumar, M. (2022). CFD Based Investigation of Thermophoresis Effect on Microparticles in Micro Channel. In *Advances in mechanical and energy technology*, Lecture Notes in Mechanical Engineering (pp. 285–296). DOI: https://doi.org/10.1007/978-981-19-1618-2_31.

89. Ahmad, A., Singh, A., Singh, D. K. & Yadav, A. K. (2022). An experimental investigation of the effect of diethyl ether as an additive on the performance of a single cylinder diesel engine. In *Recent trends in thermal and fluid sciences*, Lecture Notes in Mechanical Engineering (pp. 69–75). DOI: 10.1007/978-981-19-3498-8_7.
90. Singh, S., Yadav, A. K., Kumar, N., Choudhury, U.K. & Kumar, M. (2023). Investigation on lead free ferroelectric [(Ba0.825+xCa0.175-x)(Ti1-xSnx)O3] ceramics for energy storage density and thermal energy harvesting capacity. In *Recent Advances in Manufacturing and Thermal 2023: Engineering*, Lecture Notes in Mechanical Engineering (pp. 497–504). DOI: 10.1007/978-981-19-8517-1_38
91. Gupta, G., Agarwal, K., Yadav, A., Yadav, A. K., & Sinha, D. K. (2023). Design and fabrication of PLA-printed wearable exoskeleton with 7 DOF for upper limb physiotherapy training and rehabilitation. *Advances in Engineering Design*. Lecture Notes in Mechanical Engineering. Springer, https://doi.org/10.1007/978-981-99-3033-3_5.
92. Gupta, G., Yadav, A. K., & Sinha, D. K. (2023). Studying current safety systems for accident prevention and wellbeing of powered two-wheeler community: prevalence of safety components. *Advances in Engineering Design*. Lecture Notes in Mechanical Engineering. Springer. https://doi.org/10.1007/978-981-99-3033-3_64.
93. Dewangan, A., Singh, B., Srivastava, A., Srivastava, A., & Yadav, A. K. (2022). A broad review of biodiesel feedstocks with competency to replace diesel. In *Advances in Mechanical and Energy Technology*, Lecture Notes in Mechanical Engineering (pp. 323–331). Springer. https://doi.org/10.1007/978-981-19-1618-2_28.
94. Khan, O., Parvez, M., Kumari, P., Yadav, A. K., Akram, W., Ahmad, S., Parvez, S., & Idrisi, M. J. (2023). Modelling of compression ignition engine by soft computing techniques (ANFIS-NSGA-II and RSM) to enhance the performance characteristics for leachate blends with nano-additives. *Scientific Reports*, *13*, 15429.
95. Khan, O., Yadav, A. K., Khan, M. E., & Parvez, M. (2019). Characterization of bioethanol obtained from Eichhornia Crassipes plant; its emission and performance analysis on CI engine. *Energy Sources, Part A: Recovery, Utilization, and Environmental Effects*, *43*, 1–11.
96. Singh, D., Singh, S., Yadav, A. K., Khan, O., Dewangan, A., Alkahtani, M. Q., & Islam, S. (2023). From theory to practice: A sustainable solution to water scarcity by using a hybrid solar distiller with a heat exchanger and aluminum oxide nanoparticles. *ACS Omega*, *8*(37), 33543–33553.
97. Khan, O., Alsaduni, I., Equbal, A., Parvez, M., & Yadav, A. K. (2024). Performance and emission analysis of biodiesel blends enriched with biohydrogen and biogas in internal combustion engines. *Process Safety and Environmental Protection*, *183*(2).
98. Khan, O., Khan, M. E., Yadav, A. K., & Sharma, D. (2017). The ultrasonic-assisted optimization of biodiesel production from eucalyptus oil. *Energy Sources, Part A: Recovery, Utilization, and Environmental Effects*, *39*(13), 1323–1331.

Chapter 4

Life Cycle Assessment of sustainable building materials

Jitendra Yadav, Varun Pratap Singh, and Ashwani Kumar

4.1 INTRODUCTION

The development of sustainable construction materials has been closely linked to the fundamental aspects of human civilization throughout history. The progression of construction materials, ranging from traditional adobe and timber constructions to contemporary eco-friendly alternatives, reflects our changing connection with the environment. At the brink of the 21st century, the worldwide demand for sustainable construction materials is undergoing a significant and positive change. Driven by increased environmental consciousness, strict restrictions, and a growing need for environmentally friendly alternatives, the outlook for these materials indicates exceptional expansion. This chapter provides a thorough examination, combining historical usage with current global market trends, to give a detailed understanding of how this technology is used in many sectors.

The use of sustainable construction materials is driven by several factors, including environmental stewardship, resource efficiency, and economic concerns. Nevertheless, this adoption is not without its difficulties. Obstacles to wider implementation include factors such as financial ramifications, technological constraints, and the need for collaboration throughout the whole sector. An exhaustive analysis of these factors and obstacles establishes the foundation for comprehending the delicate equilibrium necessary for the effective incorporation of sustainable materials into conventional construction methods.

To accurately predict future trends in sustainable building materials for end-users, it is necessary to thoroughly analyze the past usage of these materials and assess their industrial capabilities. The chapter explores the lasting importance of sustainable materials by examining their historical use in several industries, including residential and commercial sectors. At the same time, it predicts their ability to influence the future of construction, providing a thorough understanding of the significant impact these materials can have on the industrial and historical technological environment. Analyzing and examining key areas in sustainable building materials gives a complete overview of the technological landscape. The section on

DOI: 10.1201/9781003496656-4

technical significance highlights the inherent worth of these materials in tackling worldwide concerns. Studying Intellectual Property Rights (IPRs) provides insights into the field of innovation. Additionally, many study areas, rising trends, and future research goals contribute to a dynamic and changing discipline, providing guidance to researchers in exploring unexplored potential. This synthesis establishes the foundation for a thorough investigation of Life Cycle Assessment (LCA) approaches, which is the main focus of this chapter. The purpose is to make a substantial contribution to the discussion on sustainable building practices and the development of a resilient built environment [1].

4.1.1 Significance of sustainable building materials

The importance of sustainable building materials in modern construction methods is complex and extensive, impacting several aspects of the built environment. Comprehending and accepting the significance of sustainable materials is crucial for tackling urgent global issues and promoting a more robust and ecologically conscious approach to construction [2]. The following are crucial factors that emphasize the need for sustainable building materials:

1. **Environmental Impact Reduction:** Sustainable building materials are designed and selected with a focus on minimizing their environmental footprint. This includes considerations such as resource depletion, energy consumption, and emissions during production and transportation. By opting for materials with lower environmental impact, construction projects contribute to mitigating climate change, reducing pollution, and conserving natural resources.
2. **Energy Efficiency and Conservation:** Sustainable materials often exhibit superior energy efficiency characteristics. This includes better insulation properties, reduced energy requirements during manufacturing, and improved thermal performance in buildings. By incorporating such materials, construction projects can contribute to energy conservation, leading to lower operational costs for buildings and decreased reliance on non-renewable energy sources.
3. **Waste Reduction and Recycling:** Sustainable building materials promote waste reduction and recycling throughout their life cycle. Many of these materials are designed to be recyclable or have a high recycled content, minimizing the amount of construction waste sent to landfills. This emphasis on circular economy principles not only conserves resources but also reduces the environmental impact associated with waste disposal.
4. **Health and Well-being of Occupants:** The use of sustainable building materials contributes to healthier indoor environments. Materials with low or no volatile organic compounds (VOCs) and other harmful

substances enhance indoor air quality, promoting the well-being of building occupants. This emphasis on health aligns with the growing awareness of the impact of the built environment on human health and productivity.

5. **Longevity and Durability:** Sustainable building materials often exhibit enhanced durability and longevity. This characteristic reduces the need for frequent replacements or renovations, extending the lifespan of buildings and infrastructure. Long-lasting materials contribute to lower life cycle costs and reduced environmental impact associated with maintenance and replacement activities.
6. **Compliance with Regulatory Standards:** As governments and regulatory bodies worldwide intensify their focus on sustainable and green building practices, the use of sustainable building materials becomes imperative for compliance. Meeting and exceeding environmental standards not only aligns construction projects with legal requirements but also positions them favorably in the context of evolving regulatory frameworks.
7. **Market Demand and Brand Image:** There is a growing market demand for sustainable and eco-friendly buildings. Adopting sustainable building materials aligns with consumer preferences and market trends, enhancing the marketability of construction projects. Moreover, companies that prioritize sustainability often enjoy a positive brand image, attracting environmentally conscious clients and investors.
8. **Resilience to Climate Change:** Sustainable building materials contribute to the resilience of structures in the face of climate change impacts. Materials that can withstand extreme weather events, resist deterioration, and adapt to changing environmental conditions enhance the overall resilience of buildings and infrastructure, ensuring their longevity and functionality.

The significance of sustainable building materials extends beyond individual construction projects; it plays a pivotal role in shaping the future of the built environment. By prioritizing materials that are environmentally responsible, energy-efficient, and health-conscious, the construction industry contributes to a sustainable, resilient, and harmonious coexistence with the natural world.

4.1.2 Purpose of Life Cycle Assessment

The purpose of Life Cycle Assessment (LCA) in the context of sustainable building materials is multifaceted, aiming to provide a comprehensive and systematic framework for evaluating the environmental, social, and economic impacts of these materials throughout their entire life cycle. The primary objectives of Life Cycle Assessment include [3]:

1. **Holistic Assessment:** Life Cycle Assessment seeks to offer a holistic and integrated assessment of sustainable building materials from cradle to grave. It goes beyond a narrow focus on individual stages of the material's life cycle and considers the cumulative impacts, ensuring that the entire life span of the material is taken into account.
2. **Environmental Performance Evaluation:** One of the key purposes of LCA is to evaluate the environmental performance of sustainable building materials. This includes analyzing the resource extraction, manufacturing processes, transportation, installation, use, maintenance, and eventual disposal or recycling of materials. By quantifying environmental impacts such as carbon footprint, energy consumption, and resource depletion, LCA provides valuable insights for decision-makers.
3. **Identification of Hotspots and Improvement Opportunities:** Life Cycle Assessment helps identify environmental "hotspots" or stages in the life cycle where the material has a significant impact. This enables stakeholders to pinpoint areas for improvement and innovation, guiding efforts to enhance the overall sustainability of building materials. By focusing on improvement opportunities, LCA supports the evolution of materials toward eco-friendly alternatives.
4. **Decision Support for Material Selection:** LCA serves as a critical decision support tool for architects, builders, and decision-makers in the construction industry. By providing comprehensive data on the environmental, social, and economic aspects of different materials, it enables informed decision-making during the material selection process. This promotes the use of materials that align with sustainability goals and regulatory requirements.
5. **Resource Efficiency and Circular Economy Promotion:** Life Cycle Assessment contributes to the promotion of resource efficiency and the principles of the circular economy. By assessing the potential for recycling, reusing, or repurposing building materials at the end of their life, LCA guides the industry toward practices that minimize waste generation and contribute to a more sustainable and circular construction model.
6. **Social and Economic Considerations:** Beyond environmental aspects, LCA also encompasses social and economic considerations. It evaluates the social impacts of material production, including factors such as labor conditions and community engagement. Additionally, LCA examines economic factors, providing insights into the financial aspects of material choices, including life cycle costs and economic feasibility.
7. **Regulatory Compliance and Standards Adherence:** Life Cycle Assessment facilitates compliance with evolving environmental standards and regulations in the construction industry. By providing a systematic approach to assessing sustainability metrics, LCA helps

ensure that building materials meet or exceed established environmental standards, positioning construction projects favorably in the regulatory landscape.

8. **Continuous Improvement and Innovation:** An essential purpose of Life Cycle Assessment is to foster a culture of continuous improvement and innovation within the construction industry. By regularly assessing and reevaluating materials in light of evolving knowledge and technological advancements, LCA contributes to the ongoing development of more sustainable and resilient building materials.

The primary objective of Life Cycle Assessment (LCA) is to provide the construction industry with a robust evaluation framework that encompasses environmental, social, and economic considerations. LCA aims to steer the industry toward a more sustainable and accountable future by assessing the whole life cycle of sustainable building materials.

4.2 LIFE CYCLE ASSESSMENT (LCA) METHODOLOGIES

Life Cycle Assessment (LCA) techniques serve as structured and standardized frameworks designed to comprehensively evaluate the ecological, societal, and economic impacts of products or systems throughout their entire life cycle, from inception to disposal. The overarching purpose of Life Cycle Assessment is to provide a thorough and unbiased evaluation, fostering well-informed decision-making that advances sustainability goals. LCA encompasses various methodologies, and the selection of a specific technique depends on the unique circumstances of the evaluation. The essential elements of LCA methodologies include goal definition and scope, life cycle inventory (LCI), life cycle impact assessment (LCIA), interpretation, improvement analysis, reporting and communication, and peer review [4].

In the initial stages of an LCA, goal definition, and scope establishment are critical components. This involves clearly identifying the goals and objectives of the assessment, such as comparing different materials, optimizing processes, or informing product design. System boundaries are then defined to establish the scope of the assessment, identifying the stages of the life cycle to be included, from raw material extraction to production, use, maintenance, and disposal. The subsequent steps involve the systematic gathering of data through life cycle inventory (LCI), which entails collecting information on inputs and outputs at each stage of the product's life cycle [5]. This data is then broken down into specific unit processes, standardizing the information for consistency and comparability. Following the LCI, the life cycle impact assessment (LCIA) identifies and quantifies potential environmental, social, and economic impacts based on the collected inventory data. The impacts are characterized and normalized to facilitate comparisons among different impact categories. The interpretation phase involves analyzing the results, identifying significant impacts or "hotspots," and conducting sensitivity analyses to assess the robustness of the findings [6].

In the final stages of an LCA, improvement analysis explores potential enhancements in environmental performance through scenario analysis and trade-off analysis. Results are presented in a clear format, emphasizing key findings, significant impacts, and areas for improvement [7]. Stakeholder engagement is integral throughout the process, ensuring transparency and credibility, while peer review provides external validation to enhance the overall quality and reliability of the LCA study. Overall, the systematic and comprehensive nature of LCA methodologies positions them as valuable tools for guiding sustainable decision-making in various industries.

4.2.1 Key components of LCA

Life Cycle Assessment (LCA) is a systematic and holistic approach to evaluating the environmental, social, and economic impacts of a product, process, or system throughout its entire life cycle. The process involves several key components, each playing a crucial role in providing a comprehensive and objective analysis. Figure 4.1 shows the key components and process flow diagram of LCA as discussed below [8, 9]:

1. **Goal and Scope Definition:**
    - **Purpose:** Clearly articulate the purpose and goals of the LCA. This involves specifying what the study aims to achieve, such as informing decision-making, comparing alternatives, or identifying areas for improvement.
    - **Scope:** Define the boundaries and limitations of the assessment, outlining the product or system being studied and specifying the life cycle stages to be included.
2. **Life Cycle Inventory (LCI):**
    - **Data Collection:** Collect comprehensive data on all inputs and outputs associated with each stage of the life cycle, from raw material extraction and production to use, maintenance, and disposal.
    - **Unit Processes:** Break down the life cycle into unit processes, identifying specific activities and subprocesses. Quantify the resource inputs, emissions, and waste associated with each unit process.
3. **Life Cycle Impact Assessment (LCIA):**
    - **Impact Categories:** Identify and characterize potential environmental, social, and economic impacts associated with the life cycle stages. Common impact categories include climate change, resource depletion, human toxicity, and biodiversity loss.

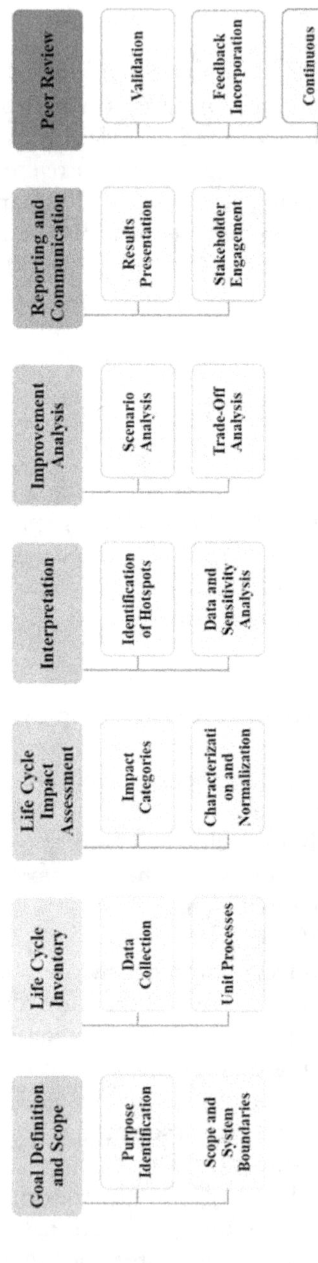

Figure 4.1 Key components and process flow diagram of LCA

   - **Characterization:** Convert raw LCI data into impact scores for each category, providing a quantitative measure of the potential effects.
4. **Interpretation:**
   - **Data Analysis:** Analyze the LCA results to draw conclusions and insights. This involves identifying significant impacts, hotspots, and areas for improvement.
   - **Sensitivity Analysis:** Assess the sensitivity of the results to variations in assumptions or data inputs. This helps evaluate the robustness of the findings.
5. **Improvement Analysis:**
   - **Scenario Analysis:** Explore different scenarios or alternatives to understand how changes in processes, materials, or technologies could impact the overall life cycle performance.
   - **Recommendations:** Develop recommendations for improving the environmental, social, and economic aspects of the product or system based on the LCA results.
6. **Reporting and Communication:**
   - **Results Presentation:** Present the LCA results in a clear and understandable manner, using visualizations and key performance indicators to communicate the findings.
   - **Stakeholder Engagement:** Engage stakeholders in the reporting process, providing opportunities for feedback and ensuring transparency in communication.
7. **Peer Review:**
   - **External Validation:** Subject the LCA study to external validation and peer review by independent experts. This helps ensure the quality, credibility, and objectivity of the assessment.
   - **Feedback Incorporation:** Incorporate feedback from peer review into the final LCA report, addressing any concerns or recommendations raised by external reviewers.
8. **Continuous Improvement:**
   - **Learning and Adaptation:** Promote a culture of continuous improvement by incorporating lessons learned from the LCA process into future assessments.
   - **Methodological Updates:** Stay informed about developments in LCA methodologies and adapt approaches to reflect advancements in the field.

These key components collectively guide practitioners through the LCA process, providing a structured framework for assessing and improving the sustainability performance of products, processes, or systems. The iterative nature of LCA encourages ongoing refinement and application of best practices in sustainability analysis.

4.2.3 Application of LCA to sustainable building materials

The application of Life Cycle Assessment (LCA) to sustainable building materials plays a pivotal role in comprehensively evaluating and enhancing the environmental, social, and economic performance of construction materials throughout their entire life cycle. LCA brings valuable insights to material selection, enabling a comparative analysis of different building materials and guiding architects, builders, and decision-makers toward choices with lower overall environmental footprints [10]. In the design and innovation phase, LCA facilitates scenario analysis, empowering designers and engineers to explore the environmental implications of various design choices and encouraging innovation in material design, construction methodologies, and architectural practices. Construction practices benefit from LCA as well, with the assessment and optimization of on-site activities and assembly processes, refining resource consumption and waste generation to minimize environmental impacts [11].

The energy efficiency and performance of sustainable building materials are evaluated through LCA, specifically in thermal analysis, aiding the selection of materials that contribute to enhanced energy efficiency. Life cycle inventory (LCI) data collection for materials covers the entire spectrum from raw material extraction to end-of-life scenarios, forming a comprehensive basis for assessing environmental impacts [12]. LCA extends its influence to waste management and recycling considerations, assessing the environmental consequences of different end-of-life scenarios and encouraging the use of materials aligned with circular economy principles. Furthermore, LCA quantifies the carbon footprint of building materials by analyzing greenhouse gas emissions associated with their life cycle, contributing crucial information for selecting materials with lower embodied carbon. Figure 4.2 discusses the application of LCA in sustainable building materials [13].

LCA supports policy compliance and certification by aiding in adherence to environmental regulations and standards related to sustainable construction. It provides a quantitative basis for certification programs, emphasizing the environmental performance of building materials. Beyond environmental considerations, LCA assesses the impact of building materials on indoor air quality and occupant health, contributing to the creation of healthier indoor environments [14]. The transparent reporting of LCA results to various stakeholder's fosters trust and accountability in the construction industry [15]. Lastly, LCA supports a continuous improvement cycle by providing feedback on the environmental performance of building materials, encouraging the industry to evolve, adopt more sustainable practices, and embrace emerging technologies.

The application of LCA to sustainable building materials is a powerful tool for promoting environmentally responsible construction practices. It

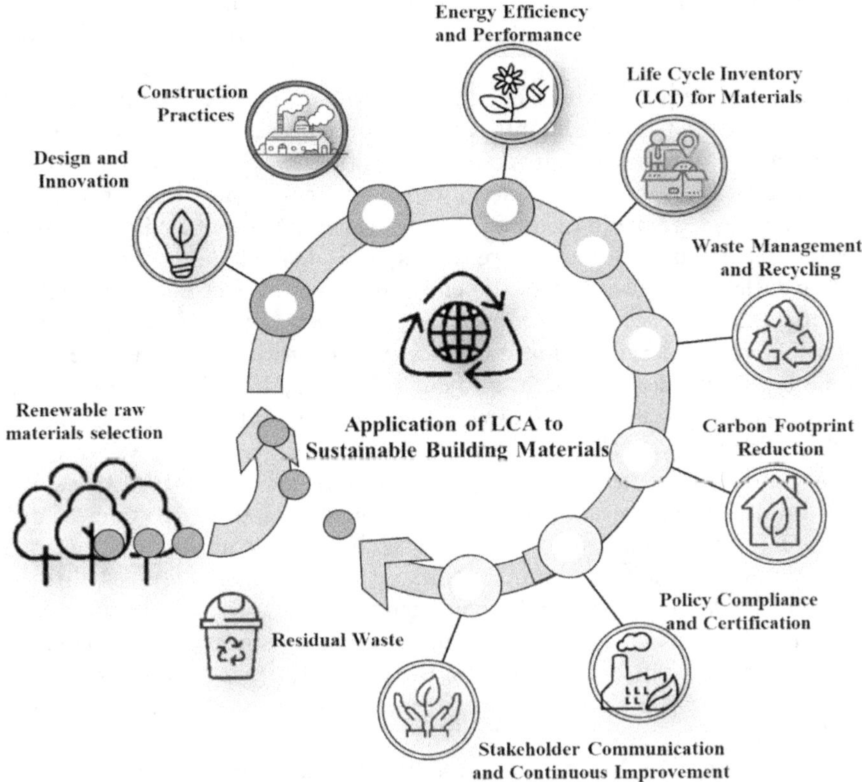

Figure 4.2 Application of LCA into sustainable building materials

guides decision-makers in selecting materials, informs design choices, and encourages a holistic approach to sustainability in the construction industry. Through a comprehensive life cycle perspective, LCA contributes to building a more resilient and sustainable built environment.

4.3 ENVIRONMENTAL IMPACT ANALYSIS OF LCA

The environmental impact analysis within the framework of Life Cycle Assessment (LCA) constitutes a critical and systematic evaluation of the potential environmental effects associated with a product, process, or system throughout its entire life cycle. This multifaceted analysis spans various stages, commencing with raw material extraction and manufacturing and extending through use, maintenance, and eventual disposal [16]. Key aspects of the environmental impact analysis in LCA include the assessment of resource depletion during raw material extraction, considering both renewable and non-renewable resources, and acknowledging potential ecosystem disruptions. In production processes, a detailed analysis involves evaluating energy consumption, including the sources of energy used and their environmental implications, as well as examining emissions and pollutants generated during manufacturing, encompassing considerations of greenhouse gas emissions, air pollutants, and water discharges [17].

Transportation is a critical stage in the environmental impact analysis, requiring the evaluation of the carbon footprint associated with transporting raw materials, intermediate products, and finished goods. Construction and installation phases assess the environmental impact of on-site assembly, including energy and resource use, emissions, and waste generation. Maintenance and renovation activities are scrutinized for their resource intensity, energy consumption, and emissions [18, 19]. The end-of-life and disposal stage involves evaluating the environmental impact of scenarios such as disposal, recycling, or reuse, considering emissions, energy requirements, and waste generation. Cumulative impacts across all life cycle stages are considered to understand the overall contribution to environmental degradation. Furthermore, specific aspects such as biotic and abiotic resource depletion, eco-toxicity and human toxicity, climate change impact through greenhouse gas emissions, and the impact on water and air quality are comprehensively assessed to provide a holistic understanding of the environmental ramifications associated with the product, process, or system under scrutiny. Figure 4.3 shows the various environmental impact analyses of LCA [20].

The environmental impact analysis in LCA aims to provide a comprehensive and quantitative understanding of the environmental implications associated with a product or process. This information is valuable for decision-makers seeking to minimize negative environmental effects and promote more sustainable practices.

4.4 SOCIAL AND ECONOMIC CONSIDERATIONS

Life Cycle Assessment for sustainable building materials entails a conscientious consideration of social and economic dimensions throughout each stage of a material's life cycle. In the extraction phase, emphasis is placed on fair labor practices and positive community engagement. During production, worker well-being and diversity are prioritized, ensuring ethical practices. Transportation processes are scrutinized for their social impact, from supply chain ethics to engaging with communities along routes. Construction and installation involve stringent safety measures for workers and active participation from local communities. Maintenance and renovation activities focus on skill development opportunities and collaborative efforts with communities. End-of-life considerations include creating

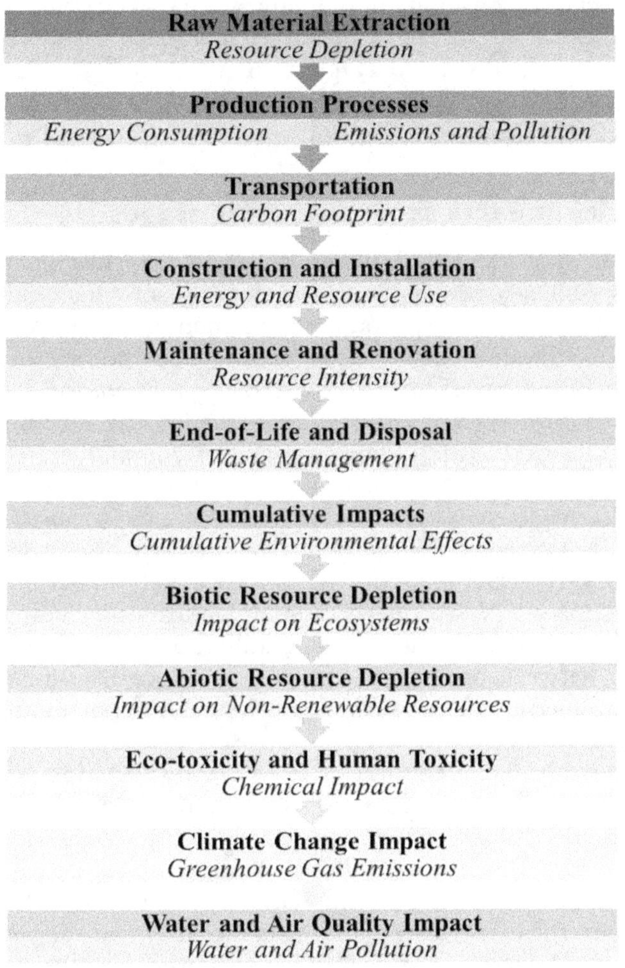

Figure 4.3 The environmental impact analysis of Life Cycle Assessment (LCA)

employment opportunities in waste management and addressing potential health impacts on nearby communities [21]. In terms of economic factors, comprehensive life cycle costing and highlighting the long-term economic benefits and return on investment associated with sustainable building materials are essential. Job creation opportunities and positive impacts on local economies further underscore the economic benefits. The market competitiveness of sustainable materials is enhanced by aligning with consumer preferences and positioning them as resilient choices in evolving market conditions. Stakeholder engagement strategies involve participatory decision-making and transparent communication, fostering positive community contributions and mitigating potential negative impacts. Integrating corporate social responsibility into organizational values ensures an ethical approach to decision-making. This holistic approach ensures that sustainable building materials not only meet environmental standards but also contribute significantly to social equity, economic prosperity, and community well-being [22].

4.4.1 Social impacts across life cycle stages

In the realm of sustainable building materials, the social aspects of the life cycle, specifically pertaining to the stages of raw material extraction, production processes, transportation, construction and installation, maintenance and renovation, and end-of-life and disposal, are of paramount importance. In the context of raw material extraction, the social impacts are evaluated through the lens of labor conditions, encompassing considerations of worker safety, fair wages, and community relations [23]. Additionally, potential community displacement due to resource extraction activities is scrutinized, with a focus on analyzing the socio-economic consequences for affected communities. As the life cycle progresses to production processes, the examination extends to occupational health and safety implications for workers involved in manufacturing, including an evaluation of health and safety conditions and adherence to labor standards. The relationship between production facilities and local communities is also assessed, considering factors such as employment opportunities and community development [24].

Moving further along the life cycle, the transportation stage involves investigating social impacts along the supply chain, particularly focusing on the labor practices of logistics and transportation providers. Concurrently, potential social consequences of transportation-related activities on communities are analyzed, with an emphasis on health and safety considerations [25]. In the construction and installation phase, the social impacts are examined in terms of labor practices, worker well-being, and community engagement during construction activities. The involvement and participation of local communities in construction processes are assessed, emphasizing inclusivity and empowerment [26]. As the life cycle advances to maintenance and renovation, the evaluation extends to the social benefits related to skilled labor opportunities during these activities, considering the potential for skill development and employment. Community involvement in maintenance and renovation projects is also scrutinized for social cohesiveness and collaborative efforts. Finally, in the end-of-life and disposal stage, the social impacts are examined concerning waste management activities, including an evaluation of employment opportunities in waste management and potential social benefits [25]. Additionally, the potential health impacts on communities surrounding disposal sites are considered, taking into account factors such as air and water quality. Overall, this comprehensive social analysis throughout the life cycle contributes to a more nuanced understanding of the sustainable building material's broader societal implications [27].

4.4.2 Economic factors and life cycle performance of sustainable building

The economic factors and life cycle performance of sustainable building materials play a pivotal role in shaping the viability and success of environmentally conscious construction practices. Economic considerations extend beyond initial costs to encompass the entire life cycle, including maintenance and end-of-life stages. Comprehensive life cycle costing, assessing return on investment, and evaluating market competitiveness are key components [13], [28].

1. **Life Cycle Cost Analysis:** Sustainable building materials undergo rigorous life cycle cost analyses, evaluating expenses across production, transportation, installation, maintenance, and disposal phases. This comprehensive approach ensures that the economic benefits of sustainable materials are accurately assessed over time, guiding decision-makers toward financially prudent choices.
2. **Return on Investment (ROI):** The economic performance of sustainable building materials is gauged through the lens of return on investment. By demonstrating the long-term benefits, including energy savings, reduced maintenance costs, and potential increases in property value, these materials justify their upfront investments and position themselves as financially sound choices for builders and investors.
3. **Employment Opportunities:** Sustainable building practices contribute to job creation throughout the life cycle. From the extraction of raw materials to manufacturing, construction, and maintenance, these practices stimulate employment, fostering economic development and promoting socially responsible construction practices.
4. **Local Economic Impact:** The use of sustainable building materials has a direct impact on local economies. By supporting local industries involved in the production and distribution of these materials, construction projects contribute to the economic resilience and sustainability of nearby communities.
5. **Market Competitiveness:** Sustainable building materials enhance market competitiveness by aligning with consumer preferences for environmentally conscious choices. As green building practices become more prevalent, materials that prioritize sustainability gain a competitive edge, attracting environmentally conscious consumers and meeting evolving market demands.
6. **Long-Term Economic Resilience:** The economic resilience of sustainable building materials is a critical consideration. These materials are designed to withstand changing market conditions, regulatory

landscapes, and consumer preferences, ensuring a long-term positive economic impact and contributing to the industry's adaptability.

In conclusion, the economic factors associated with sustainable building materials extend beyond mere cost considerations. They encompass a strategic and comprehensive assessment of life cycle costs, return on investment, employment generation, local economic impact, market competitiveness, and long-term economic resilience. By aligning economic benefits with sustainable practices, the construction industry can foster a built environment that is not only environmentally responsible but also economically robust and socially equitable.

4.4.3 Stakeholder engagement and community impact

The economic factors and life cycle performance of sustainable building materials are instrumental in shaping the viability and success of environmentally conscious construction practices. Going beyond initial costs, a comprehensive life cycle cost analysis assesses expenses across production, transportation, installation, maintenance, and disposal phases, ensuring that the economic benefits of sustainable materials are accurately evaluated over time. Return on investment (ROI) becomes a crucial metric, demonstrating the long-term advantages, including energy savings, reduced maintenance costs, and potential increases in property value, positioning sustainable materials as financially sound choices [29]. Sustainable building practices contribute to job creation, stimulate employment, and foster economic development throughout the life cycle, while also positively impacting local economies by supporting local industries. Market competitiveness is enhanced as sustainable materials align with consumer preferences for environmentally conscious choices, gaining a competitive edge and meeting evolving market demands. The long-term economic resilience of sustainable building materials is critical, designed to withstand changing market conditions, regulatory landscapes, and consumer preferences, contributing to the industry's adaptability.

Stakeholder engagement and community impact are integral components of sustainable building practices, reflecting a commitment to inclusive decision-making, transparent communication, and positive contributions to local communities. Effective stakeholder engagement involves actively involving individuals and groups with an interest in the project, employing participatory decision-making, transparent communication, inclusivity, and feedback mechanisms. Community impact considerations focus on maximizing positive contributions, implementing mitigation strategies for potential negative effects during construction and maintenance activities, prioritizing social responsibility, and integrating ethical decision-making processes. Social responsibility in sustainable building practices includes corporate social responsibility (CSR), active contribution to community development, and ethical considerations in decision-making, ensuring that construction practices align with social and environmental goals. In conclusion, stakeholder engagement and community impact form the cornerstone of sustainable building practices, fostering not only environmental sustainability but also social and economic resilience in the built environment [30].

4.5 TECHNOLOGICAL INNOVATIONS IN SUSTAINABLE MATERIALS

Technological innovations in sustainable materials have played a pivotal role in advancing environmentally friendly construction practices, aiming to mitigate the environmental impact of building materials across their entire life cycle. Noteworthy advancements include the development of highly durable and sustainable bamboo composites achieved through treating and compressing bamboo fibers, providing strength comparable to traditional materials while being renewable and biodegradable [31]. The utilization of recycled plastics in creating building blocks and construction components addresses plastic waste concerns, offering a durable alternative and promoting circular economy principles. Engineered wood products, such as cross-laminated timber (CLT) and laminated veneer lumber (LVL), showcase advancements in wood engineering, presenting sustainable alternatives with increased strength and versatility [32]. Transparent solar panels, integrating solar cells into building materials like windows and facades, enable the generation of solar energy without compromising aesthetics, enhancing energy efficiency and sustainability. Self-healing concrete, incorporating microorganisms or encapsulated healing agents, autonomously repairs cracks, extending the lifespan of structures, reducing maintenance needs, and minimizing environmental impact. Green roofs and walls, featuring vegetation integration, enhance insulation, mitigate urban heat island effects, improve air quality, and provide wildlife habitat. 3D printing with recycled materials reduces waste, allows complex designs, and promotes a circular economy [33]. Light-transmitting concrete, with embedded optical fibers or translucent materials, improves natural lighting, reducing the need for artificial lighting and enhancing energy efficiency [34–38]. Biodegradable building materials, including bioplastics and composites, minimize environmental impact during disposal, particularly for temporary structures. Integration of smart materials and nanotechnology enhances the performance of building materials, improving energy efficiency, durability, and functionality. Hybrid solar tiles, seamlessly blending solar panels into roofing tiles, enable solar energy generation while integrating with building design [39]. These innovations collectively contribute to a sustainable and resilient paradigm in construction, emphasizing efficiency, circularity, and environmental consciousness. A detailed description of innovative sustainable materials is discussed in Figure 4.4 and Table 4.1.

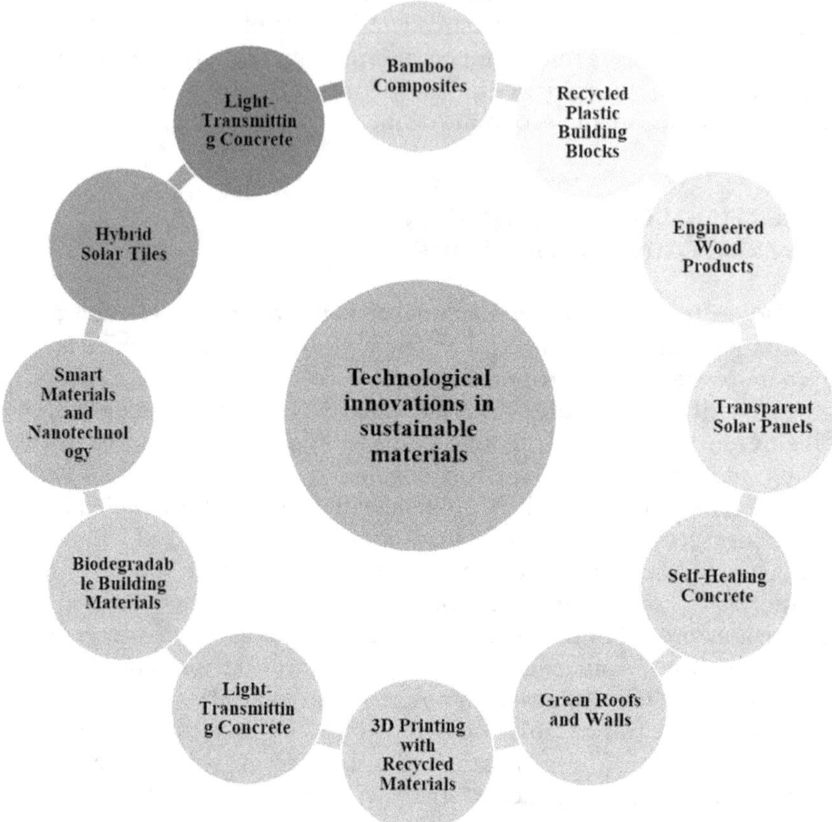

Figure 4.4 Technological innovations in sustainable materials

These technological innovations demonstrate a commitment to sustainability in the construction industry, offering alternatives that reduce environmental impact, enhance energy efficiency, and contribute to the overall resilience of the built environment.

This table provides a general comparative analysis and its importance to consider project-specific requirements and regional factors when selecting materials. Each material has its unique advantages and considerations, and the appropriateness of a particular material depends on the specific goals and context of a construction project.

4.5.1 Advancements in material science

Advancements in material science have significantly contributed to the development of sustainable building materials, addressing environmental

Table 4.1 Comparative analysis of the sustainable building materials

S. No.	*Material*	*Renewability*	*Strength/ Durability*	*Insulation/ Recyclability*	*Cost*	*Aesthetics*	*Installation Flexibility*	*Environmental Impact*
1	Bamboo composites	High	High/high	Moderate/ high	Moderate	Natural appearance	Flexible	Low
2	Recycled plastic building blocks	Low (depends on source)	Variable/ high	Low/variable	Variable	Variable	Flexible	Medium
3	Engineered wood products	High	High/high	High/variable	Moderate	Variable	Flexible	Medium
4	Transparent solar panels	Medium (depends on production)	Variable/ medium	Low/low	High	Transparent	Requires specific design	Medium
5	Self-healing concrete	Low	High/high	Variable/low	High	Variable	Standard	Medium
6	Green roofs and walls	High	Variable/ variable	High/variable	Variable	Green aesthetic	Variable	High
7	3D printing with recycled materials	Medium	Variable/ variable	Variable/high	Variable	Variable	Flexible	Medium
8	Light-transmitting concrete	Low (depends on source)	High/high	Low/low	High	Variable	Variable	Low
9	Biodegradable building materials	High	Variable/ variable	Variable/high	Variable	Variable	Flexible	High
10	Smart materials and nanotechnology	Variable	Variable/ variable	High/variable	Variable	Variable	Variable	Medium
11	Hybrid solar tiles	Variable	Variable/ high	Variable/low	High	Integration with roofing	Variable	High

concerns and promoting energy efficiency. Here are some key advancements in material science applied to sustainable construction [40]:

1. **High-Performance Concrete with Supplementary Cementitious Materials (SCMs):** The use of supplementary cementitious materials (SCMs) such as fly ash, slag, and silica fume has elevated high-performance concrete to a prominent position among sustainable construction materials. These innovations not only bolster the robustness and longevity of concrete but also lead to a substantial decrease in carbon emissions. High-performance concrete with supplementary cementitious materials (SCMs) plays a crucial role in contemporary building methods by reducing environmental impact and boosting sustainability [41].
2. **Engineered Wood Products:** Engineered wood products, such as cross-laminated timber (CLT) and laminated veneer lumber (LVL), are sustainable alternatives to conventional wood, made possible by innovative processing processes. These products demonstrate enhanced durability and adaptability, while also significantly contributing to the reduction of the need for natural wood resources. The incorporation of engineered wood into construction methods represents a notable advancement toward the use of ecologically sustainable and durable building materials [42].
3. **Recycled Aggregates in Concrete:** The use of recycled concrete aggregates in the creation of fresh concrete is a significant progress in sustainable building. This approach not only decreases the need for natural aggregates but also eliminates waste and decreases the overall carbon footprint of concrete. Construction projects contribute to circular economy concepts and promote a more ecologically responsible approach to material consumption by integrating recycled resources into the concrete mix [43].
4. **Bamboo Composite Materials:** Bamboo composite materials are created by applying treatment and compression processes to bamboo, resulting in a renewable and biodegradable alternative that possesses strength equivalent to conventional construction materials. This sustainable building materials innovation is in line with the increasing focus on environmentally friendly techniques in the construction industry. Bamboo composites offer both structural integrity and emphasize the significance of carefully using natural resources [44].
5. **Green Roofs and Walls with Advanced Planting Systems:** Green roofs and walls equipped with sophisticated planting systems: Green roofs and walls, along with enhanced planting systems and substrate technology, have become essential elements of sustainable building. These advancements lead to improved insulation, reducing the impact of the urban heat island effect and promoting biodiversity. Green roofs and walls exemplify the integration of nature into the built environment,

demonstrating a harmonious approach to building that effectively tackles ecological problems and fosters sustainable urban development [45].

6. **Self-Healing Concrete:** The introduction of microbes or encapsulated healing agents into concrete allows it to repair itself, which is a significant breakthrough in building materials. This invention enhances the durability of concrete structures, diminishes the requirement for maintenance, and mitigates the ecological consequences of repairs. Self-healing concrete demonstrates a forward-thinking and environmentally friendly approach to building infrastructure [46].
7. **Transparent Solar Panels:** Transparent solar panels, which are effortlessly incorporated into windows and facades, demonstrate the merging of sustainable energy with architectural design. This innovation enables buildings to harness solar energy while preserving natural lighting and beauty. Transparent solar panels enhance energy efficiency and showcase the potential of multifunctional, environmentally friendly building materials [47].
8. **Biodegradable Building Materials:** The advancement of biodegradable substitutes for conventional building materials represents a notable step toward the principles of a circular economy. These materials effectively reduce the negative effects on the environment when they are disposed of, providing a sustainable solution to the difficulties associated with waste management. Biodegradable construction materials demonstrate a devotion to responsible material usage and emphasize the industry's commitment to reducing its ecological impact [48].
9. **High-Performance Insulation Materials:** Nanotechnology-driven innovations in insulation materials have ushered in a new era of high-performance solutions. These advanced insulation materials exhibit enhanced thermal performance, reducing energy consumption for heating and cooling. The integration of nanotechnology into insulation materials represents a key strategy for achieving energy efficiency in buildings while addressing the environmental impact of traditional insulation [49].
10. **Smart Building Materials:** The integration of sensors, actuators, and responsive materials has given rise to a new generation of smart building materials. These materials enable adaptive building systems, optimizing energy usage and enhancing occupant comfort. By responding dynamically to environmental conditions, smart building materials exemplify a forward-thinking approach to construction that prioritizes both sustainability and technological innovation [50].
11. **Recycled Plastic Building Blocks:** The utilization of recycled plastics for constructing building blocks is a pioneering solution to address the global challenge of plastic waste. This innovation not only reduces plastic pollution but also provides a durable alternative to traditional construction materials. Recycled plastic building blocks contribute to

a circular economy by repurposing waste materials for constructive and sustainable building practices [51].

12. **Light-Transmitting Concrete:** Light-transmitting concrete, achieved through the embedding of optical fibers or translucent materials, represents a revolutionary approach to construction materials. This innovation improves natural lighting within structures, reducing the need for artificial lighting and enhancing overall energy efficiency. Light-transmitting concrete exemplifies the fusion of aesthetics and sustainability in modern architecture [52].
13. **Low-Carbon Cement Alternatives:** The development of low-carbon cement alternatives, such as geo-polymer and calcium aluminate cements, addresses the environmental impact associated with traditional Portland cement. These alternatives reduce carbon emissions, contributing to sustainable construction practices. The exploration and adoption of low-carbon cement alternatives underscore the industry's commitment to mitigating climate change through innovative material science [53].

These advancements showcase the commitment of material scientists and researchers to create sustainable alternatives that contribute to environmentally friendly and energy-efficient construction practices. Continued research and innovation in material science will likely yield further breakthroughs, shaping the future of sustainable building materials [72–76].

4.5.2 Integration of smart technologies

The integration of smart technologies in Life Cycle Assessment (LCA) processes for sustainable building materials represents a significant advancement in the construction industry. Smart technologies enhance the accuracy, efficiency, and depth of LCA by providing real-time data, monitoring, and analytics throughout the entire life cycle of building materials. The details of smart technologies integrated into LCA for sustainable building materials are discussed in this section [54, 55]:

1. **Sensors for Data Collection:** Smart sensors embedded within sustainable building materials serve as the frontline data collectors in the Life Cycle Assessment (LCA) process. These sensors continually monitor and gather real-time information on critical parameters such as energy consumption, structural conditions, and environmental influences. The data collected provides a dynamic and accurate representation of the material's performance at different stages of its life cycle.
2. **Internet of Things (IoT) for Monitoring:** The Internet of Things (IoT) plays a pivotal role in extending connectivity and monitoring capabilities. IoT devices are employed to transmit data related to energy usage, temperature variations, and other relevant parameters. This

continuous stream of information enables stakeholders to monitor sustainable building materials in real-time, identifying inefficiencies or opportunities for improvement promptly.

3. **Blockchain for Transparency:** Blockchain technology is integrated into the LCA framework to establish a transparent and tamper-proof record of the entire life cycle of sustainable building materials. This decentralized ledger ensures traceability, accountability, and transparency in the supply chain, allowing stakeholders to confidently assess the material's environmental impact.
4. **Big Data Analytics for Decision-Making:** Big data analytics processes the vast amount of data collected from sensors and other sources. This analytical approach offers insights into material performance, energy consumption patterns, and environmental impacts. Stakeholders can leverage these analytics to make informed decisions and optimize strategies throughout the material's life cycle.
5. **Machine Learning for Predictive Analysis:** Machine learning algorithms are employed to analyze historical data, enabling predictive analysis for future trends and potential environmental impacts. By learning from past patterns, machine learning contributes to proactive decision-making, allowing stakeholders to anticipate material behavior and optimize LCA strategies accordingly.
6. **Augmented Reality (AR) for Visualization:** Augmented reality (AR) technologies provide stakeholders with a visual representation of the environmental impacts of sustainable building materials. This immersive experience allows for a deeper understanding of the material's effects in real-world scenarios, facilitating better-informed decision-making.
7. **Digital Twins for Simulation:** Digital twins create virtual replicas of sustainable building materials, enabling simulation and analysis of various scenarios. This technology allows stakeholders to assess material performance under different conditions, contributing to the optimization of life cycle strategies and the identification of potential improvements.
8. **Mobile Apps for User Engagement:** Mobile applications engage end-users in the sustainability efforts of building materials. These apps provide information on material composition, guidelines for usage, and proper disposal methods. By fostering user awareness and responsibility, mobile apps contribute to a more sustainable approach to building material usage.
9. **Remote Monitoring and Control Systems:** Remote monitoring systems, integrated into sustainable building materials, allow centralized control and adjustment of material parameters. This capability enhances operational efficiency by enabling real-time adjustments based on environmental conditions, optimizing energy usage, and minimizing environmental impact.

10. **Wireless Communication for Connectivity:** Wireless communication technologies facilitate seamless connectivity between smart devices and sustainable building materials. This interconnected network ensures efficient data exchange and collaboration, promoting real-time monitoring and control. The integration of wireless communication enhances the overall efficiency and accuracy of LCA processes for sustainable building materials.

By integrating smart technologies into LCA processes, the construction industry can gain deeper insights into the environmental performance of sustainable building materials. This integration supports more informed decision-making, improves the efficiency of life cycle assessments, and contributes to the overall advancement of sustainable construction practices.

4.5.3 Role of innovation in Life Cycle improvement

In the realm of sustainable building materials, innovation plays a pivotal role in driving continuous improvement throughout the life cycle [56]. This encompasses the extraction of raw materials, manufacturing processes, construction, usage, and eventual disposal or recycling. The multifaceted role of innovation in life cycle improvement of sustainable building materials can be outlined as follows:

1. **Material Efficiency and Optimization:** Innovation contributes to the development of materials that are inherently more efficient in terms of resource use. This involves optimizing the composition of materials to reduce waste during production and construction, ensuring a more sustainable life cycle from the outset.
2. **Alternative and Renewable Materials:** Innovations introduce alternative and renewable materials that replace traditional, resource-intensive options. For example, the exploration of bamboo composites, recycled plastics, or agricultural waste-based materials as substitutes for conventional building materials reduces environmental impact and enhances sustainability.
3. **Energy-Efficient Manufacturing Processes:** Innovations in manufacturing processes focus on energy efficiency and emissions reduction. This includes the implementation of advanced technologies, such as 3D printing or automated manufacturing, to decrease energy consumption and minimize the carbon footprint associated with production.
4. **Smart Technologies and Monitoring:** Integration of smart technologies, such as sensors and IoT devices, allows for real-time monitoring of building materials' performance. This innovation enables data-driven decision-making throughout the life cycle, identifying areas for improvement, optimizing energy usage, and enhancing overall efficiency.

5. **Circular Economy Practices:** Innovations promote circular economy practices by developing building materials that are easily recyclable or biodegradable. This shift toward a circular approach mitigates waste generation, encourages material reuse, and minimizes the environmental impact associated with disposal.
6. **Durability and Longevity:** Innovations in material science lead to the development of durable and long-lasting building materials. Materials with extended lifespans reduce the frequency of replacements and repairs, contributing to resource conservation and overall life cycle improvement.
7. **Low-Impact Construction Techniques:** Innovation extends to construction techniques that minimize environmental impact. Prefabrication, modular construction, and other low-impact methodologies reduce the consumption of resources, decrease construction waste, and enhance the sustainability of building projects.
8. **Energy-Positive Building Materials:** Advancements in sustainable build ing materials include those that generate or store energy. Transparent solar panels, for instance, not only contribute to energy generation but also integrate seamlessly into architectural designs, showcasing the potential of combining functionality with sustainability.
9. **Digitalization and Building Information Modeling (BIM):** Innovation in digitalization and BIM enables more efficient planning, design, and construction processes. This technology facilitates accurate simulations, optimizing material use, energy efficiency, and overall sustainability throughout the life cycle.
10. **Community and Stakeholder Engagement:** Innovative approaches include involving communities and stakeholders in the decision-making process. This ensures that the selection and use of sustainable building materials align with local preferences, cultural considerations, and environmental goals, fostering a more holistic approach to life cycle improvement.
11. **Carbon Capture and Utilization:** Emerging innovations explore the integration of carbon capture and utilization technologies directly into building materials. This approach sequesters carbon dioxide, mitigating greenhouse gas emissions and contributing to the overall life cycle improvement by transforming a potential pollutant into a valuable resource.
12. **Adaptive and Responsive Materials:** Innovations in adaptive and responsive materials allow buildings to dynamically respond to environmental conditions. These materials can regulate temperature, lighting, and energy usage, further optimizing resource consumption and enhancing the overall sustainability of the built environment.

Innovation is the driving force behind the evolution of sustainable building materials. By continually pushing the boundaries of what is possible, researchers, engineers, and architects contribute to a built environment that is not only environmentally friendly but also economically viable and socially responsible. The role of innovation is dynamic, adapting to new challenges and opportunities, and it remains a key catalyst for advancing sustainability in construction practices.

4.6 CHALLENGES AND OPPORTUNITIES

The Life Cycle Assessment (LCA) of sustainable building materials presents both challenges and opportunities, reflecting the complexities of assessing environmental impacts across the entire life cycle of construction materials. Understanding and addressing these factors is crucial for promoting sustainable practices in the construction industry. Here are some key challenges and opportunities in the LCA of sustainable building materials [57, 58]:

4.6.1 Obstacles in implementing LCA

Implementing Life Cycle Assessment (LCA) in sustainable building technology encounters specific obstacles that are crucial to address for its effective integration. The complexity of building processes, involving various materials and intricate life cycle stages, poses a challenge to conducting comprehensive LCAs [59]. Limited data availability for sustainable building materials and technologies can hinder accurate assessments, as can the absence of standardized methodologies tailored to the unique characteristics of the construction industry. The upfront costs associated with detailed LCA studies may deter some organizations from embracing this approach, especially smaller businesses. Additionally, the lack of uniform regulations and standards across the sustainable building sector can create inconsistencies and impede the widespread adoption of LCA practices [60]. Overcoming these challenges necessitates collaborative efforts to develop industry-specific LCA standards, improve data availability, and educate stakeholders on the long-term benefits of integrating life cycle assessments into sustainable building practices.

1. **Data Availability and Quality:** Gathering comprehensive and accurate data at every stage of a sustainable building material's life cycle is a persistent challenge. The availability and reliability of data can vary, impacting the precision of Life Cycle Assessments (LCAs). To address this challenge, efforts must be directed toward investing in standardized databases, refining data collection methodologies, and fostering transparency in reporting to enhance the overall reliability of LCA data [61].
2. **System Boundaries and Scope Definition:** The challenge lies in defining the boundaries of the system and determining the scope of the

assessment. These aspects can be subjective and influence the outcomes of an LCA. However, this challenge presents an opportunity to meticulously define system boundaries, adopt a cradle-to-cradle approach, and engage stakeholders in a collaborative effort to ensure a comprehensive assessment that encapsulates all relevant environmental impacts.

3. **Dynamic Nature of Building Designs:** Building designs and functionalities may evolve over time, rendering initial LCAs less relevant. The challenge is to accommodate the dynamic nature of construction projects. The opportunity here lies in conducting periodic LCAs and integrating flexibility into designs to anticipate and adapt to future changes, ensuring that the assessment remains pertinent throughout the building's entire life cycle.
4. **Interactions and Trade-offs:** Addressing trade-offs between different environmental impact categories and understanding the intricate interactions between various life cycle stages present a considerable challenge. However, this challenge can be turned into an opportunity by employing multi-criteria decision analysis. This approach allows for the evaluation of trade-offs and consideration of cumulative impacts, fostering a more holistic decision-making process.
5. **Lack of Harmonized Methodologies:** The absence of harmonized methodologies across different LCAs poses a challenge, as varying approaches hinder the comparability of results. This challenge underscores the opportunity to encourage the widespread adoption of standardized methodologies, such as those outlined in ISO 14040 and ISO 14044, promoting consistency and facilitating meaningful comparisons across assessments [62].

4.6.2 Opportunities for improvement

1. **Innovative Material Development:** There exists a substantial opportunity in the development and integration of innovative materials with lower environmental impacts into construction practices. By exploring and adopting alternatives, the construction industry can align with sustainability goals and contribute to the ongoing evolution of environmentally friendly building materials.
2. **Advancements in Technology:** The rapid advancements in technology, including machine learning and big data analytics, present an opportunity to enhance LCA methodologies. Leveraging these technologies can improve data accuracy, streamline the assessment process, and contribute to a more sophisticated understanding of the environmental impacts associated with sustainable building materials.
3. **Circular Economy Practices:** Embracing circular economy principles represents a significant opportunity in the quest for sustainable

building materials. Designing materials with reuse, recycling, or repurposing in mind can substantially reduce the overall environmental impact, promote resource efficiency, and contribute to a circular and regenerative approach to construction.

4. **Stakeholder Engagement:** Involving stakeholders, ranging from manufacturers to end-users, in the LCA process offers a unique opportunity to ensure a comprehensive understanding of environmental impacts. Engaging stakeholders fosters collective responsibility and encourages the adoption of sustainable practices throughout the construction industry.
5. **Policy and Regulation Support:** The development and implementation of supportive policies and regulations present a pivotal opportunity to incentivize the use of sustainable building materials. By creating an enabling environment, policymakers can drive industry-wide adoption of environmentally friendly practices, aligning construction activities with broader sustainability goals.
6. **Education and Awareness:** There is a substantial opportunity to increase awareness and education within the construction industry about the significance of LCA and sustainable practices. By fostering a culture of responsibility and informed decision-making, education and awareness initiatives can contribute to the widespread adoption of sustainable building materials.
7. **Life Cycle Costing Integration:** Integrating life cycle costing alongside LCA provides a comprehensive assessment that considers both economic and environmental aspects. This opportunity allows decision-makers to make informed choices, aligning financial considerations with sustainability goals and ensuring the selection of building materials that are not only environmentally friendly but economically viable.
8. **Collaboration and Knowledge Sharing:** The opportunity lies in fostering collaboration among researchers, industry professionals, and policymakers. By encouraging knowledge sharing, best practices, and innovations in LCA methodologies, stakeholders can collectively contribute to the refinement and widespread adoption of sustainable building materials, promoting a collaborative and informed approach to sustainability in construction.

4.6.3 Regulatory landscape and compliance

Understanding the regulatory landscape and ensuring compliance play a crucial role in the Life Cycle Assessment (LCA) of sustainable building materials. Regulatory frameworks are instrumental in shaping environmental standards, guiding manufacturers, and influencing the adoption of sustainable practices in the construction industry.

Regulatory Frameworks: The regulatory landscape governing the LCA of sustainable building materials encompasses a diverse range of international, national, and regional standards. Organizations like the International Organization for Standardization (ISO) provide guidelines, such as ISO 14040 and ISO 14044, outlining principles and requirements for conducting credible LCAs. Additionally, many countries have established specific regulations addressing environmental impact assessments and sustainable construction practices.

Environmental Labelling and Certification Programs: Several regions have introduced environmental labeling and certification programs that mandate or encourage the use of sustainable building materials. These programs often require adherence to specific environmental performance criteria, verified through LCAs. Compliance with these criteria allows manufacturers to label their products as environmentally certified, providing transparency to consumers and stakeholders.

Energy Efficiency Standards: Regulations related to energy efficiency also intersect with the LCA of building materials. Governments worldwide are increasingly adopting energy performance standards that impact the manufacturing and use of construction materials. Compliance with these standards may necessitate improvements in energy efficiency throughout the life cycle of materials, influencing material selection and manufacturing processes.

Waste Management Regulations: Effective waste management is a critical aspect of sustainable building practices. Regulatory frameworks addressing waste reduction, recycling, and disposal impact the end-of-life phase of building materials. Compliance with waste management regulations requires a thorough consideration of the recyclability and biodegradability of materials, emphasizing a cradle-to-cradle approach.

Carbon Emission Targets: Many jurisdictions have established carbon emission reduction targets to mitigate climate change. Compliance with these targets involves assessing the carbon footprint of building materials through LCAs. Sustainable building materials should align with these emission reduction goals, influencing choices in material selection, manufacturing processes, and transportation [70, 71].

Market Access Requirements: To access certain markets, manufacturers may need to comply with specific environmental requirements. This could include demonstrating adherence to sustainability standards or providing evidence of LCA compliance. Understanding and meeting these market access requirements are essential for manufacturers seeking to position their sustainable building materials in global markets.

Challenges in Compliance: While regulatory frameworks provide a necessary structure for promoting sustainable building practices, challenges in compliance may arise. These challenges include variations in regulatory requirements across regions, the need for standardized assessment methodologies, and the dynamic nature of regulatory updates. Navigating these challenges requires a proactive approach from manufacturers and a commitment to staying informed about evolving regulatory landscapes.

Opportunities for Industry Leadership: Proactively embracing and exceeding regulatory requirements can position companies as leaders in sustainable construction. By integrating rigorous LCAs into their practices and ensuring compliance with emerging standards, manufacturers can not only meet regulatory expectations but also contribute to the advancement of environmentally responsible building materials.

In summary, the regulatory landscape and compliance requirements significantly influence the Life Cycle Assessment of sustainable building materials. Manufacturers, policymakers, and stakeholders must collaborate to navigate these regulations effectively, fostering a construction industry that prioritizes environmental sustainability and meets the evolving expectations of a global regulatory framework.

4.7 FUTURE OUTLOOK OF LCA IN SUSTAINABLE BUILDING MATERIALS

As the global focus on sustainable practices intensifies, the future outlook of Life Cycle Assessment (LCA) in the realm of sustainable building materials promises a dynamic landscape characterized by emerging trends, anticipated technological developments, and an evolving role for LCA.

4.7.1 Trends in sustainable building materials

The future of sustainable building materials is poised to witness several transformative trends. Among these, the integration of circular economy principles stands out as a key driver. Sustainable materials will increasingly be designed with reusability, recycling, and repurposing in mind, aligning construction practices with a circular model to minimize waste and optimize resource utilization. Additionally, a surge in the use of bio-based materials derived from renewable sources is anticipated [63]. Innovations in biomimicry and bio-fabrication are likely to contribute to the development of materials with enhanced sustainability profiles. The incorporation of smart technologies into building materials, enabling them to respond dynamically to environmental conditions, is another noteworthy trend. Lastly, there will be a growing emphasis on transparent supply chains, with consumers and stakeholders demanding clear information about the environmental impacts of materials, thereby fostering sustainable sourcing and production practices [64].

4.7.2 Anticipated technological developments

Anticipated technological developments are poised to reshape the landscape of sustainable building materials. Advanced material recycling processes are expected to take center stage, with innovations in sorting, processing, and reusing construction materials playing a pivotal role in reducing the environmental footprint. The integration of digital twin technologies is set to enhance monitoring and simulation capabilities, providing real-time insights into material performance for predictive maintenance [65]. Nanotechnology is anticipated to contribute significantly, offering materials with improved strength, durability, and energy efficiency. Machine learning algorithms are likely to optimize Life Cycle Assessment (LCA) processes, enhancing the accuracy of environmental impact predictions. Furthermore, advancements in 3D printing technology are expected, enabling the creation of intricate and sustainable designs while minimizing material waste and promoting sustainable manufacturing [67–69].

4.7.3 The evolving role of LCA

The evolving role of Life Cycle Assessment (LCA) in sustainable building materials points toward increased integration with Building Information Modeling (BIM). This integration is set to provide more accurate and real-time assessments during the design and construction phases, influencing decision-making at every stage. Efforts toward standardization and harmonization of LCA methodologies are anticipated to continue, with globally accepted standards enhancing comparability and consistency across assessments [65]. The scope of LCAs is expected to broaden to include social and economic considerations, providing a more comprehensive understanding of the overall sustainability of building materials. LCA outcomes are likely to play a more direct role in policy development, influencing regulations and incentivizing the adoption of sustainable materials [66]. The development of user-friendly LCA tools will further simplify the assessment process, making sustainability considerations accessible to a broader range of stakeholders within the construction industry.

4.8 CONCLUSION

In conclusion, the role of Life Cycle Assessment (LCA) in the realm of sustainable building materials is pivotal, serving as a compass for environmentally conscious construction practices. As highlighted in the preceding sections, the integration of LCA methodologies offers a holistic understanding of the environmental, social, and economic impacts associated with building materials throughout their life cycle. The examination of trends, technological advancements, and the evolving role of LCA illuminates a path toward a more sustainable built environment. Embracing circular economy principles, incorporating smart technologies, and navigating complex regulatory landscapes are integral components of the future landscape shaped by LCA. The insights garnered from this comprehensive approach empower stakeholders to make informed decisions, promoting the adoption of materials that align with global sustainability goals.

4.8.1 Key findings

The major key findings of this chapter can be highlighted as follows:

1. **Circular Economy Integration:** The adoption of circular economy principles is a key trend, influencing the design and use of building materials for reusability and recycling.
2. **Smart Technologies:** The incorporation of smart technologies in building materials is on the rise, enabling dynamic responses to environmental conditions for optimized energy usage.
3. **Bio-based Materials:** Anticipated trends include increased utilization of bio-based materials derived from renewable sources, contributing to enhanced sustainability profiles.
4. **Technological Advancements:** Advanced material recycling processes, digital twin integration, and nanotechnology are anticipated technological developments, shaping the landscape of sustainable building materials.
5. **Standardization and Harmonization:** Efforts toward standardization and harmonization of LCA methodologies are crucial for improving comparability and consistency across assessments.
6. **Policy Influence:** LCA outcomes are expected to play a more direct role in policy development, influencing regulations and incentivizing the adoption of sustainable materials.
7. **User-Friendly LCA Tools:** The development of user-friendly LCA tools is essential for making sustainability considerations accessible to a broader range of stakeholders within the construction industry.

4.8.2 Implications for industry practices

The implications for industry practices in the LCA of sustainable building materials are profound. Manufacturers, designers, and policymakers need to align their practices with circular economy principles, integrating smart technologies and adopting innovative materials. Transparency in supply chains, adherence to energy efficiency standards, and compliance with waste management regulations are imperative for industry stakeholders. The role of LCA in policy development emphasizes the need for a proactive approach to meet and exceed regulatory expectations. Inclusive LCAs that consider social and economic dimensions should become integral to industry practices, fostering a more comprehensive understanding of sustainability.

4.8.3 Recommendations for future research

As we look to the future, several avenues for future research emerge. It is recommended that researchers explore the dynamic interplay between circular economy principles and LCA, seeking ways to optimize reusability and recycling of building materials. Technological advancements, especially in digital twin integration and nanotechnology, warrant in-depth investigations to understand their full potential in enhancing the sustainability of building materials. Continued efforts toward standardization and harmonization of LCA methodologies are crucial for advancing the reliability and comparability of assessments. Additionally, research should focus on the development of more user-friendly LCA tools to democratize sustainable decision-making across diverse stakeholders. Finally, further studies on the social and economic dimensions of sustainable building materials are recommended to contribute to a more inclusive understanding of their overall impact.

REFERENCES

1. G. K. C. Ding, "3 – Life cycle assessment (LCA) of sustainable building materials: An overview," in *Eco-efficient Construction and Building Materials*, F. Pacheco-Torgal, L. F. Cabeza, J. Labrincha, and A. de Magalhães, Eds. Woodhead Publishing, 2014, pp. 38–62.
2. P. O. Akadiri and P. O. Olomolaiye, "Development of sustainable assessment criteria for building materials selection," *Eng. Constr. Archit. Manag.*, vol. 19, no. 6, pp. 666–687, Jan. 2012. doi: 10.1108/09699981211277568.
3. A. F. Abd Rashid and S. Yusoff, "A review of life cycle assessment method for building industry," *Renew. Sustain. Energy Rev.*, vol. 45, pp. 244–248, 2015. doi: 10.1016/j.rser.2015.01.043.
4. S. A. Hosseinijou, S. Mansour, and M. A. Shirazi, "Social life cycle assessment for material selection: A case study of building materials," *Int. J. Life Cycle Assess.*, vol. 19, no. 3, pp. 620–645, 2014. doi: 10.1007/s11367-013-0658-1.
5. V. P. Singh *et al.*, "Heat transfer and friction factor correlations development for double pass solar air heater artificially roughened with perforated multi-V ribs," *Case Stud. Therm. Eng.*, vol. 39, no. September, p. 102461, 2022. doi: 10.1016/j.csite.2022.102461.
6. D. G. Sahlol, E. Elbeltagi, M. Elzoughiby, and M. Abd Elrahman, "Sustainable building materials assessment and selection using system dynamics," *J. Build. Eng.*, vol. 35, p. 101978, 2021. doi: 10.1016/j.jobe.2020.101978.
7. A. Kundu, A. Kumar, N. Dutt, V. P. Singh, and C. S. Meena, "Chapter 7: Modelling and simulation of thermal energy system for design optimization," in *Thermal Energy Systems: Design, Computational Techniques, and Applications*, A. Kumar, N. Dutt, V. Singh, and C. Meena, Eds. CRC Press, 2023, pp. 103–137.

8. I. Zabalza Bribián, A. Valero Capilla, and A. Aranda Usón, "Life cycle assessment of building materials: Comparative analysis of energy and environmental impacts and evaluation of the eco-efficiency improvement potential," *Build. Environ.*, vol. 46, no. 5, pp. 1133–1140, 2011. doi: 10.1016/j.buildenv.2010.12.002.
9. A. Sharma, A. Saxena, M. Sethi, V. Shree, and Varun, "Life cycle assessment of buildings: A review," *Renew. Sustain. Energy Rev.*, vol. 15, no. 1, pp. 871–875, 2011. doi: 10.1016/j.rser.2010.09.008.
10. T. Sivasakthivel, V. Verma, R. Tarodiya, C. S. Meena, V. P. Singh, and R. Kumar, "Chapter 11: Analysis of optimum operating parameters for ground source heat pump system for different cases of building heating and cooling mode operations," in *Thermal Energy Systems: Design, Computational Techniques, and Applications*, 1st ed., A. Kumar, N. Dutt, V. Singh, and C. Meena, Eds. CRC Press, 2023, pp. 183–207.
11. A. Vigovskaya, O. Aleksandrova, and B. Bulgakov, "Life Cycle Assessment (LCA) in building materials industry," *MATEC Web Conf.*, vol. 106, 2017, [Online], doi: 10.1051/matecconf/201710608059.
12. S. Bawankar *et al.*, "Environmental impact assessment of lithium ion battery employing cradle to grave," *Sustain. Energy Technol. Assessments*, vol. 60, no. May, 2023. doi: 10.1016/j.seta.2023.103530.
13. L. F. Cabeza, L. Rincón, V. Vilariño, G. Pérez, and A. Castell, "Life cycle assessment (LCA) and life cycle energy analysis (LCEA) of buildings and the building sector: A review," *Renew. Sustain. Energy Rev.*, vol. 29, pp. 394–416, 2014. doi: 10.1016/j.rser.2013.08.037.
14. G. S. Rathore, S. Rathor, A. Karn, and V. P. Singh, "Recent progress in solar dryers using phase changing material: A review," *Int. J. Energy Resour. Appl.*, vol. 2, no. 1, pp. 57–77, 2023. doi: 10.56896/IJERA.2023.2.1.005.
15. A. Kumar, V. P. Singh, C. S. Meena, and N. Dutt, "Preface of thermal energy systems," in *Thermal Energy Systems: Design, Computational Techniques, and Applications*, Technical Education Department, Kanpur, CRC Press, 2023, pp. x–xiii.
16. A. Kumar, V. P. Singh, C. S. Meena, and N. Dutt, "Aim and scope of thermal energy systems: Design, computational techniques, and applications," in *Thermal Energy Systems: Design, Computational Techniques, and Applications*, A. Kumar, V. P. Singh, C. S. Meena, and N. Dutt, Eds. CRC Press, 2022, pp. 9–10.
17. K. Chauhan and V. P. Singh, "Prospect of biomass to bioenergy in India: An overview," *Mater. Today Proc.*, 2023. doi: 10.1016/j.matpr.2023.01.419.
18. M. Nayal, A. K. Sharma, S. Jain, and V. P. Singh, "Chapter 11 green hydrogen production methods, designs and applications," in *Highly Efficient Thermal Renewable Energy Systems Design, Optimization and Applications*, 1st ed., R. W. V. Verma, S. Thangavel, N. Dutt, and A. Kumar, Eds. CRC Press, 2024, pp. 178–193.
19. N. Dutt, A. Jageshwar, H. Ashwani, K. Mukesh, K. Awasthi, and V. Pratap, "Thermo – Hydraulic performance of solar air heater having discrete D – Shaped ribs as artificial roughness," *Environ. Sci. Pollut. Res.*, 2023. doi: 10.1007/s11356-023-28247-9.
20. S. Baharetha, A. A. Al-Hammad, and H. Alshuwaikhat, "Towards a unified set of sustainable building materials criteria," *ICSDEC 2012*, pp. 732–740, Jan. 08, 2013. doi: 10.1061/9780784412688.088.

21. N. Dutt, A. Binjola, A. J. Hedau, A. Kumar, V. P. Singh, and C. S. Meena, "Comparison of CFD results of smooth air duct with experimental and available equations in literature," *Int. J. Energy Resour. Appl.*, vol. 1, no. 1, pp. 40–47, 2022. doi: 10.56896/IJERA.2022.1.1.006.
22. S. Shams, K. Mahmud, and M. Al-Amin, "A comparative analysis of building materials for sustainable construction with emphasis on CO2 reduction," *Int. J. Environ. Sustain. Dev.*, vol. 10, no. 4, pp. 364–374, Jan. 2011. doi: 10.1504/IJESD.2011.047767.
23. V. P. Singh, S. Jain, and J. M. L. Gupta, "Analysis of the effect of perforation in multi-v rib artificial roughened single pass solar air heater: – Part A," *Exp. Heat Transf.*, pp. 1–20, Oct. 2021. doi: 10.1080/08916152.2021.1988761.
24. S. Amanjeet, B. George, J. Satish, and S. Matt, "Review of life-cycle assessment applications in building construction," *J. Archit. Eng.*, vol. 17, no. 1, pp. 15–23, Mar. 2011. doi: 10.1061/(ASCE)AE.1943-5568.0000026.
25. A. Kumar, V. P. Singh, C. S. Meena, and N. Dutt, *Thermal Energy Systems: Design, Computational Techniques, and Applications*, 1st ed. CRC Press, 2023.
26. R. Kumar *et al.*, "Experimental and RSM-based process-parameters optimisation for turning operation of EN36B steel," *Materials (Basel)*, vol. 16, no. 1, pp. 1–18, 2023. doi: 10.3390/ma16010339.
27. N. Kumar Sharma, "Sustainable building material for green building construction, conservation and refurbishing," *Int. J. Adv. Sci. Technol.*, vol. 29, no. 10S, pp. 5343–5350, 2020, [Online]. Available: https://www.researchgate.net/publication/342946652.
28. R. Meglin, S. Kytzia, and G. Habert, "Regional circular economy of building materials: Environmental and economic assessment combining material flow analysis, input-output analyses, and life cycle assessment," *J. Ind. Ecol.*, vol. 26, no. 2, pp. 562–576, Apr. 2022. doi: 10.1111/jiec.13205.
29. A. K. Sharma, M. Nayal, S. Jain, and V. P. Singh, "Chapter 07 optimization techniques of solar thermal and hybrid energy systems," in *Highly Efficient Thermal Renewable Energy Systems Design, Optimization and Applications*, 1st ed., R. W. V. Verma, S. Thangavel, N. Dutt, and A. Kumar, Eds. CRC Press, 2024, pp. 1–14.
30. R. Heijungs, G. Huppes, and J. B. Guinée, "Life cycle assessment and sustainability analysis of products, materials and technologies. Toward a scientific framework for sustainability life cycle analysis," *Polym. Degrad. Stab.*, vol. 95, no. 3, pp. 422–428, 2010. doi: 10.1016/j.polymdegradstab.2009.11.010.
31. V. P. Singh and G. Dwivedi, "Technical analysis of a large-scale solar updraft tower power plant," *Energies*, vol. 16, no. 1, p. 103325, 2023. doi: 10.3390/en16010494.
32. P. Dewick and M. Miozzo, "Sustainable technologies and the innovation–regulation paradox," *Futures*, vol. 34, no. 9, pp. 823–840, 2002. doi: 10.1016/S0016-3287(02)00029-0.
33. M. Apep and B. Dwi, "Uncertain supply chain management technological innovation and the environmentally friendly building material supply chain: Implications for sustainable environment," vol. 11, pp. 1405–1416, 2023. doi: 10.5267/j.uscm.2023.8.006.
34. R. K. Shubham Srivastava, Deepti Verma, Shreya Thusoo, Ashwani Kumar, and Varun Pratap Singh, "Nanomanufacturing for energy conversion

and storage devices," in *Nanomanufacturing and Nanomaterials Design: Principles and Applications*, 2022, pp. 165–174.
35. V. P. Singh *et al.*, "Nanomanufacturing and design of high-performance piezoelectric nanogenerator for energy harvesting," in *Nanomanufacturing and Nanomaterials Design: Principles and Applications*, 1st ed., S. Singh, S. K. Behura, A Kumar, K. Verma, Eds. Boca Raton, CRC Press, 2022, pp. 241–272, doi: 10.1201/9781003220602.
36. Y. G. Ashwani Kumar, Arun Kumar Singh Gangwar, Avinash Kumar, Chandan Swaroop Meena, Varun Pratap Singh, Nitesh Dutt, and Arbind Prasad, "Biomedical study of femur bone fracture and healing," in *Advanced Materials for Biomedical Applications*, 1st ed., A. Kumar, Y. Gori, A. Kumar, C. S. Meena, N. Dutt, Eds. Boca Raton, CRC Press, 2022, pp. 212–235.
37. A. Datta, A. Kumar, A. Kumar, A. Kumar, and V. P. Singh, "Advanced materials in biological implants and surgical tools," in *Advanced Materials for Biomedical Applications*, 1st ed., N. D. Ashwani Kumar, Yatika Gori, Avinash Kumar, and Chandan Swaroop Meena, Eds. Boca Raton, CRC Press, 2022, pp. 283–298.
38. A. Kumar, Y. Gori, C. S. Meena, V. P. Singh, and V. K. Sharma, "Design and analysis of heavy vehicle medium duty transmission gearbox system," *Addit. Manuf. Ind. 4.0 Methods, Tech. Model. Nano Asp.*, pp. 67–86, 2022. doi: 10.1201/9781003360001-4.
39. M. Nayal, A. K. Sharma, S. Jain, and V. P. Singh, "Chapter 06 design and modelling of solar, geothermal and hybrid energy systems.pdf," in *Highly Efficient Thermal Renewable Energy Systems Design, Optimization and Applications*, 1st ed., R. W. V. Verma, S. Thangavel, N. Dutt, and A. Kumar, Eds. CRC Press, 2024, pp. 1–15.
40. L. F. Cabeza, C. Barreneche, L. Miró, M. Martínez, A. I. Fernández, and D. Urge-Vorsatz, "Affordable construction towards sustainable buildings: Review on embodied energy in building materials," *Curr. Opin. Environ. Sustain.*, vol. 5, no. 2, pp. 229–236, 2013. doi: 10.1016/j.cosust.2013.05.005.
41. S. Park, S. Wu, Z. Liu, and S. Pyo, "The role of Supplementary Cementitious Materials (SCMs) in Ultra High Performance Concrete (UHPC): A review," *Materials*, vol. 14, no. 6. 2021. doi: 10.3390/ma14061472.
42. H. R. Milner and A. C. Woodard, "8 – Sustainability of engineered wood products," in *Woodhead Publishing Series in Civil and Structural Engineering*, J. M. B. T.-S. of C. M. Second E. Khatib, Ed. Woodhead Publishing, 2016, pp. 159–180.
43. A. Mistri, S. K. Bhattacharyya, N. Dhami, A. Mukherjee, and S. V Barai, "A review on different treatment methods for enhancing the properties of recycled aggregates for sustainable construction materials," *Constr. Build. Mater.*, vol. 233, p. 117894, 2020. doi: 10.1016/j.conbuildmat.2019.117894.
44. K. Chaowana, S. Wisadsatorn, and P. Chaowana, "Bamboo as a sustainable building material—Culm characteristics and properties," *Sustainability*, vol. 13, no. 13. 2021. doi: 10.3390/su13137376.
45. K. L. Getter and D. B. Rowe, "The role of extensive green roofs in sustainable development," *HortScience HortSci*, vol. 41, no. 5, pp. 1276–1285, 2006. doi: 10.21273/HORTSCI.41.5.1276.
46. M. R. Hossain, R. Sultana, M. M. Patwary, N. Khunga, P. Sharma, and S. J. Shaker, "Self-healing concrete for sustainable buildings. A review,"

Environ. Chem. Lett., vol. 20, no. 2, pp. 1265–1273, 2022. doi: 10.1007/s10311-021-01375-9.

47. A. A. F. Husain, W. Z. W. Hasan, S. Shafie, M. N. Hamidon, and S. S. Pandey, "A review of transparent solar photovoltaic technologies," *Renew. Sustain. Energy Rev.*, vol. 94, pp. 779–791, 2018. doi: 10.1016/j.rser.2018.06.031.
48. A. Hussain and M. A. Kamal, "Energy efficient sustainable building materials: An overview," *Key Eng. Mater.*, vol. 650, pp. 38–50, 2015. doi: 10.4028/www.scientific.net/KEM.650.38.
49. D. Kumar, M. Alam, P. X. W. Zou, J. G. Sanjayan, and R. A. Memon, "Comparative analysis of building insulation material properties and performance," *Renew. Sustain. Energy Rev.*, vol. 131, p. 110038, 2020. doi: 10.1016/j.rser.2020.110038.
50. L. A. Dobrescu, "From traditional to smart building materials in architecture," *IOP Conf. Ser. Mater. Sci. Eng.*, vol. 1203, no. 3, p. 32113, 2021. doi: 10.1088/1757-899X/1203/3/032113.
51. P. Lamba, D. P. Kaur, S. Raj, and J. Sorout, "Recycling/reuse of plastic waste as construction material for sustainable development: A review," *Environ. Sci. Pollut. Res.*, vol. 29, no. 57, pp. 86156–86179, 2022. doi: 10.1007/s11356-021-16980-y.
52. S. M. Chiew, I. S. Ibrahim, M. A. Mohd Ariffin, H.-S. Lee, and J. K. Singh, "Development and properties of light-transmitting concrete (LTC) – A review," *J. Clean. Prod.*, vol. 284, p. 124780, 2021. doi: 10.1016/j.jclepro.2020.124780.
53. A. A. Shubbar, M. Sadique, H. K. Shanbara, and K. Hashim, "The development of a new low carbon binder for construction as an alternative to cement BT – Advances in sustainable construction materials and geotechnical engineering," 35, 2020, pp. 205–213.
54. A. Fnais *et al.*, "The application of life cycle assessment in buildings: Challenges, and directions for future research," *Int. J. Life Cycle Assess.*, vol. 27, no. 5, pp. 627–654, 2022. doi: 10.1007/s11367-022-02058-5.
55. M. Hu, "Building impact assessment—A combined life cycle assessment and multi-criteria decision analysis framework," *Resour. Conserv. Recycl.*, vol. 150, p. 104410, 2019. doi: 10.1016/j.resconrec.2019.104410.
56. M. Saini, A. Sharma, V. P. Singh, G. Dwivedi, and S. Jain, "Solar thermal receivers – A review," *Adv. Mater. Manuf. Energy Eng. Vol. II*, vol. II, pp. 310–325, 2022. doi: 10.1007/978-981-16-8341-1.
57. P. O. Akadiri, "Understanding barriers affecting the selection of sustainable materials in building projects," *J. Build. Eng.*, vol. 4, pp. 86–93, 2015. doi: 10.1016/j.jobe.2015.08.006.
58. S. Jain, N. Kumar, V. P. Singh, S. Mishra, and N. K. Sharma, "Transesterification of algae oil and little amount of waste cooking oil blend at low temperature in the presence of NaOH," *Energies*, vol. 16, no. 1, pp. 1–12, 2023. doi: 10.3390/ en16031293.
59. V. P. Singh, C. S. Meena, A. Kumar, and N. Dutt, "Double pass solar air heater: A review," *Int. J. Energy Resour. Appl.*, vol. 1, no. 2, pp. 22–43, 2022. doi: 10.56896/IJERA.2022.1.2.009.
60. A. R. Singh, S. K. Singh, and V. P. Singh, "Process parameters optimization of carbon nano tube based catalytic transesterification of algal oil," *Mater. Today Proc.*, no. xxxx, 2023. doi: 10.1016/j.matpr.2023.01.418.

61. J. Ayarkwa, D.-G. Joe Opoku, P. Antwi-Afari, and R. Y. M. Li, "Sustainable building processes' challenges and strategies: The relative important index approach," *Clean. Eng. Technol.*, vol. 7, p. 100455, 2022. doi: 10.1016/j.clet.2022.100455.
62. S. Gounder, A. Hasan, A. Shrestha, and A. Elmualim, "Barriers to the use of sustainable materials in Australian building projects," *Eng. Constr. Archit. Manag.*, vol. 30, no. 1, pp. 189–209, 2023. doi: 10.1108/ECAM-10-2020-0854.
63. V. P. Singh *et al.*, "Recent developments and advancements in solar air heaters: A detailed review," *Sustain.*, vol. 14, no. 19, pp. 1–57, Sep. 2022. doi: 10.3390/su141912149.
64. A. Kundu, A. Kumar, V. P. Singh, C. S. Meena, and N. Dutt, "Chapter 1: Introduction to thermal energy resources and their smart applications," in *Thermal Energy Systems: Design, Computational Techniques, and Applications*, 1st ed., A. Kumar, V. P. Singh, C. S. Meena, N. Dutt, Eds. Boca Raton, CRC Press, 2023, pp. 1–15.
65. V. P. Singh, S. Jain, and A. Kumar, "Establishment of correlations for the Thermo-Hydraulic parameters due to perforation in a multi-V rib roughened single pass solar air heater," *Exp. Heat Transf.*, vol. 35, no. 5, pp. 1–20, 2022. doi: 10.1080/08916152.2022.2064940.
66. V. P. Singh, S. Jain, A. Karn, A. Kumar, and G. Dwivedi, "Mathematical modeling of efficiency evaluation of double pass parallel flow solar air heater," *Sustainability*, vol. 14, no. 17, pp. 1–22, 2022. doi: 10.3390/su141710535.
67. A. Kumar, Y. Singla, and T. Namboodri, "Globalization and international issues in sustainable manufacturing," in *Sustainability in Smart Manufacturing: Trends, Scope, and Challenges*, S. Shah, H. Nautiyal, G. Gugliani, A. Kumar, T. Namboodri, and Y. K. Singla, Eds. CRC Press, Boca Raton, FL, 2024; Chapter 1; ISBN 9781032 740713.
68. R. Kumar, A. Prasad, and A. Kumar, *Sustainable Smart Manufacturing Processes in Industry 4.0.* Taylor & Francis (CRC Press), 2023. ISBN: 9781003436072, doi: 10.1201/9781003436072.
69. S. Shah, H. Nautiyal, G. Gugliani, A. Kumar, T. Namboodri, and Y. K. Singla, Editors, *Sustainability in Smart Manufacturing: Trends, Scope, and Challenges.* CRC Press, Boca Raton, FL, 2024; ISBN 9781032 740713.
70. R. K. Upadhyay, V. P. Singh, and A. Kumar, "Sustainable transportation: Policy, planning and implementation," in *Energy Efficient Vehicles: Technologies and Challenges*, V. P. Singh, A. Kumar, C. S. Meena, and G. Dwivedi, Eds. CRC Press, Boca Raton, FL, 2024; Chapter 05; ISBN 9781032548111.
71. V. P. Singh and A. Kumar, "Techno-economic and future aspects of the HEV-EV-FCV: Decarbonisation, digitalization and sustainability," in *Energy Efficient Vehicles: Technologies and Challenges*, V. P. Singh, A. Kumar, C. S. Meena, and G. Dwivedi, Eds. CRC Press, Boca Raton, FL, 2024; Chapter 11; ISBN 9781032548111.
72. V. Verma, S. Thangavel, N. Dutt, A. Kumar, and R. Weerasinghe, "Recent development of thermal energy storage: Solar, geothermal and hydrogen energy," in *Highly Efficient Thermal Renewable Energy Systems: Design, Design, Optimization and Applications*, V. Verma, S. Thangavel, N. Dutt, A. Kumar, and R. Weerasinghe, Eds. CRC Press, Boca Raton, FL, 2024; Chapter 01; ISBN 9781032595641.

73. M. Ahmadizadeh, M. Heidari, S. Thangavel, E. A. Naamani, M. Khashehchi, V. Verma, and A. Kumar, "Technological advancements in sustainable and renewable solar energy systems," in *Highly Efficient Thermal Renewable Energy Systems: Design, Design, Optimization and Applications*, V. Verma, S. Thangavel, N. Dutt, A. Kumar, and R. Weerasinghe, Eds. CRC Press, Boca Raton, FL, 2024; Chapter 02; ISBN 9781032595641.
74. M. Chitt, S. Thangavel, V. Verma, and A. Kumar, "Green hydrogen productions: Methods, designs and smart applications," in *Highly Efficient Thermal Renewable Energy Systems: Design, Design, Optimization and Applications*, V. Verma, S. Thangavel, N. Dutt, A. Kumar, and R. Weerasinghe, Eds. CRC Press, Boca Raton, FL, 2024; Chapter 16; ISBN 9781032595641.
75. M. Khashehchi, S. Thangavel, P. Rahmanivahid, M. Heidari, T. Moazzeni, V. Verma, and A. Kumar, "Solar desalination techniques: Challenges and opportunities," in *Highly Efficient Thermal Renewable Energy Systems: Design, Design, Optimization and Applications*, V. Verma, S. Thangavel, N. Dutt, A. Kumar, and R. Weerasinghe, Eds. CRC Press, Boca Raton, FL, 2024; Chapter 19; ISBN 9781032595641.
76. V. Verma, S. Thangavel, N. Dutt, A. Kumar, and R. Weerasinghe, Editors, *Highly Efficient Thermal Renewable Energy Systems: Design, Design, Optimization and Applications*. CRC Press, Boca Raton, FL, 2024; ISBN 9781032595641.

Chapter 5

Efficiency meets innovation

City gas systems in modern buildings

Bhalchandra Shingan and Adarsh Kumar Aarya

5.1 INTRODUCTION

Several challenges have evolved as a result of the worsening of urban transportation and its severe consequences for the environment and human health. Primary reliance on internal combustion engines powered by petrol or diesel has resulted in poor air quality, lost time, noise, traffic bottlenecks, and additional environmental pollutants [1]. In response to these difficulties, the need for sustainable energy solutions in cities has never been greater. This chapter investigates the causes behind the increasing need for sustainable energy in modern buildings, as well as the possible advantages of adopting cleaner, more efficient options as a gas distribution system. Urbanization is a distinguishing feature of the 21st century, with more people coming to cities in quest of greater opportunities and a higher standard of living. This increased urbanization raises energy demand to power dwellings, businesses, and transit networks designed in modern buildings. Traditional energy sources, which are mostly based on coal, petrol, and diesel, greatly contribute to pollution and greenhouse gas emissions. As cities grow in size, the demand for traditional energy systems exacerbates environmental challenges and the effects of climate change [2]. Cities are vulnerable to disruptions in energy supply induced by natural catastrophes, geopolitical conflicts, or infrastructure breakdowns. Sustainable energy solutions make energy infrastructure more robust and secure [3]. Investing in sustainable energy solutions addresses both environmental and economic problems. Significant developments in the city gas distribution (CGD) industry have resulted in the creation of employment and new business prospects. Cities that embrace sustainable technology have the potential to create a green economy, promoting innovation and driving economic growth [4]. CGD sector advancements have drastically decreased the cost of sustainable energy solutions, making them cheaper and more accessible. Solar and wind power, which were long deemed expensive, are now on par with, if not cheaper than, traditional energy sources. As these technologies grow more affordable, cities will have a stronger incentive to switch to sustainable energy sources [5].

 DOI: 10.1201/9781003496656-5

5.1.1 Natural as a sustainable energy source

Natural gas is becoming more widely recognized as a sustainable energy source due to its many uses, lower carbon emissions than traditional fossil fuels, and possible role in the transition to a more sustainable energy future. Natural gas, which is mostly composed of methane, is a cleaner-burning fuel that emits fewer pollutants and greenhouse gases when burned than coal and oil. The composition of natural gas is shown in Table 5.1. Because of this, it is a critical factor in attempts to cut carbon dioxide emissions and ameliorate climate change [6]. Natural gas is also abundant, with huge reserves all over the planet, ensuring a continuous and stable energy source. Furthermore, technological advances, such as hydraulic fracturing, have enabled the exploitation of hitherto unreachable deposits. The development of renewable natural gas, which is produced from organic waste via methods such as anaerobic digestion, improves its environmental credentials by reusing organic matter and lowering methane emissions from landfills [7]. Natural gas may act as a transitional fuel as the energy landscape changes, supporting the incorporation of intermittent renewable energy sources like wind and solar by providing a robust and responsive backup. However, concerns remain, such as the possibility of methane leaks during extraction and transportation, demanding strict laws and technical developments to reduce environmental damage. Nonetheless, with continued technological breakthroughs and ethical extraction and utilization practices, natural gas remains a critical component in the worldwide goal of a sustainable and diverse energy portfolio [8].

5.1.2 Applications of natural gas in modern buildings

Natural gas is a versatile and efficient energy source for a variety of applications that contribute to comfort, sustainability, and economic viability in modern buildings. Natural gas is one of the most cost-effective types

Table 5.1 Typical composition of natural gas [9]

Gas	*Composition*	*Range*
Methane	CH_4	70–90%
Ethane	C_2H_6	0–20%
Propane	C_3H_8	
Butane	C_4H_{10}	
Pentane and higher hydrocarbons	C_5H_{12}	0–10%
Carbon dioxide	CO_2	0–8%
Oxygen	O_2	0–0.2%
Nitrogen	N_2	0–5 %
Hydrogen sulfide, carbonyl sulfide	H_2S, COS	0–5%
Rare gases	Ar, He, Ne, Xe	Trace

of energy available to modern buildings as shown in Figure 5.1. In reality, as a source of energy in buildings, natural gas has traditionally been more affordable than electricity [10].

Heating is one of the most prevalent uses for natural gas in buildings. It is extensively used for space heating in homes, offices, and industries. Natural gas furnaces and boilers are well-known for their efficiency and cost-effectiveness, offering a steady and rapid heat source that can be readily adjusted to suit specific temperature requirements. As a result, natural gas heating systems are a popular choice for central heating in homes, workplaces, and other large buildings [11].

In addition to space heating, natural gas is frequently utilized for water heating. Gas water heaters are efficient and can create a consistent supply of hot water for domestic use right away. This is necessary for activities like bathing, cooking, and cleaning, and it contributes to the general comfort and usefulness of contemporary dwellings. Natural gas is often used in kitchen appliances. Many people choose gas burners and ovens because of their exact temperature control, fast heat response, and general dependability. Natural gas cooking appliances are popular in both professional and domestic kitchens due to their performance and simplicity [12].

Natural gas is increasingly being used in combined heat and power (CHP) systems in modern buildings, in addition to heating and cooking. CHP, also known as cogeneration, is the process of producing both electricity and usable thermal energy from the same energy source at the same time. Natural gas-fired CHP systems are extremely efficient because they use the heat created during electricity generation to heat space, water, or other industrial operations. This comprehensive method improves total energy efficiency while lowering greenhouse gas emissions [13]. Furthermore, natural gas is

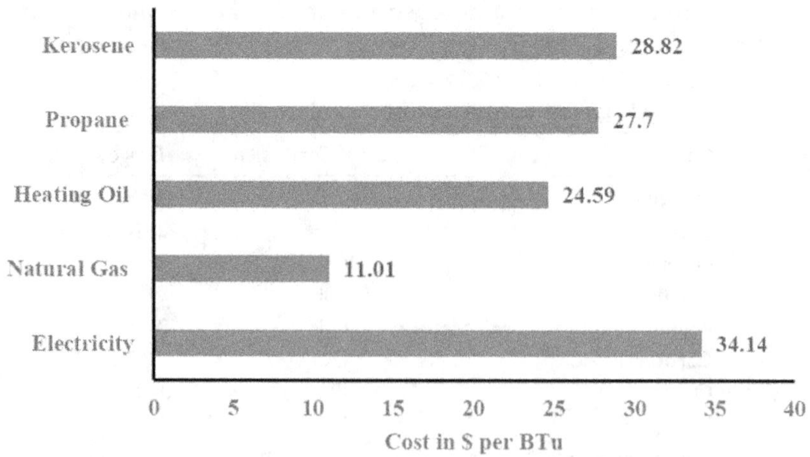

Figure 5.1 Cost comparison with different sources [10]

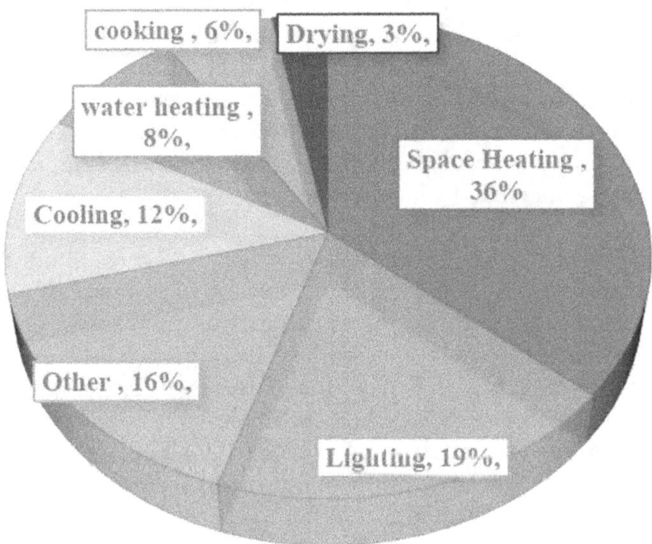

Figure 5.2 Applications of natural gas in modern buildings

becoming more popular in modern buildings for backup power production. During power outages, natural gas generators provide a dependable supply of electricity, assuring the continued functioning of important systems such as lighting, security, and necessary appliances. This is especially vital for companies, hospitals, and other institutions that require continuous electricity [14]. Different applications of natural gas in modern buildings are shown in Figure 5.2.

5.2 CITY GAS DISTRIBUTION SYSTEM

City gas distribution is a network of pipes that transports gas to residential, commercial, and industrial users. It is available for automobiles in the form of piped natural gas (PNG) (low-pressure distribution network) and compressed natural gas (CNG) (high-pressure distribution network). The city gas distribution system is a city-owned underground pipeline distribution network that provides natural gas to a number of consumers for a variety of purposes. The network of pipes and devices that transfer high-pressure gas across the primary transmission network to a medium-pressure distribution network and subsequently to service pipelines is known as city gas distribution. Following that, the gas is distributed to residential, industrial, and commercial premises, as well as CNG stations located in certain charge zones or geographical regions [14].

The history of city gas distribution networks has been a revolutionary journey highlighted by technology advances, regulatory reforms, and an increased emphasis on sustainability. Historically, the provision of produced gas sourced from coal or oil was the primary focus of city gas distribution. However, with the discovery and utilization of natural gas reserves in the mid-twentieth century, a substantial change occurred. The incorporation of natural gas into city gas distribution networks transformed the industry by providing a cleaner and more efficient energy source. Pipelines for gas transportation superseded older techniques, increasing the dependability and safety of gas supply [15].

The worldwide city gas distribution system grew throughout time, serving not just to household heating but also to a wide range of applications such as industrial operations, power production, and transportation [16]. More recently, the growth of renewable natural gas (RNG) and the incorporation of smart technology has altered the evolution of municipal gas distribution networks. RNG, which is made from organic waste, helps to provide a more sustainable and ecologically friendly gas supply. Smart meters, sensors, and automation have increased system monitoring, maintenance, and user involvement, enhancing the efficiency and responsiveness of municipal gas distribution networks. As cities seek cleaner, more robust energy systems, municipal gas distribution is expected to adopt technologies that promote sustainability, resilience, and increased energy accessibility [17].

5.3 COMPONENTS OF CGD

There are following types of networks in CGD [18]:

- **Primary Network:** Steel pipes are commonly used in the construction of medium pressure distribution systems, gas mains, and distribution mains. The working pressure ranges from 7 to 49 bar. The goal is to provide gas to the secondary distribution network.
- **Secondary Network:** The secondary network is a low-pressure distribution system, typically comprised of MDPE tubes with pressures ranging from 7 bar to 100 mbar.
- **Tertiary Network:** The tertiary network is a low-pressure distribution system made up of service lines, service regulators, and customer metering systems made of MDPE and copper/GI tubing. The operating pressure in the tertiary network is less than 100 mbar. A block diagram of CGD value chain is presented in Figure 5.3.

5.3.1 City gate station (CGS)

As an essential component of city gas distribution networks, city gate stations play a critical role in the transportation and delivery of natural gas to metropolitan regions. City gate stations serve as a vital juncture

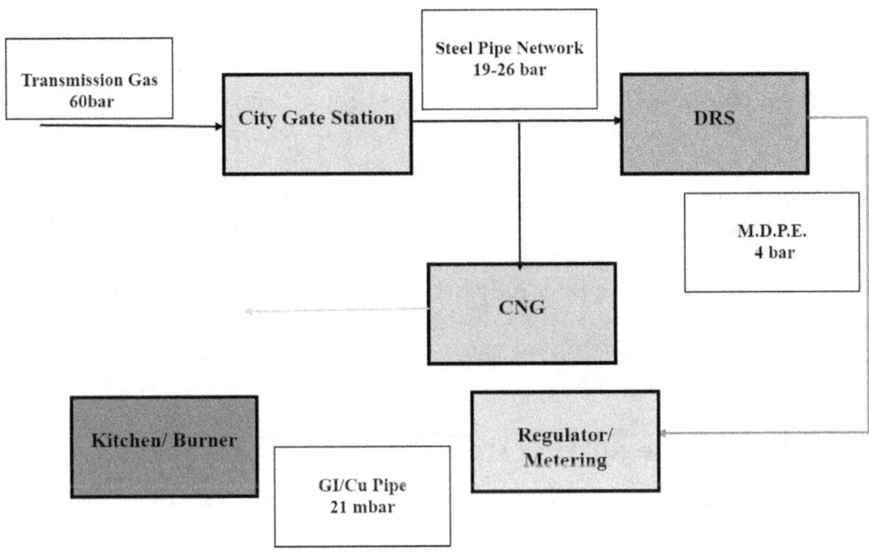

Figure 5.3 CGD value chain

when natural gas is accepted from high-pressure transmission pipes and subsequently delivered into lower-pressure municipal networks, guaranteeing a continuous and dependable gas supply to households, companies, and industries. These stations are outfitted with cutting-edge technology to manage and measure gas flow, assuring safe and efficient natural gas distribution. City gate stations ensure the quality and safety of the gas supply by reducing pressure, odorizing, and monitoring. The broad adoption of city gas distribution networks, with strategically placed city gate stations, helps environmental sustainability by encouraging the use of cleaner and more efficient natural gas as a major energy source. City gate stations are emerging as essential nodes in the drive for efficient and ecologically aware city gas distribution networks as metropolitan areas increasingly shift to sustainable and eco-friendly energy alternatives [19].

5.3.2 Industrial pressure reduction station (IPRS)

The industrial pressure reduction station is a key component in meeting the different energy demands of industries. It is a cornerstone within the complicated network of city gas distribution systems. The pressure reduction station, which is a vital element of this network, specializes in converting high-pressure natural gas supplied from transmission pipes into lower pressures suited for industrial usage. This complex procedure necessitates the use of modern pressure reduction technology and control systems to enable a smooth and safe changeover. Furthermore, these stations are critical in optimizing gas supply for diverse industrial applications, promoting efficiency and dependability. Industrial pressure reduction stations contribute greatly to risk reduction and the overall resilience of municipal gas distribution infrastructure by adhering to high safety regulations and deploying cutting-edge equipment. They are built to manage variable demand loads, ensuring that enterprises have a steady and stable supply of natural gas, which is frequently a cleaner and more sustainable alternative to traditional energy sources. In essence, the industrial pressure reduction station is a keystone in the drive to support industrial expansion while also harmonizing with the wider goals of sustainable and ecologically conscientious energy usage in urban settings [20].

5.3.3 District regulating station (DRS)

The DRS serves as a regulating station where the pressure of entering gas via higher-pressure transmission pipelines is carefully regulated to meet the needs of certain districts. Maintaining the integrity and safety of the gas distribution network requires rigorous pressure management. The DRS, which is outfitted with innovative pressure control mechanisms, safety valves, and monitoring systems, plays an important role in protecting the gas supply infrastructure from fluctuations and unanticipated incidents. Furthermore, DRS units are strategically situated to optimize natural gas distribution to diverse residential, commercial, and industrial regions, assuring a steady and dependable supply for end consumers. District regulating stations play a critical role in improving the adaptability of city gas distribution networks to the dynamic demands of different urban zones, promoting energy efficiency, and contributing to the overall resilience and sustainability of gas distribution infrastructure within cities [20].

5.3.4 CNG stations

CNG stations are an important link in the CGD network, providing an environmentally friendly alternative to conventional vehicle fuels. These stations compress natural gas, which is mostly made of methane, to a transportable pressure about 250 bar. CNG, being a cleaner and more ecologically friendly alternative to traditional fossil fuels, contributes greatly to lowering carbon emissions and reducing air pollution in cities. The strategic location of CNG stations inside city gas distribution networks not only encourages the use of greener energy sources but also allows for the smooth integration of natural gas as a viable and sustainable fuel for a variety of vehicles, including buses, taxis, and private cars. CNG stations in city gas distribution, in essence, serve as critical nodes in the transition to a greener and more sustainable transportation ecology [21].

5.3.5 Gas systems design for modern buildings

The design of gas systems for modern buildings must consider safety, energy efficiency, and environmental sustainability. Gas systems in buildings are often utilized for heating, hot water, cooking, and, in certain cases, electricity generation (Figure 5.4). Adherence to local construction rules and safety laws is mandated in gas system design [22]. The specifications for gas pipe size, ventilation, gas appliance installation, and safety features such as shut-off valves and emergency procedures are outlined by these standards. Energy-efficient gas appliances must be chosen to reduce energy consumption and operational expenses [23]. Included are high-efficiency furnaces, water heaters, and kitchen appliances that meet or exceed energy efficiency criteria. When gas systems are integrated with building automation systems, centralized control and monitoring become possible [24]. Improved gas appliance performance, regulation of heating and cooling based on occupancy, and real-time data for maintenance and energy management can be achieved with smart controls. The usage of renewable gases is encouraged by creating gas networks that can handle biogas or biomethane, which may be made from organic waste and injected into the petrol supply, thereby lowering the carbon footprint of petrol usage [25]. The support of district heating and cooling for larger projects or metropolitan regions is encouraged by constructing gas systems. Overall efficiency can be enhanced, and energy waste reduced by centralized thermal energy

Figure 5.4 PNG installation at buildings [28]

production and delivery [26]. Gas systems that are adaptable to future additions or alterations should be created, considering prospective changes to new gas compositions (e.g., hydrogen) and adapting infrastructure to upcoming technologies [27].

5.3.6 Safety measures

Safety measures in modern buildings within the context of CGD are deemed vital to ensuring the well-being of inhabitants and the proper operation of gas infrastructure. CGD safety procedures in the residential sector are specified as follows. Qualified individuals should perform all gas-related installations, including pipes and appliances. Regular maintenance inspections by skilled technicians are required to identify and solve possible concerns before they become serious [29]. While installing a residential gas system, local construction requirements, safety standards, and regulations must be followed. These standards cover aspects such as pipe size, ventilation regulations, and safety measures like shut-off valves [30]. Gas leak detection sensors should be installed in high-traffic areas like kitchens and utility rooms. Leaks can be identified quickly by these sensors, which generate warnings. Emergency shut-off systems that enable residents to turn off the gas supply in the event of a leak or other emergency should be installed [31]. Residents should be given clear instructions and training on how to utilize these devices. To prevent the accumulation of gases such as carbon monoxide, sufficient ventilation should be provided in locations where gas appliances are utilized. Adequate airflow is required for safe gas combustion and the removal of combustion byproducts [32]. Only gas appliances that have been tested and approved to fulfill safety criteria should be used. These appliances should be fitted in accordance with the manufacturer's instructions and local restrictions. Gas appliances should be inspected and maintained on a regular basis to ensure they are safe and efficient. This involves inspecting burners, valves, and connections for wear or damage. Ongoing education, regular maintenance, and adherence to safety regulations must be included in a complete gas safety program in modern buildings [33].

5.4 ENERGY EFFICIENCY WITH CITY GAS SYSTEMS

Through numerous methods, city gas systems play a critical role in improving energy efficiency in modern buildings. Here are a few ways that city gas systems help with energy efficiency [34].

i) **Cleaner Combustion:** City gas, which is frequently made of methane, burns cleaner and more effectively than certain other fossil fuels as presented in Table 5.2. It creates less pollutants and greenhouse gas emissions when used for heating or cooking in buildings. This helps

Table 5.2 Emissions from different fuels [36]

Emissions	*Natural Gas*	*Diesel*	*Furnace Oil*	*Coal*	*Petrol*
Carbon dioxide	117,000	135,250	164,000	208,000	285,700
Nitrogen Oxide	92	1,632	448	457	4,081
Sulfur oxide	0.6	1,121	1,122	2,591	204
Particulates	7	1,021	8	2,744	41

Emissions are pounds of air pollutants per BTU of energy.

to clean up the environment and lowers the overall carbon footprint connected with energy usage [35].

ii) **High Efficiency Appliances:** High-efficiency equipment like gas furnaces, water heaters, and stoves typically use city gas. These appliances frequently have innovative technology that improve energy efficiency, allowing them to operate better while utilizing less energy than their competitors [37].

iii) **Infrastructure for Renewable Gas:** City gas systems can also help to integrate renewable gases into the energy mix, such as biomethane or hydrogen. These gases can be generated from biological waste or electrolysis using renewable energy sources. Integrating these renewable gases into municipal gas infrastructure improves the sustainability of building energy usage [38].

iv) **Smart Metering and Monitoring:** Smart metering and monitoring technology may be integrated into city gas infrastructure. This enables more exact measurement and control of gas consumption in buildings, allowing for improved energy management and promoting energy-efficient practices [39].

v) **Incentives for Energy-Efficient Technologies:** City governments and utility companies frequently offer incentives to encourage the use of energy-efficient technologies in buildings. These incentives can help to promote the use of high-efficiency gas appliances and systems, resulting in total energy savings [40].

5.5 INNOVATIONS IN CITY GAS DISTRIBUTION

Emerging technologies and developments in gas distribution systems are critical to the advancement of sustainable development. These developments are aimed at increasing efficiency, lowering environmental impact, and embracing renewable energy sources. Here are a few noteworthy trends in the field [41]. Pipeline material innovations lead to better durability, less leakage, and improved safety. To extend the lifespan of gas distribution infrastructure, composite materials and corrosion-resistant metals are being investigated [42]. Predictive maintenance, issue detection, and system optimization all benefit from data analytics and artificial intelligence (AI). These technologies allow gas distribution companies to predict problems, increase efficiency, and decrease environmental impact [43]. Gas distribution system integration with smart grids and the Internet of Things (IoT) improves overall system efficiency. This comprises enhanced metering infrastructure, remote monitoring, and predictive maintenance, all of which contribute to lower energy use and environmental effect. To swiftly identify and treat gas leaks, enhanced leak detection technologies such as smart sensors and monitoring systems are being implemented. These systems give real-time data, allowing for immediate action and reducing the environmental and safety hazards connected with gas leaks. Gas distribution networks are progressing toward decentralized energy systems, which include microgrids for local energy generation and consumption. This strategy improves energy resilience, minimizes transmission losses, and promotes community-level integration of renewable energy sources. By implementing waste-to-energy technology, gas distribution networks are experimenting with circular economy practices. This entails transforming waste materials into energy, which helps to contribute to a more sustainable and resource-efficient energy model for modern buildings [44].

5.6 CONCLUSION

The increasing need for sustainable energy solutions in modern buildings is a reaction to the issues faced by rapid urbanization, environmental deterioration, and the impending threat of climate change. Cities are crucial in the worldwide fight for a sustainable future, and transitioning to cleaner, more efficient energy sources is a key step. Embracing the CGD sector solves not just environmental problems, but also improves energy security, boosts economic growth, and promotes social fairness. As cities expand, the incorporation of sustainable energy solutions will be critical in establishing resilient, flourishing, and environmentally conscious cities for future generations. Natural gas uses in modern buildings include heating, water heating, cooking, combined heat and power systems, and backup power production, with emerging choices such as renewable natural gas helping to create a more sustainable and diverse energy environment. While natural gas has significant advantages for modern buildings, environmental problems, notably the potential for methane emissions during extraction and transportation, must be addressed. Implementing modern technology and tight regulations can assist to mitigate these effects, ensuring that natural gas use in buildings is consistent with larger sustainability goals. City gas systems help in building energy efficiency by offering a cleaner and more efficient energy supply, enabling high-efficiency equipment, simplifying district heating and cooling, and integrating renewable gases. The entire impact contributes to decreased energy usage, lower emissions, and more sustainable urban settings. Designing gas systems for modern buildings involves a holistic approach that considers safety, energy efficiency, sustainability, and future adaptability. Emerging technologies in gas distribution systems are aligning with sustainable development goals by incorporating renewable gases, researching hydrogen infrastructure, applying sophisticated monitoring and analytics, embracing decentralized energy systems, and supporting circular economy practices. Continuous growth and enhancement of city gas systems in modern buildings is critical for meeting environmental goals and creating a more sustainable energy infrastructure.

REFERENCES

1. H. Al-Thani, M. Koç, R. J. Isaifan, and Y. Bicer, "A review of the integrated renewable energy systems for sustainable urban mobility," *Sustainability*, vol. 14, no. 17, p. 10517, 2022.
2. F. Cumo, D. A. Garcia, L. Calcagnini, F. Rosa, and A. S. Sferra, "Urban policies and sustainable energy management," *Sustainable Cities and Society*, vol. 4, pp. 29–34, 2012.
3. D. Chwieduk, "Towards sustainable-energy buildings," *Applied Energy*, vol. 76, no. 1–3, pp. 211–217, 2003.
4. S. Rajput, N. Sabharwal, A. Singh, B. Shingan, and B. P. Yadav, "City gas distribution incident analysis in India using Pareto principle: A comprehensive analysis," *Journal of Failure Analysis and Prevention*, pp. 1–13, 2022. doi: 10.1016/j.jenvp.2021.101671.
5. K. Yadav and A. Sircar, "Modeling parameters influencing city gas distribution sector based on factor analysis method," *Petroleum Research*, vol. 7, no. 1, pp. 144–154, 2022.
6. B. Shingan and V. Parthasarthy, "Recent progress in cold utilization of liquefied natural gas," *Chemical Engineering & Technology*, vol. 46, no. 9, pp. 1823–1837, 2023.
7. A. Al-Enazi, Y. Bicer, E. C. Okonkwo, and T. Al-Ansari, "Evaluating the utilisation of clean fuels in maritime applications: A techno-economic supply chain optimization," *Fuel*, vol. 322, p. 124195, 2022.
8. J. Pospíšil, P. Charvát, O. Arsenyeva, L. Klimeš, M. Špiláček, and J. J. Klemeš, "Energy demand of liquefaction and regasification of natural gas and the potential of LNG for operative thermal energy storage," *Renewable and Sustainable Energy Reviews*, vol. 99. Elsevier Ltd, pp. 1–15, Jan. 01, 2019. doi: 10.1016/j.rser.2018.09.027.
9. S. Mokhatab, W. A. Poe, and J. Y. Mak, *Handbook of Natural Gas Transmission and Processing: Principles and Practices*, Gulf Professional Publishing, 2018.
10. "Uses of natural gas," http://www.naturalgas.org/.
11. A. Rawat, S. Gupta, and T. J. Rao, "Risk analysis and mitigation for the city gas distribution projects," *International Journal of Energy Sector Management*, vol. 15, no. 5, pp. 1007–1029, 2021.
12. S. K. Kudaisya and S. K. Kar, "A comprehensive review of city gas distribution in India," In Kar, S., Gupta, A. Eds, *Natural Gas Markets in India: Opportunities and Challenges*, pp. 113–165, 2017.

13. M. Bianchi, A. De Pascale, and P. R. Spina, "Guidelines for residential micro-CHP systems design," *Applied Energy*, vol. 97, pp. 673–685, 2012.
14. D. Liu, W. Zhou, and X. Pan, "Risk evaluation for city gas transmission and distribution system based on information revision," *Journal of Loss Prevention in the Process Industries*, vol. 41, pp. 194–201, 2016.
15. D. J. Pandian *et al.*, "City gas India roundtable 2010: Initiatives and challenges," *Vikalpa*, vol. 35, no. 4, pp. 61–92, 2010.
16. O. Ridaliani A. Prima, A. Rinanti, B. Satiyawira, A. Hamid, M. Maulani, W. Dahani, and H. Pramadika, "*The strong growth predictors for Indonesia's city gas distribution*," *TRKU*, vol. 62, no. 03, pp. 477–487, 2020.
17. N. Bist, A. Sircar, and K. Yadav, "Recent trends in natural gas and city gas distribution network," In Anirbid Sircar, Gautami Tripathi, Namrata Bist, Kashish Ara Shakil, Mithileysh Sathiyanarayanan, Eds., *Emerging Technologies for Sustainable and Smart Energy*, CRC Press, 2022, pp. 57–72.
18. "Environmental, Health, and Safety Guidelines GAS DISTRIBUTION SYSTEMS Environmental, Health, and Safety Guidelines for Gas Distribution Systems," 2007. [Online]. Available: www.ifc.org/ifcext/enviro.nsf/Content/EnvironmentalGuidelines.
19. A. Gupta, "Preparedness to handle emergency in city gas distribution networks," 2009. Part of the proceedings of Pipeline Technology Conference 2009, Hannover Messe, Hannover, Germany.
20. T. Kiuchi, H. Izumi, and T. Huke, "An operation support system of large city gas networks based on fluid transient model," *J. Energy Resour. Technol*, vol. 117, no. 4, pp. 324–328, 1995.
21. N. Keogh, D. Corr, R. O'Shea, and R. F. D. Monaghan, "The gas grid as a vector for regional decarbonisation-a techno economic case study for biomethane injection and natural gas heavy goods vehicles," *Applied Energy*, vol. 323, p. 119590, 2022.
22. V. Bianco, F. Scarpa, and L. A. Tagliafico, "Analysis and future outlook of natural gas consumption in the Italian residential sector," *Energy Conversion and Management*, vol. 87, pp. 754–764, 2014.
23. Z. Zakaria and A. Mustafa, "Gas reticulation optimisation for domestic sector in Malaysia," *International Journal of Oil, Gas and Coal Technology*, vol. 2, no. 4, pp. 365–378, 2009.
24. D. Zhang, H. Zhu, H. Zhang, H. H. Goh, H. Liu, and T. Wu, "An optimized design of residential integrated energy system considering the power-to-gas technology with multi-functional characteristics," *Energy*, vol. 238, p. 121774, 2022.
25. H. R. Ellamla, I. Staffell, P. Bujlo, B. G. Pollet, and S. Pasupathi, "Current status of fuel cell based combined heat and power systems for residential sector," *Journal of Power Sources*, vol. 293, pp. 312–328, 2015.
26. V. Bianco, F. Scarpa, and L. A. Tagliafico, "Analysis and future outlook of natural gas consumption in the Italian residential sector," *Energy Conversion and Management*, vol. 87, pp. 754–764, 2014.
27. L. I. Lubis, I. Dincer, G. F. Naterer, and M. A. Rosen, "Utilizing hydrogen energy to reduce greenhouse gas emissions in Canada's residential sector," *International Journal of Hydrogen Energy*, vol. 34, no. 4, pp. 1631–1637, 2009.
28. U. Modi, A. Sircar, and S. Sahjpal, *City Gas Distribution – An Indian Perspective*, Technology Publications, 2017.
29. Y. Shimizu *et al.*, "Development of real-time safety control system for urban gas supply network," *Journal of Geotechnical and Geoenvironmental Engineering*, vol. 132, no. 2, pp. 237–249, 2006.

30. Y. C. Wijnia, R. J. M. Hermkens, and J. Flonk, “The safety indicator: Measuring safety in gas distribution networks,” In Amadi-Echendu, J., Brown, K., Willett, R., Mathew, J., Eds., *Definitions, Concepts and Scope of Engineering Asset Management, Engineering Asset Management Review*, vol 1. Springer, London, pp. 327–344, 2010.
31. J. S. Oh, J. G. Sung, and Y. D. Kim, “Smart wireless system for city-gas safety management,” *WIT Transactions on Ecology and the Environment*, vol. 155, pp. 1173–1183, 2012.
32. L. Y. U. Hui, W. Zhu, Y. Wang, and Q. Wang, “Study on index system of safety risk assessment for city gas industry,” *Fifth Symposium of Risk Analysis and Risk Management in Western China (WRARM 2017)*, Atlantis Press, 2017, pp. 29–33.
33. B. Tchórzewska-Cieślak, K. Pietrucha-Urbanik, M. Urbanik, and J. R. Rak, “Approaches for safety analysis of gas-pipeline functionality in terms of failure occurrence: A case study,” *Energies*, vol. 11, no. 6, p. 1589, 2018.
34. D. Schott, “Empowering European cities: Gas and electricity in the urban environment,” In Mikael Hård and Thomas J. Misa, Eds., *Urban Machinery. Inside Modern European Cities*, The MIT Press, pp. 165–186, 2008.
35. B. Shingan, P. Vijay, and K. Pandian, “Advanced design of power generation cycle with cold utilization from LNG,” *Arabian Journal for Science and Engineering*, pp. 1–16, 2023.
36. M. I. Khan, T. Yasmin, and A. Shakoor, “International experience with compressed natural gas (CNG) as environmental friendly fuel,” *Energy Systems*, vol. 6, no. 4, pp. 507–531, 2015.
37. J. Leicher *et al.*, “The impact of hydrogen admixture into natural gas on residential and commercial gas appliances,” *Energies*, vol. 15, no. 3, p. 777, 2022.
38. P. Y. Hoo, H. Hashim, and W. S. Ho, “Opportunities and challenges: Landfill gas to biomethane injection into natural gas distribution grid through pipeline,” *Journal of Cleaner Production*, vol. 175, pp. 409–419, 2018.
39. A. Guzzini, “Gas distribution: from networks’ integrity management to new smart meters’ performances technical evaluation,” [Dissertation thesis], Alma Mater Studiorum Università di Bologna. Dottorato di ricerca in Meccanica e scienze avanzate dell’ingegneria, 31 Ciclo. DOI 10.48676/unibo/amsdottorato/9001, 2019.
40. T. Ota, M. Kakinaka, and K. Kotani, “Demographic effects on residential electricity and city gas consumption in the aging society of Japan,” *Energy Policy*, vol. 115, pp. 503–513, 2018.
41. S. M. Sirin, “Energy market reforms in Turkey and their impact on innovation and R&D expenditures,” *Renewable and Sustainable Energy Reviews*, vol. 15, no. 9, pp. 4579–4585, 2011.
42. P. J. G. Pearson and S. Arapostathis, “Two centuries of innovation, transformation and transition in the UK gas industry: Where next?,” *Proceedings of the Institution of Mechanical Engineers, Part A: Journal of Power and Energy*, vol. 231, no. 6, pp. 478–497, 2017.
43. M. Arentsen, “Facilitating sustainable innovations: Sustainable innovation as a tool for regional development.”
44. A. Petersson and A. Wellinger, “Biogas upgrading technologies–developments and innovations,” *IEA Bioenergy*, vol. 20, pp. 1–19, 2009.

Chapter 6

India's future on sustainable infrastructure

A systematic review of initiatives

Jaimin Sureshbhai Parmar, Chandan Swaroop Meena, Nawal Kishor Banjara, and Jagruti Piyushkumar Shah

6.1 INTRODUCTION

The term infrastructure is widely used in today's world. There are many definitions available of the infrastructure. Infrastructure, according to the World Bank 1994, p. 2, [1] is defined as facilities and processes included in three aspects: public utility, public work, and other transport infrastructure. Public utilities include power, telecommunications, piped water supply, sanitation and sewerage, solid waste collection and disposal, and piped gas. Public works include roads and major dam and canal works (for irrigation and drainage). Other transport sectors include urban and interurban railways, urban transport, ports and waterways, and airports. A modern understanding of infrastructures cannot be derived from that definition. As new services are emerging / owing to new emerging services from recent developments in science and technology (and will continue to emerge), as the increasing population of the countries; and finding easement, comfort, safety, and speed in each aspect of human life. (Has and will govern the face of national and international infrastructure.)

Infrastructure is the basic necessity/medium on which different services of a nation, state, city, organization, or facility operate on a daily basis. For example, mobility/transportation of people and goods requires roads, railway tracks, airstrips/runways, metro lines, tram lines, etc. Water supply and sewage drainage need pipelines, pumping stations, and a system consisting of manpower and machinery to operate. Without these essential infrastructures, the services cannot work. A bus, train, or airplane needs a level, strong, and smooth surface to operate; however, only roads, railway tracks, and airstrips or runways are not the infrastructure in the totality. A bus can operate but bus service cannot operate on roads alone, a train can operate but train service cannot operate on only railway tracks, an airplane can take off and land on a runway or airstrip but air service cannot operate on only base infrastructure. Hence, a service infrastructure is also coming into the picture when we discuss the infrastructure. A bus service needs various stations and stands throughout the route, a train service needs a traffic monitoring system, stations, junctions, and terminals, and an airway

 DOI: 10.1201/9781003496656-6

service needs terminals, taxiways, hangers, tower control, etc. as service infrastructure.

A service infrastructure facilitates users of infrastructure services to access the services, for example, purchasing tickets, board and un-board safely. It helps the operators estimate the time of journey and fare of the services, monitor and regulate the quality of services, implement mobility services, and operate the services smoothly., The new technological developments has led modern services like, e-ticketing services, QR ticketing, reservation services, real-time tracking services, etc. to require another set of service infrastructure for which the understanding of the infrastructure expands. Now, the scope of the definition also includes public data acquisition, processing, monitoring, and management of websites, applications, etc. It is more feasible to define and update the attributes of the infrastructure, as the availability of a wide range of facilities and services is in a non-physical state.

Sustainability means to maintain balance, and is defined as "meeting the needs of the present without compromising the ability of future generations to meet their own needs" – United Nations Brundtland Commission (1987) [2]. We need resources to survive, and so do infrastructure services and our city, state, and nation. In this quest to utilize precious resources of the earth for our survival, we were historically short-sighted. We made technologies that utilize some non-renewable resources; now there is a threat to these resources; the scarcity of resources led us to find alternatives, control consumption, and adopt new sustainable policies in recent times. The environmental sustainability of infrastructure can be achieved by reducing dependency on non-renewable resources (conservation) and increasing the usage of renewable resources. Sustainable infrastructure is required to be resilient, and capable of sustaining for longer periods in the havoc and chaos of disasters, the constant effect of the surrounding environment, and the increasing demand of future population.

This chapter provides a structured review of the policies, schemes, missions, programs, and projects on sustainable infrastructure development, which can generally be categorized into five major sectors, i.e., transport, energy, industry, mining and quarrying, and water. Besides these, the national, state, or city infrastructure categories are more than that. As a city is a complex socio-technical element of human society, the infrastructure affects nearly all aspects of our lives. These ministries are interconnected and focused on various goals, relative to the aim and functionality of individual ministry. Sustainable development can be a primary outcome or byproduct of outcomes for each ministry as various ministries of India directly or indirectly contribute to sustainable infrastructure, for example, all the given ministries, "Ministry of Housing and Urban Affairs; Road Transport and Highway; Jal Shakti; Environment, Forest and Climate Change; Coal; Mines; Rural Development; Railways; Civil Aviation; Power; New and Renewable Energy; Petroleum and Natural Gas; Tourism," are integral components, formulating national policies, coordinates with various ministries, subordinate offices, government and private entities for the collective effort towards achieving sustainable development of backbone i.e., infrastructures in India.

6.2 MINISTRY OF ENVIRONMENT, FOREST AND CLIMATE CHANGE (MOEF & CC)

The Ministry of Environment, Forest and Climate Change is an essential ministry for a sustainable environment both for humans and wildlife through planning, promotion, coordination, and successful implementation of environmental and forestry policies, projects, and programs. This ministry plays a crucial role in controlling pollution, conserving forests, protecting of environment and wildlife, etc. Integration of the built environment with the natural environment is an essential requirement for the existence of today's world, as both environments are codependent for survival. Therefore, the ministry has taken several initiatives as below [3].

6.2.1 Lifestyle for Environment (LiFE)

- **Aim:** LiFE aims to bring sustainable change in the cities through a change in habits, mindset, and ultimately lifestyle of the 1 billion people (we call it P3—Pro Planet People), 5.15 lakh villages, 3700 urban local bodies, and 766 districts within the time of 5 years.
- **Objective:** To involve citizens at the individual level in the climate change movement.
- **Key provisions/features:** The approach focuses on individual behavior, co-creates globally, and leverages local cultures.
- **Implementation:** Three phases, i.e., change in demand, change in supply, and change in policy.
- **Progress/Achievements:** Introduced on November 1, 2021. Pesticide-free Punukula, Sikkim Organic Mission, Sattvik Food Festival, Traditional Wisdom of Conserving Indigenous Seeds, Auroville Earth Institute, Ecocab Punjab, Ranchi's Cycle Revolution, GoCoop Weaving Technology and Fashion Together, Black Cotton Initiative, Puducherry Mahila Housing Trust, Ecozen Solutions, Waste to Energy Tamil Nadu, Goonj New Delhi, Conserving wetlands Jammu and Kashmir, Climate Adaptation in Wetlands along the Mahanadi River Catchment Area, Coastal habitat rehabilitation and management. Rehabilitation and management of grasslands, coastal regions, and an arid island between salt flats—Kutch, and Enhancing Climate Resistance of Forest and Forest Dependent Communities—Jharkhand.
- **Impacts:** Saving in consumption of energy (22.5 billion kWh) and water (9 trillion liters). Reduced waste including food waste, e-waste,

and single-used plastics (375 million tons) by 2027–2028. People adopt a healthy and sustainable lifestyle.

- **Challenges/limitations:** Difficult to monitor change and practicality constraints in sustaining behavior. [4]

6.2.2 National Action Plan on Climate Change (NAPCC)

- **Aim:** To reduce the climate change impacts by adapting ecologically sustainable growth.
- **Objective:** To identify, categorize, monitor, and resolve the impacts. To promote eco-friendly alternatives. Make a strategy to alleviate climate change by reducing GHGs.
- **Key provisions/features:** Principles like Protection, National Sustainable Growth, Demand Site Management, Mitigation of GHGs, Sustainable Development of Market Place, Public-Private Partnership, and International Cooperation. The approach focuses on the urgency and the critical nature of climate change concerns. Comprises eight missions (National Solar Mission, National Mission for Enhanced Energy Efficiency, National Mission on Sustainable Habitat, National Water Mission, National Mission for Sustaining the Himalayan Ecosystem, National Mission for a Green India, National Mission for Sustainable Agriculture, and National Mission on Strategic Knowledge for Climate Change).
- **Implementation:** Through eight missions (National Solar Mission, National Mission for Enhanced Energy Efficiency, National Mission on Sustainable Habitat, National Water Mission, National Mission for Sustaining the Himalayan Ecosystem, National Mission for a Green India, National Mission for Sustainable Agriculture, National Mission on Strategic Knowledge for Climate Change) carried out by respective ministries as given in the respective clause of the chapter.
- **Progress/Achievement:** Released in June 2008. Achievements given in respective ministries of eight missions.
- **Impacts:** Attention of various ministries, developing strategies, plans, schemes, policies, and taking initiatives regarding Technological Research, Promotion of Sustainable Development, Use of Renewable Sources, and Sustainable Alternatives.
- **Challenges/limitations:** Coordination of various ministries, private-public entities, and international bodies. [5]

6.3 MINISTRY OF HOUSING AND URBAN AFFAIRS (MOHUA)

The Ministry of Housing and Urban Affairs with the help of five Attached and three Subordinate Offices (Central Public Works

Department—CPWD, Directorate of Printing, Directorate of Estates, Land and Development Office—L&DO, and National Building Organization—NBO), three Public Sector Undertaking (NBCC India Ltd., Housing and Urban Development Corporation—HUDCO, and Hindustan Prefab Limited—HPL), and eight Statutory/Autonomous Bodies, including one non-statutory registered society and a government company (Delhi Urban Arts Commission—DUAC, The National Capital Region—NCR, Delhi Development Authority—DDA, The National Institute of Urban Affairs—NIUA, Rajghat Samadhi Committee, Building Material Technology Promotion Council—BMTPC, National Cooperative Housing Federation—NCHF, Central Government Employees Welfare Housing Organization—CGEWHO), develops the strategy and implements comprehensive strategies and initiatives for the planned development of housing and the urban areas in India [6].

Perhaps, it is one of the most foundational ministries, owing to the increasing need for housing and service infrastructure due to high population, urbanization, urban poverty, enhanced service delivery, and low capacity of aging, polluting, and unserviceable infrastructure. The contribution of housing and urban infrastructure in a comprehensive development of physical, institutional, social, environmental, economic, and resilience sustainability is essential for sustainable cities that are human settlement inclusive, safe, resilient, and sustainable, according to the 2030 development agenda of the United Nations [7]. Cities contribute approximately 75% of the global CO_2 emissions with transport and buildings as the largest contributors, according to the United Nations, and 70% of the global greenhouse gases emission. "Cities are the forefront of climate fight" according to World Bank press released on May 18, 2023 [8]. The need for policy changes is recommended by the World Bank report on "Thriving: Making Cities Green, Resilient, and Inclusive in a Changing Climate." [9]

The ministry is taking several actions and initiatives, for example, Missions of National Action Plan on Climate Change (NAPCC); National Heritage City Development and Augmentation Yojana (HRIDAY); National Urban Transport Policy (NUTP), 2006; Metro Rail Policy, 2017; National Common Mobility Card (NCMC); Integrated Sustainable Urban Transport Systems for Smart Cities (SMART-SUT); North Eastern Region Urban Development Programme (NERUDP).

6.3.1 National Mission on Sustainable Habitat on Climate Change (NMSH 2.0)

- **Aim:** To build a framework for implementation of enabling climate actions and facilitate adoption.
- **Objective:** To manage the carbon footprint and disaster resilience of urban habitats. To promote low-carbon urban growth, and reduce GHG emission. To build climate change and disaster-resilient cities.

- **Key provisions/features:** Five thematic areas, i.e., Energy and Green Building, Urban Planning, Mobility and Air Quality, Green Cover and Biodiversity, Water Management, and Waste Management.
- **Implementation:** Through commitments, roadmaps, and sector-wise action plans on Adaptation, Mitigation, and Monitoring.
- **Progress/Achievements:** As a part of the NAPCC, launched in 2010. NMSH 2.0 launch in 2020–2021 with a timeline of 2030.
- **Impacts:** Development of Standards, City Development and Mobility Plans, and Capacity Building.
- **Challenges/Limitations:** Lack of Public Participation due to inadequate or unavailable alternatives. Poor enforcement, coordination, integration, and awareness in communities and ministries.

6.3.2 Swachh Bharat Mission Urban 2.0 (SBM-U 2.0)

- **Aim:** To make cities garbage-free with sustainable waste management.
- **Objective:** To sustain ODF+ or ODF++ (Open Defecation Free), 3-star garbage-free, water + status. To change Behavior through IEC (Information, Education and Communication).
- **Key provisions/features:** Mass behavior change approach. Four-tier mission management structure. ₹1.41 lakh crore budget.
- **Implementation:** Collection through Individual Household Latrines (IHHL), Community Toilets (CTs)/Public Toilets (PTs) and urinals, Municipal Solid Waste Management, and Used water management. Conveyance through Zone-wise Sewer Network, Home Collection of Municipal Waste. Treatment through stand-alone FSTP/STP.
- **Progress/Achievement:** 5-year initiative started from October 1, 2021 to October 1, 2026.
- **Impacts:** Improving urban landscape by reduction of pollution, healthy lifestyle, improved aesthetic, sustainable communities, private sector participation, innovation, and cost-effective solutions. 85 lakh sewer connections, benefiting more than four crore people.
- **Challenges/limitations:** Relies on third-party evaluation.
- **Review:** It is a comprehensive approach to address the pollution and waste in urban areas contributing to sustainable infrastructure development.

6.3.3 Atal Mission for Rejuvenation and Urban Transformation 2.0 (AMRUT 2.0)

- **Aim:** Drinking water and water sanitation infrastructure connected to every household aiming towards "Making Cities Water Secure."
- **Objective:** To plan, construct, operate, maintain, and repair the infrastructure, providing 2.68 crore tap water connections and 2.64 crore sewer/septage connections.

- **Key provisions/features:** Adopting approach focuses on community participation, and principles of conservation, circular economy, and rejuvenation of water bodies. ₹2.87 lakh crore budget.
- **Implementation:** A three-tier institutional mechanism, Monitoring through Indicative Annexures, "Pey Jal Survekshan," "Technology Sub-Mission," "Central Water Balance Plan," and "Central Water Action Plan."
- **Progress/Achievement:** Five-year initiative launched on October 1, 2021, providing 1.1 crore water tap connections, benefiting more than four crore people.
- **Impacts:** Data-led governance, water security, improved living standards, conservation through metering, competitiveness between ULBs, and GIS (Geographic Information System) and SCADA (Supervisory Control and Data Acquisition)-based operation and maintenance. Promotion of Public-Private Partnership (PPP/3Ps) projects.
- **Challenges/limitations:** Addressing disparities in infrastructure distribution, and overcoming technical hurdles.

6.3.4 Smart Cities Mission (SCM)

- **Aim:** Driving economic growth and improving quality of life with comprehensive work on physical, social, institutional, and economic, pillars of the city.
- **Objective:** To encourage the development of cities that use "smart solutions." To offer basic infrastructure, a healthy and sustainable environment, and a respectable standard of living for their people.
- **Key provisions/features:** Government support up to Rs. 48,000 crores over the span of five years. Six principles, i.e., focused on community, more from less, cooperative and competitive federalism, integration, innovation and sustainability, technology as a tool instead of goals, and convergence. Area-based development approach.
- **Implementation:** Through instruments like CITIES—City Investment to Innovate Integrate and Sustain, NULP—National Urban Learning Platform, MPI—Municipal Performance Index, ISAC—India Smart City Award Contest, IMAF—ICCC Maturity Assessment Framework, DMAF—Data Maturity Assessment Framework, Climate Smart Cities, City INX, Maturity Assessment Framework, and EOI—Ease of Living Index to Selected 100 cities.
- **Progress/Achievement:** Launched on June 25, 2015. 75+ case studies on innovative projects on Urban Infrastructure, Urban Management and Climate Change, and Resilient Cities. Overall, 6,237 out of 7,959 projects worth ₹114,934 crore were completed with private and public investments.
- **Impacts:** PPP, ICT, and IOT-based high living standards, ease of business, sustainable alternatives, efficient and productive work culture, and innovation in urban infrastructure development.

- **Challenges/limitations:** First-time "Challenge" or competition method, planning-based investment projects rather than conventional DPR-based approach, financial uncertainty, participation of smart people.

6.3.5 National Heritage City Development and Augmentation Yojana (HRIDAY)

- **Aim:** Improvement in overall quality of life with a specific focus on sanitation, security, tourism, heritage revitalization, and livelihoods retaining the city's cultural identity.
- **Objective:** To preserve the unique character of the city. To plan, develop, monitor, and implement heritage-sensitive infrastructure. To promote inclusive heritage-based communities. To improve tourism in surrounding areas. Three-component scheme, Heritage Documentation and Mapping, Heritage Revitalization, City information/knowledge management and skill development.
- **Key provisions/features:** Four years duration of the scheme from December 2014. Focuses on physical, social, institutional, and economic infrastructure.
- **Implementation:** Through project preparation, appraisal, and implementation. Heritage Management Plans (HMPs) in Detailed Project Report (DPR).
- **Progress/Achievement:** Initiated on January 21, 2015, and concluded on March 31, 2019. A total of 77 projects were implemented with an allocation of ₹453.90 crore.
- **Impacts:** Stakeholder/public participation, uniqueness of community, improved infrastructure, tourism experience, services, and economy of the city.
- **Challenges/limitations:** Ensuring long-term management of sustainability in balancing modernization with heritage preservation

6.3.6 National Urban Housing and Habitat Policy, 2007

- **Aim:** Promoting partnership of public, private, and cooperative sectors for sustainable development of habitat.
- **Objective:** Ensuring Sustainability and Affordability of Habitat. Encouraging public-private partnership. Enhancing supply of land and housing stock. Promoting rental housing and in situ slum rehabilitation.
- **Key provisions/features:** Regional planning approach, government as facilitator and regulator, and involvement of multiple stakeholders. Deendayal Antyodaya Yojana-National Urban Livelihoods Mission (DAY-NULM), Pradhan Mantri Awas Yojana (Urban), and Pradhan Mantri Awas Yojana (Rural).

- **Progress/achievement:** According to schemes and missions under the policy.
- **Impacts:** Improved social sustainability, reduced poverty, enhanced quality standards of living.
- **Challenges/limitations:** Inter-agency coordination and communication, addressing urbanization pressures, and adapting to evolving urban challenges and demographics.

6.3.7 Deendayal Antyodaya Yojana-National Urban Livelihoods Mission (DAY-NULM)

- **Aim:** Sustainable development of habitat (land, shelter, and services) in an equitable and affordable manner.
- **Objective:** To reduce poverty and vulnerability. To provide social and financial sustainability access to employment/business/vendor opportunities. To improve skill through training.
- **Key provisions/features:** Strategies like capacity building, poverty alleviation, training, and support. Social Mobilization and Institution Development (SM&ID). Capacity Building and Training (CB&T). Employment through Skills Training and Placement (EST&P). Self-Employment Programme (SEP). Scheme of Shelter for Urban Homeless (SUH). Innovative and Special Projects involving Public, Private, and Community Partnerships (PPCP). PM Street Vendor's AtmaNirbhar Nidhi (PM SVANidhi).
- **Implementation:** Monitoring and evaluation through three-level mission management units (MMUs), i.e., national, state, and city level.
- **Progress/achievement:** Launched on September 24, 2013. Nearly 5.26 lakh persons established individual or group businesses as of March 2020, borrowing a total of Rs. 4417.86 crore. MoUs with International, Government, and Private entities.
- **Impacts:** Involvement of urban poor in the transparent, accountable, and inclusive community self-reliability process of sustainable urban infrastructure development.
- **Challenges/limitations:** Addressing potential bureaucratic hurdles, socio-political drift, and financial risks.

6.3.8 Pradhan Mantri Awas Yojana (Urban)

- **Aim:** Providing all-weather pucca houses in urban areas for the urban poor.
- **Objective:** To address residential vulnerability issues. To improve the living standards of the urban poor.
- **Key provisions/features:** Security of tenure, woman empowerment, better quality of life, all-weather housing units, adequate physical and social infrastructure, securing UN-SDGs. Beneficiary-Led

Construction/Enhancement (BLC), Affordable Housing in Partnership (AHP), Affordable Rental Housing Complexes (ARHCs), and In-situ Slum Redevelopment (ISSR).

- **Implementation:** Involvement of stakeholders, Monitoring, and Evaluation through a robust and comprehensive Management Information System (MIS). Innovative solutions through the Global Housing Technology Challenge (GHTC).
- **Progress/Achievement:** Launched on June 25, 2015. 78.27 Lakhs out of 118.63 Lakhs Houses Completed. 8.11 lakh crore total investment.
- **Impacts:** Sustainable communities, affordable housing, social security for minority and poor.
- **Challenges/limitations:** Addressing urban migration challenges, and targeting vulnerable populations effectively.

6.3.9 National Urban Transport Policy, 2014

- **Aim:** To improve urban mobility with respect to urban problems.
- **Objective:** To assure safety, affordability, speed, comfort, dependability, and sustainability in public access. Introducing Intelligent Transport Systems, Safety, and Emergency Response, Urban Transport Planning, Public Transport, Infrastructure Design, Non-Motorized Transport, Financing, Traffic Management, Governance and Capacity Building.
- **Key provisions/features:** Integrating land use and transport planning, equitable allocation of road space, public transport oriented, utilizing technologies for public transport, low-carbon transportation. "Avoid, Shift and Improve" approach in transportation planning.
- **Implementation:** Inclusive planning and holistic approach in design. Establishing institutional framework, comprehensive mobility plans, integrated land use, and financial support.
- **Progress/Achievement:** Rs. 88,552 crore investment through private and public funds. Projects worth 795 crores out of 7626 crores were completed in 21 cities out of 54.
- **Impacts:** Universal accessibility, improved road safety, reduced adverse environmental impacts, highly reliable infrastructure, economic growth, and innovation in sustainable transportation. Promoting public-private partnerships, involvement of users, innovative technological advancements, and alternative modes of transportation.
- **Challenges/Limitations:** Resistance to behavior change, funding constraints, and coordination issues.

6.3.10 Metro Rail Policy, 2017

- **Aim:** High-capacity public transit for urban areas.
- **Objective:** To promote public transport. To provide a framework for metro rail projects. To ensure quality, standards, and safety in metro projects.

- **Key provisions/features:** Flexibility of diverse financial models, namely, Centre, State, Centre & State and PPPs, Land Value Capture and Transit Oriented Development to improve financial viability Constitution of Unified Metropolitan Transport Authority (UMTA), Urban Transport Fund, Multi-modal integration, and Last Mile connectivity. Involvement of unbundled Public-Private Partnerships Projects which are evaluated for economic returns and social viability. Avenues of mobilizing capital at a reasonable cost through the issuance of bonds. Theme of "Intelligent, Inclusive & Sustainable Mobility."
- **Implementation:** System Approach, Comprehensive Mobility Plan (CMP), MoU for metro operation, and other modes/government/stakeholders.
- **Progress/Achievement:** 363 km constructed out of 683 km metro and RRTS network approved after Metro Rail Policy. New urban transit modes, for instance, RRTS, MetroLite, MetroNeo, and WaterMetro.
- **Impacts:** Reduced traffic on road and pollution; rapid, safe, and economical transport. Higher quality ride and service regularity.
- **Challenges/limitations:** High capital investments, long gestation periods, land acquisition issues, potential cost overruns, and financial sustainability.

6.3.11 Integrated Sustainable Urban Transport Systems for Smart Cities (SMART-SUT)

- **Aim:** Improving the planning and implementation of SUT transport projects and pilot actions for the selected Indian cities.
- **Objective:** To support key institutions that have a transport-planning mandate for sustainable urban transport.
- **Key provisions/features:** Multi-level and multi-actor approach. To encourage the planning of low-carbon mobility in Kochi, Bhubaneshwar, and Coimbatore.
- **Implementation:** Consultation, Data Collection and Assessment, Action Plan, Implementation, Monitoring, and Evaluation. Capacity Building and Framework Building. City Level Action Plan.
- **Progress/Achievement:** Launched in August 2017 and ended in June 2022. Launching E-autos in Kochi. Institutional Assessment and Strengthening of KMC. Integrated Planning Document for transportation projects.
- **Impacts:** Strengthening institutions, improvement of urban transport services, parking policy and management, capacity building for individuals and communities, and developing integrated planning documents.
- **Challenges/limitations:** Sustained funding, and land acquisition in the urban context.

6.3.12 North Eastern Region Urban Development Programme (NERUDP)

- **Aim:** Enabling states to meet urban development challenges.
- **Objective:** To provide essential urban services, like water supply, waste management, and sewer connection. To strengthen institutions and financial reforms for managing projects and capacity development of ULBs.
- **Key provisions/features:** Supranational Institutional Investment through the North Eastern Region Capital Cities Development Investment Programme (NERCCDIP). Multi-tranche financing facility.
- **Implementation:** In states of Arunachal Pradesh, Assam, Manipur, Meghalaya, Mizoram, Nagaland, Sikkim, and Tripura. Phase 1 covers five capital cities: Agartala (Tripura), Aizawl (Mizoram), Gangtok (Sikkim), Kohima (Nagaland), and Shillong (Meghalaya). Phase 2 covers three capital cities of Guwahati (Assam), Imphal (Manipur) and Itanagar (Arunachal), and the city of Dibrugarh (Assam).
- **Progress/Achievement:** Phase 1 was approved on February 2, 2009. Total estimated budget of Rs 3,019.2 crores. The timeline was extended to June 22, 2021.
- **Impacts:** Improved living conditions and sustainable management of urban services.
- **Challenges/limitations:** Timely project execution, managing diverse state contexts, addressing unique urban challenges in NER.

6.4 MINISTRY OF RURAL DEVELOPMENT

The Ministry, comprising the Department of Rural Development and the Department of Land Resource, serves as the nodal Ministry for a myriad of rural development and welfare activities and plays a pivotal role in the sustainable development of rural infrastructures. Socio-economic transformation of rural areas through social pension and skill development initiatives leads to sustainable communities in rural areas. A sustainable modern rural area can help in reducing the population bursting in urban areas due to the migration of large populations owing to economic opportunities and improved living conditions in rural areas, thereby fostering a balanced distribution of population and alleviating the strain on urban infrastructure [10].

6.4.1 Pradhan Mantri Awaas Yojana—Grameen (PMAY-G)

- **Aim:** To provide pucca houses with basic amenities.
- **Objective:** Housing for everyone (2.95 crore houses by 2021–2022)

- **Key provisions/features:** Minimum unit size 25 sq.m. with hygienic and safe cooking area. Classification of plain and hilly regions for addressing special needs. Identification of beneficiaries. National Technical Support Agency (NTSA). Adopting saturation approach.
- **Implementation:** Through target population study, financial allocation, action plan, construction, beneficiary identification, support, and service. Reporting, Monitoring, Participation and Auditing of the program.
- **Progress/Achievement:** In effect since April 1, 2016. Completed 2.10 crore house units.
- **Impacts:** Improving living conditions, social sustainability, and security of poor/minority communities.
- **Challenges/Limitations:** Addressing regional variations, ensuring quality construction, and effectively reaching to remote vulnerable populations.

6.4.2 Pradhan Mantri Gram Sadak Yojana (PMGSY)

- **Aim:** To provide rural road connectivity.
- **Objective:** Plan, fund, construct, repair, and maintain the all-weather roads with essential support structures.
- **Key provisions/features:** National Rural Road Development Agency (NRRDA), Three-tier control mechanism, Quality Coordinators, Capacity Building, Manuals and Guidelines for technical support. Decentralized and evidence-based planning, standards and specifications according to IRC—Indian Road Congress.
- **Implementation:** District Rural Road Plan, Identification of districts with road connectivity needs. Allocation of resources with three priority levels. Monitoring Evaluation and Approval through Programme Implementation Units.
- **Progress/Achievement:** First launched in 2000, second in 2013, and third in 2016. A total of 1,25,000 Km of roads were constructed and upgraded with an estimated Rs 80,250 crore budget.
- **Impacts:** Aligned to SDGs. Improving Public Access, Travel Time, and promoting local businesses.
- **Challenges/Limitations:** Remote location access, condition assessment, public participation, and effective resource allocation.

6.4.3 Shyama Prasad Mukherjee Rurban Mission (SPMRM)

- **Aim:** Making rural areas socially, economically, and physically sustainable regions.
- **Objective:** Stimulating local economic development, and enhancing the basic service as of urban region to 300 clusters.

- **Key provisions/features:** Integrated Cluster Action Plans (ICAPs) and Detailed Project Reports (DPRs), Mission Management Units on the national and state level, Project Management Units at the District level, and Cluster Development and Management Unit (CDMU).
- **Implementation:** Selection of Cluster, Delineation & notification of Planning area, Cluster Profiling, Decency Analysis and identification of needs, Identifying and Detailing of Mission Components Scheme Convergence Investment/Phasing Arriving at CGF, Estimate Implementation Strategy, O&M Strategy, Obtaining Gram Sabha Resolutions, Submission of ICAP to MoRD, Revision of ICAP based on approved DPR, and Five Yearly Iteration to ICAP
- **Progress/achievement:** Launched in 2016, 248 clusters planned, 177,613 work out of 186,496 projected works completed or near completion with expenditure of Rs. 15,072.34 crore. Road length of 177,613 out of 749,257 km completed.
- **Impacts:** Improving living standards, sustainable development of infrastructure, and effective implementation.
- **Challenges/Limitations:** Public participation, and reliability on local resource.

6.4.4 Saansad Adarsh Gram Yojana (SAGY)

- **Aim:** Sustainable and self-dependent communities with high values.
- **Objective:** To make holistic development, high standard of living and quality of life, public participation, and model development for others.
- **Key provisions/features:** Leveraging the leadership, capacity, commitments, and energy of MPs. Focus on local-level participatory development and sustainability outcomes. Values as "nucleus of health, cleanliness, greenery and cordiality."
- **Implementation:** Identification of Adarsh Gram, Planning, Timelines, and Capacity Building.
- **Progress/Achievement:** 3390 Gram Panchayat are Identified. 127 Central Sector and Centrally Sponsored, and 1806 State Scheme. 209,499 out of 253,063 projects completed.
- **Impacts:** Sustainable development, high quality of life, social and economic growth.
- **Challenges/Limitations:** Addressing regional disparities, enhancing financial allocation, and refining the model for scalable and replicable development.

6.5 MINISTRY OF ROAD TRANSPORT AND HIGHWAYS (MORTH)

The Ministry of Road Transport and Highways, with the help of respective departments and entities like National Highway Authority of India

(NHAI), National Highways and Infrastructure Development Corporation Ltd. (NHIDCL), Indian Academy of Highway Engineers (IAHE), Indian Roads Congress (IRC), National Highway Logistics Management Limited (NHLML), etc., develops, plans, constructs, monitors, repairs, and maintains the most critical and largest infrastructure of this nation, i.e., road infrastructure; responsible for the economic development of a country. It is also important in terms of sustainable development as India has the second largest road network in the world about 63.32 lakh km. The road network comprises National Highways of a total stretch of 1,44,955 km, Expressways, State Highways, Major District Roads, Other District Roads, and Village Roads [11].

6.5.1 Green Highway Policy, 2015

- **Aim:** To develop an eco-friendly highway with the involvement of all stakeholders for economic and sustainable growth.
- **Objective:** To reduce soil erosion, pollution, glare, and wind flow and provide shades through "Plantation, Transplantation, Beautification & Maintenance."
- **Key provisions/features:** Technical—Data-based guideline for implementation. Emphasis on transplantation.
- **Implementation:** Selection of plantation type and arrangement, selection of species, following technical specifications based on scientific and evident studies. Protection measures.
- **Progress/achievement:** 244.68 lakh plants in 869 NH projects of length of 51,178 km.
- **Impacts:** Equilibrium in the development of highways and protection of the environment. Enhancing road safety and lifespan. Alignment of UN-SDGs.
- **Challenges/Limitations:** Regular maintenance needs, finances, and resources like fertilizer and water requirements.

6.6 MINISTRY OF RAILWAYS

The Ministry of Railways is at the forefront of comprehensive planning and execution concerning development (construction of railway infrastructure, i.e., railway tracks, stations, support facilities, etc., manufacturing/acquiring the trains, maintaining economic and environmental balance, operations of trains, and services like ticketing, ticketless travel, time management, and safety auditing) and management of functioning of world's fourth largest railway infrastructure by size and one of most commonly used mode of transportation, which provides safe, ,economical and rapid transportation of people and goods in India [12].

6.6.1 Mission 100% electrification moving toward net zero carbon emission

- **Aim:** Sustainable railways with no pollution direct or indirect for eco-friendly; faster and energy-efficient modes of transportation.
- **Objective:** To plan, construct, and maintain the electric lines and support facilities. To include renewable sources with electrification.
- **Key provisions/features:** Effective water management, paperless work, green certification, clean trains and railway stations, low-carbon green transportation network, and use of technologies like LED lighting, waste to energy, and bio-toilets for passenger coaches.
- **Implementation:** Third-party auditing. MoU between IR and Indian industry confederation.
- **Progress/Achievement:** Konkan Railway with 100% rail electrification. 600 railway stations implementing Environment Management System to ISO: 14001. 142 MW solar rooftop capacity and 103.4 MW of Wind energy installation. 52,508 out of 65,141 RKM BG Network.
- **Impacts:** Net zero carbon emitter by 2030.
- **Challenges/Limitations:** Technical complexities and ensuring uniform electrification progress.

6.6.2 Green railways initiatives

- **Aim:** Transforming the Indian Railways into Green Railways by 2030.
- **Objective:** Optimization of energy demand, monitoring energy consumption, and shifting to renewable sources of energy.
- **Key provisions/features:** NOC for Open Access, Wheeling and Banking provision, Merger of Solar Purchase Obligation & Non-Solar Purchase Obligation, and Unrestricted Net Metering Regulations.
- **Implementation:** Three phases of implementation, Phase 1—tendering, Phase 2—developing capacities in railway plots, and Phase 3—installing 1 GW capacity in railway plots.
- **Progress/Achievement:** Seven Production Units (PUs), 39 Workshops, 6 Diesel Sheds, 1 Stores Depot, 14 Railway Stations, and 21 Other Buildings/Campuses received Green Certification from CIII. There are 215 stations accredited by ISO 14001, the Environment Management System (EMS). Rs 450 crore on an annual basis, saving through Head of Generation.
- **Impacts:** Achievement of UN-SDGs.
- **Challenges/limitations:** Ensuring uniform implementation across diverse railway facilities, addressing financial constraints, and evolving technologies.

6.7 OTHER PASSIVELY IMPACTING MINISTRIES:

6.7.1 Ministry of Civil Aviation

The ministry oversees the development and regulation of air transport and service in a sustainable, safe and economical manner. Air transport/travel can greatly influence the development of tourism, and contribute significantly to economic growth by fostering cultural exchange, creating employment opportunities, and enhancing global connectivity. It helps sustainable infrastructure development by catalyzing investments in airports, roads, and related facilities, further boosting the overall accessibility and appeal of tourist destinations. The ministry is actively invested in the sustainable development of infrastructures like airport buildings, support facilities, pavement, aircraft, etc. [13].

6.7.2 Ministry of Jal Shakti

The ministry aims to manage water resources within the territory of India and provide clean and safe water regularly to all the citizens of India. Usable water is an essential resource for a nation and its citizens of a nation because of high scarcity, increased pollution, and the need for survival. The Department of Water Resources, River Development and Ganga Rejuvenation (DoWR, RD & GR) is responsible for policy guidelines and programs for the development, conservation, and management of water as a national resource. The resource in itself is insufficient for the people of the nation to thrive in growth as accessibility is an issue that coexists with resource management [14]. The Department of Drinking Water and Sanitation is responsible for policies, initiatives, and programs concerning potable water in sufficient quality and quantity for each citizen on a regular basis. This department also deals with the development and maintenance of sanitation infrastructure in rural areas under Swachh Bharat Mission—Gramin [15].

6.7.3 Ministry of Tourism

The ministry, with public sector undertaking and autonomous institutions, develops the tourism infrastructures and promotes tourism across the nation and provides financial assistance to city, state, or central agencies with the aim of promoting local culture, heritage, and art and catalyzing private sector investment, tourist revenue, and local businesses. Domestic Promotion, Publicity and Hospitality (DPPH) scheme, Viability Gap Scheme, Swadesh Darshan Scheme, PRASHAD Scheme, Capacity Building for Service Providers, Champion Services Sector Scheme for providing social, economic and environmentally sustainable development of tourism in India [16].

6.7.4 Ministry of New and Renewable Energy

The ministry is playing a crucial role in sustainable infrastructure development and India's commitment to achieving 50% dependence on renewable energy sources by the year 2030. Currently standing in fourth position in the world in installed capacity of renewable energy generation infrastructure, according to REN21 Renewables 2022 Global Status Report. Technological development is a key to achieving efficient resource utilization and replacing conventional non-renewable-fossil-based sources. Thus, research and development plays a crucial part in this ministry [17].

6.7.5 Ministry of Coal

The ministry is instrumental in overseeing the exploration and development of coal and lignite storage, authorization for major projects, and further developments through Public Sector Undertakings like Coal India Ltd. and Neyveli Lignite Corporation India Ltd., and joint ventures in Singareni Collieries Company Ltd. It is aimed at modern, sustainable, economic, and competitive coal sector with highly productive energy security. The ministry has sustainable objectives like infrastructure development, technology integration, research and development, and efficient and safe utilization of resources in an eco-friendly and economic manner, with initiatives like Satellite Surveillance for land reclamation, Sustainability Conclave 2022, Policy Initiatives and Reform Measures regarding Corporate Social Responsibility (CSR), the Sustainable Development Policy of Coal and Lignite PSUs, and Establishing Sustainable Development Cells (SDC) [18].

6.7.6 Ministry of Mines

The ministry is in charge of surveying and exploring all minerals other than natural gases, petroleum, and atomic minerals, as well as mining and metallurgy of non-ferrous metals such as aluminum, copper, zinc, lead, gold, nickel, and so on, as well as administration of the Mines and Minerals (Regulation and Development) Act, 1957, for all mines and minerals other than coal, natural gas, and petroleum. The sustainable utilization of natural resources of a nation is essential as it provides the foundation for long-term economic development, environmental conservation, and social well-being. A balance between resource exploitation and preservation contributes to the overall prosperity and ensures the resilience of the country [19].

6.7.7 Ministry of Power

The ministry is concerned with the development of energy infrastructure from the generation of electricity from different sources to end user supply, the construction, operation and maintenance of plants, operation facilities, transmission and distribution infrastructure in a safe, economical and equitable manner. This ministry involved in the sustainable use of electricity by reducing losses, promoting conservation of energy, using new technology, and adopting new renewable energy resources, standards and labelling, Energy Conservation Building Code (ECBC), implementation of Perform Achieve and Trade (PAT), Demand Side Management (DSM), enhancing energy efficiency in Industries, Small and Medium Enterprises (SMEs), Transport Sector, Strengthening of State Designated Agencies (SDA), promoting efficient utilization of energy and energy conservation [20].

6.7.8 Ministry of Petroleum and Natural Gas

This is an important ministry that concerns fuel infrastructure development for exploration, production, transportation, and distribution facilities in an accessible, affordable, secure, efficient, and sustainable manner. It directly impacts the security and conservation of national resources and indirectly impacts environmental pollution and economic growth. Initiatives like biofuel and biodiesel, my LPG (Liquid Petroleum Gas), EBP (Ethanol Blended Petroleum) program, etc. are helping the government in accomplishing environmentally sustainable fuel throughout the nation [21].

REFERENCES

1. World Development Report on "Infrastructure for Development" 1994, p. 2, https://documents1.worldbank.org/curated/en/535851468336642118/pdf/131840REPLACEMENT0WDR01994.pdf
2. World Commission on Environment and Development – United Nations Report On, "Our Common Future" 1987, https://www.are.admin.ch/dam/are/en/dokumente/nachhaltige_entwicklung/dokumente/bericht/our_common_futurebrundtlandreport1987.pdf.download.pdf/our_common_futurebrundtlandreport1987.pdf
3. Ministry of Environment, Forest, and Climate Change – Annual Report 2022–23; Government of India, moef.gov.in/wp-content/uploads/2023/05/Annual-Report-English-2022-23.pdf
4. MoEFCC and UNDP. 2022. *Prayaas Se Prabhaav Tak: From Mindless Consumption to Mindful Utilization*, https://missionlife-moefcc.nic.in/assets/pdf/Prayaas-Se-Prabhaav-Tak.pdf
5. Press Information Bureau; National Action Plan on Climate Change; RU-49-03-0004-011221/FAQ, https://static.pib.gov.in/WriteReadData/specificdocs/documents/2021/dec/doc202112101.pdf
6. Ministry of Housing and Urban Planning – Annual Report 2022–23, Government of India, https://mohua.gov.in/upload/uploadfiles/files/2688HUA-ENGLISH-19-4-2023.pdf
7. United Nations Environment Programme; Article on "Cities and Climate Change", https://www.unep.org/explore-topics/resource-efficiency/what-we-do/cities/cities-and-climate-change

8. The World Bank; Press Release on "Cities Key to Solving Climate Crisis"; 2023, https://www.worldbank.org/en/news/press-release/2023/05/18/cities-key-to-solving-climate-crisis
9. Mukim, Megha and Roberts, Mark (ed.). 2023. *Thriving: Making Cities Green, Resilient, and Inclusive in a Changing Climate.* © Washington, DC: World Bank, http://hdl.handle.net/10986/38295 License: CC BY 3.0 IGO.
10. Ministry of Rural Development – Annual Report 2022–23, Government of India, https://rural.nic.in/sites/default/files/AnnualReport2022_23_English_0.pdf
11. Ministry of Road Transport and Highways – Annual Report 2022–23, Government of India, https://morth.nic.in/sites/default/files/MoRTH%20Annual%20Report%20for%20the%20Year%202022-23%20in%20English.pdf
12. Ministry of Railways – Annual Report 2022–23, Government of India, https://indianrailways.gov.in/railwayboard/uploads/directorate/stat_econ/pdf/Indian%20Railways%20Annual%20Report%20%26%20Accounts%20English%202021-22_web_Final.pdf
13. Ministry of Civil Aviation – Annual Report 2022–23, Government of India, https://www.civilaviation.gov.in/ministry-documents/annual-reports/annual-report-2022-2023
14. Ministry of Jal Shakti (The Department of Water Resources, River Development and Ganga Rejuvenation) – Annual Report 2022–23, Government of India, https://cdnbbsr.s3waas.gov.in/s3a70dc40477bc2adceef4d2c90f47eb82/uploads/2023/04/2023041116.pdf
15. Ministry of Jal Shakti (The Department of Drinking Water and Sanitation) – Annual Report 2022–23, Government of India, https://jalshakti-ddws.gov.in/documents/annual-report
16. Ministry of Tourism – Annual Report 2022–23, Government of India, https://tourism.gov.in/sites/default/files/2023-02/MOT%20Annual%20Report_2022-23_English.pdf
17. Ministry of New and Renewable Energy – Annual Report 2022–23, Government of India, https://cdnbbsr.s3waas.gov.in/s3716e1b8c6cd17b771da77391355749f3/uploads/2023/08/2023080211.pdf
18. Ministry of Coal – Annual Report 2023, Government of India, https://coal.gov.in/sites/default/files/2023-03/chap3AnnualReport2023en.pdf
19. Ministry of Mines – Annual Report 2022–23, Government of India, https://mines.gov.in/admin/storage/app/uploads/6433da09a9f741681119753.pdf
20. Ministry of Power – Annual Report 2022–23, Government of India, https://powermin.gov.in/en/content/annual-report-2022-23
21. Ministry of Petroleum and Natural Gas – Annual Report 2019–20, Government of India, mopng.gov.in/files/TableManagements/2020-12-08-115045-xyd7b-AR_2019-20E.pdf

Chapter 7

Net zero energy buildings

Design considerations, challenges, and applications

Morteza Khashehchi, Sivasakthivel Thangavel, Pooyan Rahmanivahid, Milad Heidari, and Ashwani Kumar

7.1 INTRODUCTION

In the ever-evolving landscape of sustainable architecture, net zero energy buildings (NZEBs) stand as beacons of innovation and environmental responsibility. This chapter delves into the intricate world of NZEBs, exploring the key design considerations, challenges faced in their implementation, and the diverse applications that showcase their potential to reshape the future of the built environment (C. S. Meena et al. 2022a).

7.1.1 Understanding net zero energy buildings

NZEBs represent a change in thinking in the realm of sustainable architecture, challenging traditional notions of energy consumption and production within the built environment. This section delves into the fundamental principles that underpin NZEBs, offering a comprehensive understanding of the concepts guiding their design and operation (Jeongyoon Oh et al. 2017 & Matteo Bilardo, 2023).

Definition and Principles: At the core of the Net Zero Energy concept is the ambition to create buildings that, over the course of a year, generate as much energy as they consume. This involves a delicate balance between minimizing energy demand through efficiency measures and harnessing renewable sources to meet the remaining energy needs (C.S. Meena, 2022b). The section provides a nuanced definition of NZEBs, emphasizing their role in mitigating the environmental impact of buildings and contributing to broader sustainability goals (Figure 7.1).

Principles guiding NZEB design encompass a comprehensive approach that extends beyond energy production and consumption. The integration of passive design strategies, energy-efficient technologies, and renewable energy sources forms the cornerstone of these principles (V.P. Singh, 2022). Emphasizing the importance of a systems-thinking approach, designers and architects must consider the interplay of various components to optimize overall building performance (Liana Müller & Thomas Berker, 2013).

DOI: 10.1201/9781003496656-7

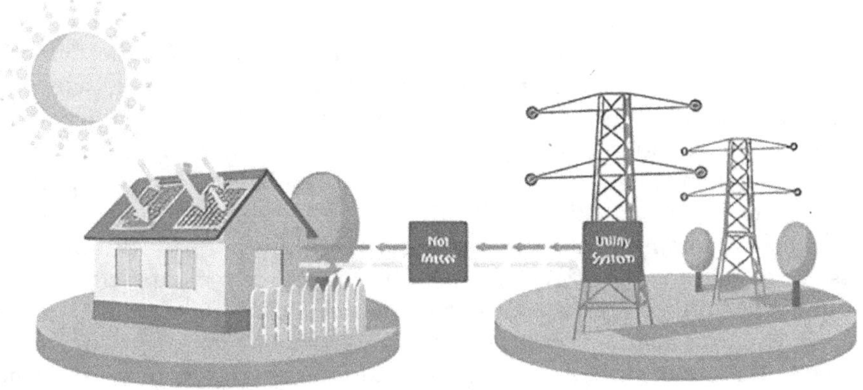

Figure 7.1 What is net zero energy building (Muhamnad Irfan et al., 2018)

7.1.2 Energy balance and metrics

Achieving a balance between energy consumption and production is essential for the success of NZEBs. This subsection delves into the intricacies of energy balance, emphasizing the need for a meticulous understanding of a building's energy requirements throughout different seasons. Key metrics such as Energy Use Intensity (EUI) and the Zero Energy Performance Index (ZEPI) are introduced as crucial tools for quantifying and evaluating the performance of NZEBs.

EUI serves as a quantitative measure of a building's energy efficiency, expressed as energy consumption per square foot or square meter. A lower EUI indicates higher energy efficiency, a vital aspect of NZEB success. ZEPI, on the other hand, provides a comprehensive assessment by considering not only energy consumption but also the renewable energy generated. This holistic metric offers a more nuanced view of a building's sustainability performance (Martínez Arias & Andrea, 2016). Understanding these metrics is pivotal for designers and stakeholders to assess the effectiveness of NZEB strategies and make informed decisions throughout the design and operation phases (Muhamnad Irfan et al.; Nsilulu T. Mbungu et al., 2020).

7.1.3 Design considerations for net zero energy buildings

The concept of NZEBs emerges as a pivotal response to the urgent global need for sustainable and environmentally conscious architecture. Against the backdrop of escalating concerns about climate change, resource depletion, and the environmental impact of conventional building practices, NZEBs stand out as a transformative solution. The significance of NZEBs lies in their fundamental departure from the traditional model of energy-intensive construction and operation. These buildings aim not only to reduce energy consumption through cutting-edge design strategies but also to generate the energy they require through renewable sources, achieving a delicate equilibrium (Martínez Arias & Andrea, 2016). This departure from the conventional paradigm is rooted in the understanding that the built environment plays a substantial role in global energy consumption and carbon emissions (Figure 7.2). By striving for net zero energy, NZEBs exemplify a commitment to mitigating the carbon footprint of buildings, fostering environmental stewardship, and highlighting the potential for a sustainable future. As the global community increasingly recognizes the imperative to transition toward cleaner and more efficient energy practices, the significance of NZEBs extends beyond individual structures, influencing industry standards, governmental policies, and societal perceptions about the possibilities of harmonizing human habitation with the natural environment. The evolution of NZEBs represents a change in thinking in architectural philosophy, one that not only addresses the pressing challenges of today but also sets a visionary course for a resilient and sustainable built environment in the years to come.

7.1.4 The evolution of sustainable design and the emergence of NZEBs

The evolution of sustainable design has been marked by a profound shift in architectural thinking, transitioning from a focus on aesthetic considerations alone to an integrative approach that prioritizes environmental responsibility and resource efficiency. Over the past few decades, the architectural and construction industry has witnessed a growing recognition of the environmental impact of buildings and the imperative to reduce their carbon footprint. This change in thinking has been underscored by a confluence of factors, including heightened awareness of climate change, advancements in technology, and a global push toward sustainability. Net zero energy buildings (NZEBs) have emerged as a pinnacle in this evolution, representing a milestone where the aspirations of sustainable design are translated into tangible, energy-efficient structures (Hoda Homayouni et al., 2020). The emergence of NZEBs reflects a maturation of the sustainable design movement, acknowledging that energy efficiency alone is insufficient, and that true sustainability requires a comprehensive approach that integrates renewable energy sources seamlessly into the built environment. The evolution from conventional design to sustainable design, and to NZEBs, is emblematic of an industry in constant pursuit of innovative solutions to address the environmental challenges of our time, while also highlighting a commitment to creating spaces that harmonize with the natural world.

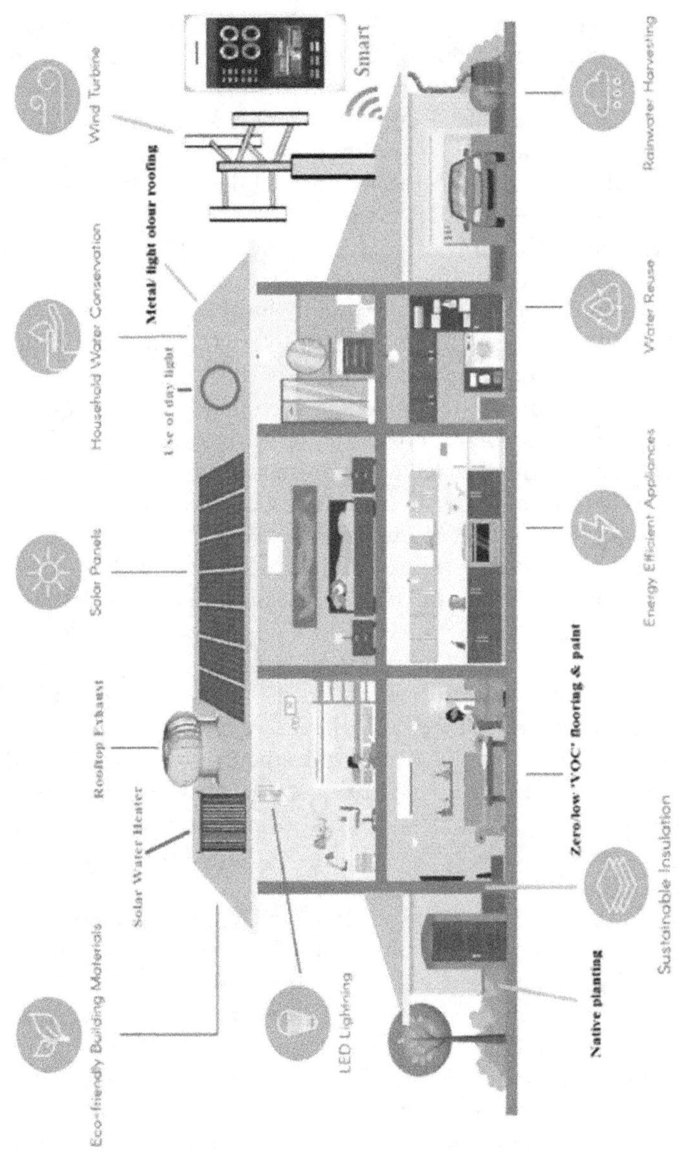

Figure 7.2 NZEB toward smart sustainable building (Rajan Kumar Jaysawal et al., 2022)

7.1.5 Passive design strategies

Passive design, a cornerstone of sustainable architecture, plays a pivotal role in minimizing energy demand within the context of net zero energy buildings (NZEBs) (Abed Al Waheed Hawila et al., 2022). This approach relies on harnessing natural elements and the building's intrinsic features to optimize energy efficiency, reducing dependence on active, energy-intensive systems. One of the key tenets of passive design is the strategic harnessing of solar energy (Alexandra R. Rempel et al. 2013). By understanding the sun's path, architects can orient buildings to maximize exposure during colder months, allowing for natural heating, while carefully designed shading elements prevent overheating in the summer (G Stauskis et al., 2015). This thoughtful manipulation of solar gain not only minimizes the need for artificial heating and cooling but also enhances the comfort of occupants. Natural ventilation is another fundamental aspect of passive design, promoting the circulation of fresh air through a building. By capitalizing on prevailing winds and temperature differentials, this strategy not only improves indoor air quality but also reduces reliance on mechanical ventilation systems (Behrang Chenari et al. 2016 & N. Dutt, 2024). Additionally, incorporating thermal mass, such as high-mass materials that absorb and store heat, helps stabilize indoor temperatures, contributing to a more comfortable and energy-efficient building (Figure 7.3).

Passive design is not merely about individual strategies but represents a comprehensive approach to architectural planning that considers the entire built environment. The importance of this approach lies in its proactive nature—it addresses energy demand at its source, during the design and construction phase, rather than relying solely on technological solutions. By minimizing the need for mechanical interventions, passive design significantly reduces the overall energy requirements of a building, aligning seamlessly with the core principles of sustainability. It not only lessens the environmental impact but also results in long-term economic benefits for building owners and occupants (Nikolaos Papadakis & Dimitrios Al. Katsaprakakis, 2023).

Moreover, the integration of passive design principles is integral to achieving the objectives of NZEBs. As these buildings strive to balance energy consumption with on-site renewable energy generation, the emphasis on minimizing energy demand becomes paramount. Passive design becomes a foundation upon which other technologies, such as solar panels and energy storage systems, can be more effectively employed (Marshahida Mat Yashim et al., 2021). In the pursuit of NZEBs, passive design thus emerges as a linchpin strategy, not only reducing operational energy demand but also providing a resilient and sustainable foundation for the future of architectural practice.

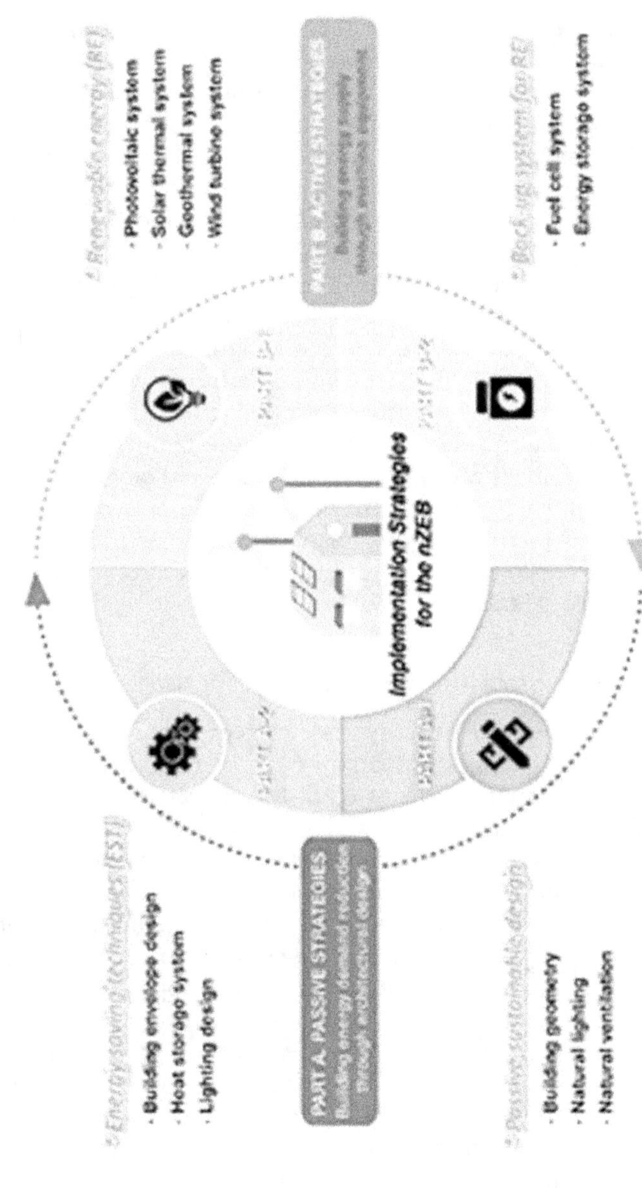

Figure 7.3 Passive and active strategies for implementing net zero energy building (Jeongyoon Ch, 2022)

7.1.6 Advanced building envelope technologies

The building envelope, a critical interface between the interior and exterior environments, plays a pivotal role in determining a structure's energy efficiency and environmental impact. Advanced building envelope technologies represent a change in thinking in architectural design, leveraging cutting-edge materials and innovative techniques to enhance thermal performance, reduce energy consumption, and contribute to the overall sustainability of buildings. High-performance insulation materials, such as aerogels and vacuum-insulated panels, challenge traditional insulation norms by providing superior thermal resistance in thinner profiles. Advanced glazing technologies, including smart glass and dynamic windows, dynamically respond to changing environmental conditions, optimizing daylighting and solar heat gain (W. J. Hee et al., 2015). These technologies not only promote energy efficiency but also enhance occupant comfort and well-being by providing ample natural light and minimizing temperature fluctuations. Furthermore, the integration of phase-change materials within the building envelope introduces an additional layer of innovation, allowing structures to store and release heat to maintain optimal indoor temperatures. The continuous evolution of advanced building envelope technologies underscores a commitment to pushing the boundaries of sustainable design, offering architects and builders an array of tools to create structures that are not only aesthetically pleasing but also resilient, energy-efficient, and environmentally responsible (Hatice Sozer, 2010).

7.1.7 Integration of renewable energy systems

The integration of renewable energy systems represents a pivotal shift in contemporary building design, reshaping the way structures interact with the power grid and the environment (C.S. Meena, 2022b). This transformative approach recognizes the imperative to reduce reliance on finite fossil fuels and embraces sustainable alternatives to meet energy demands. Central to this paradigm is the incorporation of renewable energy sources such as solar, wind, and geothermal power within the built environment. Solar photovoltaic (PV) systems, leveraging sunlight to generate electricity, have become increasingly prevalent, adorning rooftops and facades with panels that convert sunlight into a clean, renewable energy source (P.K. Kushwaha, 2023). Wind turbines, whether as part of large-scale wind farms or smaller-scale installations, harness the kinetic energy of the wind to produce electricity. Geothermal systems tap into the Earth's natural heat reservoirs to provide both heating and cooling for buildings. The integration of these renewable energy systems is not only a response to environmental imperatives but also a strategic move to enhance the resilience and self-sufficiency of buildings. In the context of net zero energy buildings (NZEBs), the integration of renewable energy sources is particularly vital, as these structures strive to balance energy consumption with on-site energy generation, aiming for a net zero energy footprint. This integration not only reduces the environmental impact of buildings but also positions them as active contributors to the broader goal of transitioning toward a sustainable and low-carbon energy future (V. Verma, 2023).

7.2 ENERGY STORAGE AND MANAGEMENT

Energy storage and management play pivotal roles in shaping a sustainable and reliable energy infrastructure. As the integration of renewable energy sources, such as solar and wind, becomes more prevalent, the need for effective energy storage solutions becomes paramount. Energy storage systems, including advanced batteries, compressed air energy storage, and pumped hydro storage, address the intermittent nature of renewable generation by storing excess energy during periods of high production for use when demand is high, or generation is low (M. Ahmadizadeh, 2024). Beyond addressing intermittency, energy storage contributes to grid stability, allowing for a more balanced supply-demand relationship. Smart energy management systems further optimize energy use by intelligently controlling the flow of electricity, coordinating energy production and consumption, and integrating demand response strategies. This comprehensive approach to energy storage and management not only enhances the efficiency and reliability of power systems but also supports the transition toward a more sustainable and resilient energy future (M. Chitt, 2024).

One of the primary challenges associated with renewable energy sources, such as solar and wind, is their inherent intermittent. Unlike traditional fossil fuel-based power generation, where output can be more easily controlled, renewables are subject to variations in weather conditions and natural elements. Addressing the intermittent challenges is crucial for ensuring a stable and reliable energy supply from these sources (M. Khashehchi, 2024).

7.3 SMART TECHNOLOGIES FOR OPTIMAL ENERGY MANAGEMENT

The advent of smart technologies has ushered in a new era of efficiency and sustainability in energy management, offering innovative solutions for optimizing resource utilization and reducing environmental impact. In the context of building design and operation, these smart technologies play a pivotal role in achieving optimal energy management, contributing to the goals of energy efficiency and sustainability.

Building Automation Systems (BAS): Building Automation Systems (BAS) form the backbone of smart energy management, providing a centralized platform to monitor, control, and optimize various building systems.

These systems integrate sensors, actuators, and controllers to collect real-time data on factors such as temperature, occupancy, and lighting conditions. By analyzing this data, BAS can dynamically adjust HVAC systems, lighting, and other building components to ensure optimal energy efficiency while maintaining occupant comfort.

Energy Management Software: Energy Management Software (EMS) complements BAS by offering advanced analytics and insights into energy consumption patterns. These software solutions leverage data from various sensors and meters to identify energy-intensive processes, inefficiencies, and opportunities for improvement. Through data visualization and reporting, EMS empowers building managers to make informed decisions, implement energy-saving strategies, and track the impact of efficiency measures over time.

Smart Metering and Monitoring: Smart metering technologies provide granular insights into energy consumption at a device or appliance level. By deploying smart meters, building operators can monitor real-time energy usage, identify energy hogs, and implement load-shifting strategies to optimize consumption during periods of lower demand or lower energy costs. Smart metering fosters transparency, enabling building occupants to actively participate in energy conservation efforts through awareness and behavioral adjustments.

Demand Response Systems: Demand Response (DR) systems enable buildings to participate in grid-balancing initiatives by adjusting their energy consumption in response to grid conditions. Smart technologies facilitate seamless communication between the building and the grid operator, allowing for load shedding during peak demand periods or grid stress. This not only contributes to grid stability but also opens opportunities for financial incentives through demand response programs.

Artificial Intelligence (AI) and Machine Learning (ML): Artificial Intelligence and Machine Learning technologies bring a predictive and adaptive dimension to energy management. These systems analyze historical data, weather forecasts, and occupancy patterns to predict future energy demand and dynamically adjust building systems for optimal efficiency. AI and ML algorithms continually learn and adapt, refining their recommendations and actions over time to achieve maximum energy savings (Suresh B. Sadineni et al., 2011).

7.4 LIFE CYCLE ASSESSMENT (LCA) IN DESIGN

Life Cycle Assessment (LCA) stands as a powerful methodology within the realm of sustainable design, offering a holistic and comprehensive approach to evaluating the environmental impacts of products, processes, or buildings throughout their entire life cycle. In the context of design, LCA has become an invaluable tool for architects, engineers, and decision-makers seeking to create environmentally responsible and resource-efficient solutions (Shahana Y. Janjua et al., 2019).

LCA Methodology: At its core, LCA considers the environmental consequences of a product or system from raw material extraction through production, use, and disposal. It assesses multiple impact categories, including energy consumption, greenhouse gas emissions, water usage, and other relevant indicators. The methodology is structured into four main stages: goal and scope definition, inventory analysis, impact assessment, and interpretation. This systematic approach allows for a rigorous and standardized examination of a design's environmental footprint.

Informing Design Decisions: One of the primary advantages of integrating LCA into the design process is its capacity to inform decision-making at critical junctures. Designers can evaluate the environmental implications of various material choices, construction methods, and building systems before implementation. This proactive approach empowers design professionals to make informed decisions that minimize environmental impact while meeting functional and aesthetic requirements.

Material Selection and Specification: LCA enables a detailed assessment of the environmental impacts associated with different materials. This includes considering factors such as raw material extraction, manufacturing processes, transportation, and end-of-life considerations. Armed with this information, designers can opt for materials with lower environmental footprints, promoting the use of sustainable, recycled, or locally sourced materials. Additionally, LCA facilitates the optimization of material quantities to reduce waste during construction.

Energy Efficiency and Performance: Beyond materials, LCA addresses the energy efficiency and performance of building systems. It assesses the environmental consequences of energy consumption during the operational phase, shedding light on the long-term sustainability of the design. This insight encourages the implementation of energy-efficient technologies, renewable energy sources, and strategies that reduce overall energy demand throughout the building's life cycle.

End-of-Life Considerations: LCA accounts for the environmental impact of a design's end-of-life phase, considering aspects such as recyclability, reusability, and disposal. By understanding these implications, designers can incorporate strategies for minimizing waste and promoting circular economy principles. This may involve designing for disassembly, specifying materials with high recyclability, or considering alternative uses for components at the end of their initial lifespan.

Certification and Rating Systems: LCA has found its way into various green building certification and rating systems worldwide. Systems like LEED (Leadership in Energy and Environmental Design) and BREEAM (Building Research Establishment Environmental Assessment Method) incorporate LCA to evaluate and reward environmentally conscious design and construction practices. Achieving certification in these systems serves as a testament to a project's commitment to sustainability.

7.5 INCORPORATING LIFE CYCLE THINKING IN NZEB DESIGN PROCESSES

Net zero energy buildings (NZEBs) represent a change in thinking in the design and construction of buildings, aspiring to balance energy consumption with on-site renewable energy generation. To truly achieve sustainability, it is imperative to go beyond operational energy considerations and embrace life cycle thinking – a comprehensive approach that assesses the environmental impacts of a building from inception to demolition. Integrating life cycle thinking into NZEB design decisions is essential for maximizing environmental benefits and ensuring a comprehensive approach to sustainability.

Materials and Embodied Carbon: Life cycle thinking begins with an examination of materials and their embodied carbon – the total greenhouse gas emissions associated with their extraction, manufacturing, transportation, and construction. NZEB design decisions must prioritize materials with lower embodied carbon, such as recycled or sustainably sourced options. This approach aligns with the goal of reducing the carbon footprint of the building throughout its life cycle.

Construction and Demolition Practices: Considerations extend beyond the operational phase to construction and eventual demolition. Sustainable construction practices, including waste reduction, efficient use of resources, and environmentally responsible demolition strategies, play a pivotal role. Design decisions that prioritize modular or prefabricated components, which facilitate disassembly and recycling, contribute to a circular economy and reduce the environmental impact associated with construction and deconstruction.

Operational Energy Efficiency: While NZEBs aim to produce as much energy as they consume, life cycle thinking emphasizes the importance of energy efficiency during the operational phase. Design decisions should prioritize passive design strategies, energy-efficient technologies, and smart building systems that optimize energy consumption. By reducing the demand for energy, NZEBs can achieve greater energy independence and minimize their overall environmental impact.

Renewable Energy Integration: Life cycle thinking underscores the significance of selecting and integrating renewable energy systems in NZEB design decisions. The choice of on-site renewable sources, such as solar panels or wind turbines, should align with environmental goals and optimize the overall energy balance. Considerations should extend to the environmental impact of manufacturing and installing renewable technologies to ensure an integrated approach to sustainability.

Long-Term Performance and Adaptability: NZEB design decisions should account for the long-term performance and adaptability of the building. Life cycle thinking encourages designs that can accommodate changes in occupancy, technology upgrades, and evolving energy demands. This forward-thinking approach enhances the resilience of NZEBs, ensuring their relevance and sustainability over an extended life cycle.

Occupant Well-being and Comfort: Life cycle thinking encompasses considerations beyond environmental impacts, extending to the well-being and comfort of building occupants. Design decisions that prioritize indoor air quality, natural lighting, and thermal comfort contribute to a healthier and more sustainable built environment. A focus on occupant well-being aligns with the broader goal of creating buildings that support human health and productivity.

7.6 COMMERCIAL AND INSTITUTIONAL NZEBS: INNOVATIVE DESIGN STRATEGIES

Economic considerations play a pivotal role in shaping decisions related to sustainable design, influencing the adoption of environmentally friendly practices and technologies in the built environment. The integration of sustainability into design decisions often involves a balance between upfront costs and long-term benefits. Various economic considerations and incentives drive the transition toward sustainable design practices.

Upfront Costs vs. Life Cycle Costing: One of the primary economic considerations in sustainable design is the balance between upfront costs and long-term savings. Sustainable technologies and materials might have higher initial costs, but they often offer lower life cycle costs due to reduced energy consumption, maintenance, and operational expenses. Life cycle costing analysis helps decision-makers evaluate the economic viability of sustainable choices over the entire lifespan of a building.

Return on Investment (ROI): The concept of Return on Investment is integral to economic considerations in sustainable design. Investments in energy-efficient systems, renewable energy sources, and green building features are evaluated based on their potential returns over time. Governments, businesses, and individuals increasingly recognize the economic benefits of reduced energy bills, enhanced property value, and operational savings associated with sustainable design practices.

Government Incentives and Subsidies: Governments around the world offer various incentives and subsidies to promote sustainable design and construction. These can include tax credits, grants, and rebates for incorporating energy-efficient technologies, renewable energy systems, or meeting specific green building standards. These economic incentives not only reduce the financial burden on developers and property owners but also encourage the widespread adoption of sustainable practices.

Green Building Certification Programs: Certification programs such as LEED (Leadership in Energy and Environmental Design) and BREEAM (Building Research Establishment Environmental Assessment Method) provide a framework for recognizing and promoting sustainable building practices. Achieving certification can enhance a building's market value, attract tenants, and qualify for financial incentives. The economic benefits associated with certification serve as a strong motivator for developers to invest in sustainable design.

Energy Efficiency Financing: Financial institutions increasingly offer specialized financing options for energy-efficient and sustainable projects. Green financing and energy efficiency loans provide developers and building owners with capital to invest in sustainable technologies. These financial instruments often offer favorable terms, recognizing the long-term economic benefits associated with energy savings and reduced environmental impact.

Rising Energy Costs and Resource Scarcity: The escalating costs of traditional energy sources and growing awareness of resource scarcity contribute to economic considerations favoring sustainable design. As energy costs rise, the financial benefits of energy-efficient systems and renewable energy sources become more pronounced, making sustainable design an economically sensible choice for both short-term and long-term financial planning.

Market Demand and Tenant Preferences: Market dynamics and changing consumer preferences also influence economic considerations in sustainable design. Tenants and property buyers increasingly prioritize environmentally friendly features and green certifications when choosing properties. Meeting this demand enhances the marketability of sustainable buildings, potentially commanding higher rents, and property values.

7.7 COMMERCIAL AND INSTITUTIONAL NZEBS: INNOVATIVE DESIGN STRATEGIES

Commercial and institutional buildings present unique challenges and opportunities in the pursuit of net zero energy goals. Innovative design strategies employed in these settings not only highlight environmental responsibility but also demonstrate the potential for sustainable, energy-efficient, and high-performance structures.

Integrated Design Approach: The success of commercial and institutional NZEBs often hinges on an integrated design approach that considers various factors, including building orientation, energy-efficient HVAC systems, advanced lighting solutions, and renewable energy integration. Innovative designs prioritize collaboration among architects, engineers, and sustainability experts to create a holistic and synergistic building solution.

Advanced Building Envelope Technologies: Innovative design strategies for Commercial and Institutional NZEBs frequently involve advanced building envelope technologies. High-performance insulation materials, efficient windows, and innovative construction methods contribute to a well-insulated and airtight building envelope. This not only enhances energy efficiency but also ensures occupant comfort and minimizes heat transfer.

Energy-Efficient HVAC Systems: Innovative HVAC (Heating, Ventilation, and Air Conditioning) systems play a crucial role in the success of Commercial and Institutional NZEBs. High-efficiency heating and cooling systems, coupled with smart building controls, enable precise temperature management while minimizing energy consumption. Ground-source heat pumps and radiant heating/cooling systems are examples of innovative HVAC solutions.

Daylighting and Artificial Lighting Strategies: Daylighting strategies that maximize natural light penetration and minimize the need for artificial lighting are integral to the innovative design of Commercial and Institutional NZEBs. Advanced lighting systems, including energy-efficient LEDs and smart lighting controls, further optimize energy usage while providing a comfortable and well-lit environment for occupants.

Flexible Space Utilization: Innovative design strategies recognize the importance of flexible space utilization in Commercial and Institutional NZEBs. Adaptable spaces that can accommodate changing occupancy requirements, technological advancements, and evolving organizational needs contribute to the longevity and sustainability of the building. Flexible designs enhance the building's ability to meet net zero energy goals over time.

Renewable Energy Integration: Incorporating renewable energy sources is a hallmark of innovative design in Commercial and Institutional NZEBs. Solar photovoltaic panels, wind turbines, and other on-site renewable energy systems are strategically integrated to generate clean energy. Innovative approaches may include building-integrated solar technologies, such as solar facades or integrated photovoltaic roofing systems.

Smart Grid Integration: Commercial and Institutional NZEBs often leverage smart grid integration to optimize energy usage and contribute to grid stability. Smart technologies and grid-responsive systems allow buildings to adjust energy consumption based on real-time grid conditions, participate in demand response programs, and enhance overall energy efficiency.

Performance Monitoring and Feedback: Innovative design strategies incorporate robust performance monitoring and feedback systems. Real-time data collection on energy consumption, indoor environmental quality, and overall building performance enables continuous optimization. These systems provide valuable insights for building operators, allowing them to make informed decisions to maintain and enhance the building's net zero energy status.

7.8 CHALLENGES AND FUTURE DIRECTIONS

7.8.1 Technical challenges in NZEB design and implementation

While net zero energy buildings (NZEBs) represent a promising avenue for sustainable construction, several technical challenges persist in their design and implementation.

Energy Storage and Management: The intermittent nature of renewable energy sources poses challenges in effectively storing and managing energy for use during periods of low generation. Advanced energy storage technologies and smart management systems are needed to address this issue.

Integration of Renewable Systems: Seamless integration of diverse renewable energy systems, such as solar, wind, and geothermal, remains a technical challenge. Achieving optimal constructive collaboration among these systems to meet the energy demands of buildings requires sophisticated control strategies and efficient energy conversion technologies.

Technological Innovation: Ongoing technological innovation is crucial to overcome technical challenges. Advancements in materials, energy-efficient appliances, and building systems are necessary to enhance the overall performance and feasibility of NZEBs.

Building Envelope Performance: Ensuring the high performance of building envelopes in varying climates is a technical hurdle. Designing envelopes that effectively balance insulation, ventilation, and daylighting while responding to different climatic conditions is essential for NZEB success.

Grid Integration and Resilience: Integrating NZEBs into existing grids and ensuring their resilience during power outages or disruptions present technical challenges. Smart grid technologies and decentralized energy systems are needed to enhance the grid integration and resilience of NZEBs.

7.8.2 Economic and Policy Challenges Hindering Widespread Adoption

Widespread adoption of NZEBs faces economic and policy challenges that must be addressed to accelerate the transition to sustainable building practices.

Upfront Costs: The higher upfront costs associated with energy-efficient technologies and renewable energy systems present a significant economic barrier. Incentives, subsidies, and innovative financing options are essential to make NZEBs financially accessible to a broader range of stakeholders.

Lack of Standardization: The absence of standardized approaches and regulations hinders the widespread adoption of NZEBs. Developing and implementing clear, consistent standards for energy performance and sustainability is crucial for creating a level playing field and fostering market confidence.

Limited Awareness and Education: Lack of awareness and understanding among stakeholders, including developers, builders, and homeowners, poses a challenge. Robust educational campaigns and training programs are necessary to increase awareness and knowledge about the benefits and feasibility of NZEBs.

Policy Frameworks: Inconsistent or inadequate policy frameworks can impede the adoption of NZEBs. Governments need to implement and enforce policies that incentivize sustainable building practices, such as tax credits, grants, and mandatory energy performance standards.

Long-Term Benefits Perception: The focus on short-term costs often overshadows the long-term benefits of NZEBs. Shifting the perception to recognize the economic, environmental, and societal advantages over the life cycle of the building is essential for overcoming economic barriers.

7.8.3 Future trends and advancements in NZEB technologies

The future of net zero energy buildings is promising, with ongoing advancements and emerging trends that are shaping the landscape of sustainable design and construction.

Advanced Building Materials: The development of advanced building materials with enhanced thermal properties, durability, and sustainability is a key trend. Innovations such as smart materials and self-healing structures contribute to improved building performance.

Smart Technologies and Automation: The integration of smart technologies and automation in NZEBs is expected to grow. AI-driven building management systems, IoT-enabled devices, and real-time data analytics will play a crucial role in optimizing energy usage and enhancing occupant comfort.

Zero-Emission Technologies: Advancements in zero-emission technologies, such as hydrogen-based heating and fuel cells, are gaining attention. These technologies offer alternatives for decarbonizing building operations and achieving emissions reduction targets.

Circular Economy Principles: The adoption of circular economy principles in construction and design is a future trend. Designing buildings for disassembly, reuse of materials, and minimizing waste align with sustainability goals and contribute to a circular building economy.

Community-Scale NZEBs: The emergence of community-scale NZEBs is anticipated. Planning and designing entire neighborhoods or districts as net zero energy communities allow for shared resources, optimized energy infrastructure, and a holistic approach to sustainability.

Energy-Positive Buildings: The concept of energy-positive buildings, which generate more energy than they consume, is gaining momentum. Advancements in renewable energy technologies, energy storage, and building design are driving the feasibility of creating buildings that contribute excess energy to the grid.

Inclusive and Equitable Design: Future trends in NZEBs emphasize inclusive and equitable design practices. Ensuring that sustainable buildings are accessible to diverse communities and socio-economic groups promotes environmental justice and resilience.

7.9 CONCLUSION

The journey toward net zero energy buildings (NZEBs) represents a transformative path for the construction and design industry, offering a sustainable and resilient vision for the future built environment. Throughout this exploration, several key themes and takeaways have emerged:

Integration of Cutting-Edge Technologies: The successful realization of NZEBs relies on the seamless integration of cutting-edge technologies. Advanced building materials, smart systems, and renewable energy solutions are pivotal components in achieving the delicate balance between energy consumption and on-site generation (V. Verma, 2024).

Holistic Design Approaches: NZEB success requires a holistic design approach that not only encompasses energy efficiency during the operational phase but also considers the entire life cycle of the building. From materials selection and construction practices to adaptive designs and end-of-life considerations, a comprehensive approach is essential (Z. Razaviyn, 2024).

Economic and Policy Considerations: Overcoming economic barriers and creating supportive policy frameworks are critical for the widespread adoption of NZEBs. Initiatives such as financial incentives, standardized regulations, and educational programs play a key role in fostering a conducive environment for sustainable building practices.

Environmental Responsibility and Resilience: NZEBs embody a commitment to environmental responsibility and resilience. By reducing carbon footprints, optimizing energy usage, and promoting circular economy principles, these buildings contribute to a more sustainable, resource-efficient, and climate-resilient future (V. Verma, 2024).

Educational and Community Engagement: Building awareness, educating stakeholders, and fostering community engagement are crucial elements in the successful adoption of NZEBs. Empowering individuals with knowledge about sustainable practices and creating a sense of community around environmental stewardship are key aspects of this transformative journey.

The evolution towards net zero energy buildings is not merely a technological shift but a paradigm change that requires collaboration, innovation, and a shared commitment to building a more sustainable and equitable world. As researchers, policymakers, industry professionals, and communities continue to collaborate, the vision of a built environment characterized by NZEBs moves closer to becoming a widespread reality. The journey is ongoing, and the lessons learned along the way serve as guideposts for future endeavors in sustainable design and construction.

REFERENCES

A. Al Waheed Hawila, R. Pernetti, C. Pozza, A. Belleri, 2022, "Plus energy building: Operational definition and assessment", *Energy and Buildings*, Volume 265, Pages 112069.

A. R. Rempel, A. W. Rempel, K. V. Cashman, K. N. Gates, C. J. Page, B. Shaw, 2013, "Interpretation of passive solar field data with EnergyPlus models: Un-conventional wisdom from four sunspaces in Eugene, Oregon", *Building and Environment*, Volume 60, Pages 158–172.

B. Chenari, J. Dias Carrilho, M. Gameiro da Silva, 2016, "Towards sustainable, energy-efficient and healthy ventilation strategies in buildings: A review", *Renewable and Sustainable Energy Reviews*, Volume 59, Pages 1426–1447.

C. S. Meena, A. Kumar, S. Jain, A. U. Rehman, S. Mishra, 2022a, "Innovation in green building sector for sustainable future", *Energies*, Volume 15, Pages 6631. https://doi.org/10.3390/en15186631.

C. S. Meena, A. N. Prajapati, A. Kumar, M. Kumar, 2022b, "Utilization of solar energy for water heating application to improve building energy efficiency: An experimental study", *Buildings*, Volume 12, Pages 2166. https://doi.org/10.3390/buildings12122166.

G. Stauskis, D. Markovskaja, N. Surmilavičiūtė, G. Šerpytytė, A. Segalis, I. Gražys, 2015, "Bioclimatic principles in architectural design. A way to better buildings", *Green Architecture*, 2014/4.

H. Sozer, 2010, "Improving energy efficiency through the design of the building envelope", *Building and Environment*, Volume 45, Issue 12, Pages 2581–2593.

H. Homayouni, C. S. Dossick, G. Neff, 2020, "Three pathways to highly energy efficient buildings: Assessing combinations of teaming and technology", *Journal of Management in Engineering*, Volume 37, Issue 2, Pages 04020110. https://doi.org/10.1061/(ASCE)ME.1943-5479.000088

J. Oh, T. Hong, H. Kim, J. An, K. Jeong, C. Koo, 2017, "Advanced strategies for net-zero energy building: Focused on the early phase and usage phase of a building's life cycle", *Sustainability*, Volume 9, Issue 12, Pages 2272. https://doi.org/10.3390/su9122272

L. Müller, T. Berker, 2013, "Passive house at the crossroa: The past and the present of a voluntary standard that managed to bridge the energy efficiency gap", *Energy Policy*, Volume 60, Pages 586–593.

M. Ahmadizadeh, M. Heidari, S. Thangavel, E. A. Naamani, M. Khashehchi, V. Verma, A. Kumar, 2024, "Technological advancements in sustainable and renewable solar energy systems", in *Highly Efficient Thermal Renewable Energy Systems: Design, Design, Optimization and Applications*; V. Verma, S. Thangavel, N. Dutt, A. Kumar, R. Weerasinghe Editors; CRC Press: Boca Raton, FL; Chapter 02. ISBN 9781032595641.

M. Chitt, S. Thangavel, V. Verma, A. Kumar, 2024, "Green hydrogen productions: Methods, designs and smart applications", in *Highly Efficient Thermal Renewable Energy Systems: Design, Design, Optimization and Applications*; V. Verma, S. Thangavel, N. Dutt, A. Kumar, R. Weerasinghe Editors; CRC Press: Boca Raton, FL, Chapter 16. ISBN 9781032595641.

M. Khashehchi, S. Thangavel, P. Rahmanivahid, M. Heidari, T. Moazzeni, V. Verma, A. Kumar, 2024, "Solar desalination techniques: Challenges and opportunities", in *Highly Efficient Thermal Renewable Energy Systems: Design, Design, Optimization and Applications*; V. Verma, S. Thangavel, N. Dutt, A. Kumar, R. Weerasinghe Editors; CRC Press: Boca Raton, FL, Chapter 19. ISBN 9781032595641.

M. Andrea Arias, *Energy Facade Retrofit: Energy Performance and the Impact of Facade Changes in Non-residential Buildings in the United States*. University of Southern California ProQuest Dissertations Publishing, 2016.

M. Mat Yashim, M. Hanif Sainorudin, M. Mohammad, A. Fudholi, N. Asim, H. Razali, K. Sopian, 2021, "Recent advances on lightweight aerogel as a porous receiver layer for solar thermal technology application", *Solar Energy Materials and Solar Cells*, Volume 228, Pages 111131.

M. Bilardo, J. H. Kämpf, E. Fabrizio, 2023, "From zero energy to zero power buildings: A new paradigm for a sustainable transition of the building stock", *Sustainable Cities and Society*, Volume 101, Issue 2024, Pages 105136.

M. Irfan, N. Abas, S. Saleem, "Net Zero Energy Buildings (NZEB): A case study of net zero energy home in Pakistan", Conference: PGSRET IEEEAt: Islamabad.

N. Dutt, A. Hedau, A. Kumar, M. K. Awasthi, S. Hedau, C. S. Meena, 2024, "Thermo-hydraulic performance investigation of solar air heater duct having staggered D-shaped ribs: Numerical approach", *Heat Transfer*, pages 1–31. https://doi.org/10.1002/htj.22998.

N. Papadakis, D. A. Katsaprakakis, 2023, "A review of energy efficiency interventions in public buildings", *Energies*, Volume 16, Issue 17, Pages 6329.

N. T. Mbungu, R. M. Naidoo, R. C. Bansal, M. W. Siti, D. H. Tungadio, 2020, "An overview of renewable energy resources and grid integration for commercial building applications", *Journal of Energy Storage*, Volume 29, Pages 101385.

P. K. Kushwaha, N. K. Sharma, A. Kumar, C. S. Meena, 2023, "Recent advancements in augmentation of solar water heaters using nanocomposites with PCM: Past, present, and future", *Buildings*, Volume 13, Pages 79. https://doi.org/10.3390/buildings13010079.

R. K. Jaysawal, S. Chakraborty, D. Elangovan, S. Padmanaban, 2022, "Concept of net zero energy buildings (NZEB) - A literature review", *Cleaner Engineering and Technology*, Volume 11, Pages 100582. https://doi.org/10.1016/j.clet.2022.100582.

S. Y. Janjua, P. Sarker, W. Biswas, 2019, "A review of residential buildings' sustainability performance using a life cycle assessment approach", *Journal of Sustainability Research*, Volume 1, Issue 1, Pages 1–29. https://doi.org/10.20900/jsr20190006

S. B. Sadineni, S. Madala, R. F. Boehm, 2011, "Passive building energy savings: A review of building envelope components", *Renewable and Sustainable Energy Reviews*, Volume 15, Issue 8, Pages 3617–3631.

V. Verma, S. Thangavel, N. Dutt, A. Kumar, R. Weerasinghe, 2024, "Recent development of thermal energy storage: Solar, geothermal and hydrogen energy", in *Highly Efficient Thermal Renewable Energy Systems: Design, Design, Optimization and Applications*; V. Verma, S. Thangavel, N. Dutt, A. Kumar, R. Weerasinghe Editors; CRC Press: Boca Raton, FL, Chapter 01. ISBN 9781032595641.

V. Verma, S. Thangavel, N. Dutt, A. Kumar, R. Weerasinghe Editors, 2024, *Highly Efficient Thermal Renewable Energy Systems: Design, Design, Optimization and Applications*; CRC Press: Boca Raton, FL. ISBN 9781032595641.

V. Verma, C. S. Meena, S. Thangavel, A. Kumar, T. Choudhary, G. Dwivedi, 2023, "Ground and solar assisted heat pump systems for space heating and cooling applications in the northern region of India – A study on energy and CO2 saving potential", *Sustainable Energy Technologies and Assessments*, Volume 59, Pages 103405. ISSN 2213-1388. https://doi.org/10.1016/j.seta.2023.103405.

V. P. Singh, S. Jain, A. Karn, A. Kumar, G. Dwivedi, C. S. Meena, 2022, "Recent developments and advancements in solar air heaters: A detailed review", *Sustainability*, Volume 14, Pages 12149. https://doi.org/10.3390/su141912149.

W. J. Hee, M. A. Alghoul, B. Bakhtyar, O. Elayeb, M. A. Shameri, M. S. Alrubaih, K. Sopian, 2015, "The role of window glazing on daylighting and energy saving in buildings", *Renewable and Sustainable Energy Reviews*, Volume 42, Pages 323–343.

Z. Razaviyn, M. Heidari, V. Verma, S. Thangavel, A. Kumar, K. K. Saxena, G. Dwivedi, 2024, "Numerical simulation of a marine energy convertor based on vortex induced vibrations", *Proceedings of the Institution of Mechanical Engineers, Part E: Journal of Process Mechanical Engineering* (in press).

Chapter 8

Strategic and technical analysis of sustainable smart cities development in Oman

Sivasakthivel Thangavel, Ashwani Kumar, and Milad Heidari

8.1 INTRODUCTION

This study is trying to address a list of substantial issues confronting the renovation process in the Sultanate of Oman by resolving the research question: "Are the cities in Oman ready for the transformation toward sustainable smart cities yet?" By solving this matter, the research will demonstrate the significance of fulfilling the recognized conversion gaps in Oman with the intent of assisting cities to operate smartly their facilities and sustainably manage their resources.

8.1.1 Overview of Oman sustainable development

The sustainability attitude has been a vital principle of the consecutive visions and five-year plans implemented by the Sultanate of Oman. Oman has also been an energetic participant in all the United Nations and provincial forums and discussions that led to the announcement of the Sustainable Development Goals (SDGs) in 2015. Since then, Oman has been concentrating on incorporating the sustainable development pillars and objectives into its development plans and strategies including Oman Vision 2040 (Figure 8.1). This proves the consistency and significance which the government chasing SDGs employment by assigning funds and formulating procedures and rules that ensure their attainment in the medium and long term.

The official outline and the methodology espoused in the preparation of Oman Vision 2040 directional strategies and goals were linked with the UN SDGs and their targets. Such an arrangement confirms that the 2030 SDGs are implanted in Oman Vision 2040 and thus obtained an equal level of consideration as the goals of the national vision. Consequently, the initiative for converting urban to be sustainable smart cities is aligned with the new orientation of Oman's strategy for the coming decades (Al-Bahrani, n.d.; Al-Nasrawi et al., 2017). Figure 8.1 illustrates the targets of Oman Vision 2040 considering the development of the three dimensions of sustainability.

 DOI: 10.1201/9781003496656-8

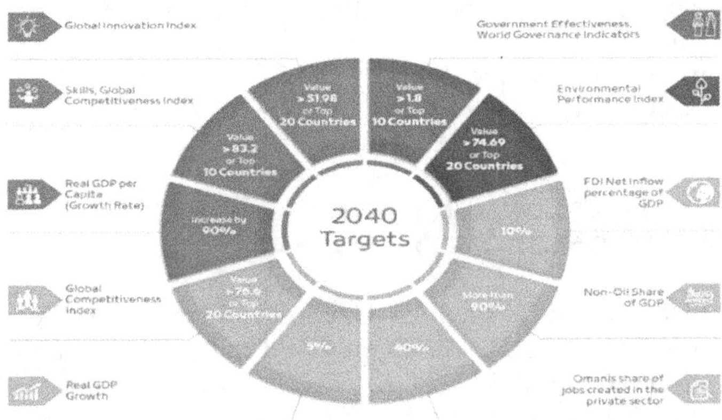

Figure 8.1 Targets of Oman Vision 2040 (Oman Observer, 2020). https://www.omanobserver.om/article/6459/Front%20Stories/vision-2040-to-improve-omans-global-standings

8.1.2 Background knowledge

Throughout the previous two decades, the Arab Gulf countries have undergone the fastest urban development in the world. The conventional communities in this area have transparently entered the market economy and modern way of life. Despite the growth achieved at various levels, these communities became targets of their capitalist style. The urbanization development in Oman particularly in its capital (Muscat) is the driving force behind the migration level from the rural places over the last decades, and it is expected to continue in the coming years. Generally, the population in the Sultanate of Oman has raised from 2,018,074 in 1993 to 4,481,042 in 2020 as per NCSI's latest statistics.

On the same hand, the global industrialization trend and growth of population in urban areas have posed global concerns in the consumption of energy natural resources, environment, and life quality in the urban area. Recent studies in urban and academic spheres concentrate on sustainability in urban planning and identify the primary development obstacles of the current condition. Numerous countries across the world are presently introducing Sustainable Smart City (SSC) schemes to tackle the fast urbanization and globalization challenges. Nevertheless, several countries are still undergoing knowledge and financial gaps in alliance with conversion toward SSCs. Thus, the smart cities notion appeared as an applicable resolution to this extraordinary urbanization and the necessity for sustainability. The International Telecommunication Union Focus Group on Smart Sustainable Cities (ITU-T FG-SSC) announced a statement:

> A Smart Sustainable City is an innovative city that uses Information and Communication Technologies (ICTs) and other means to improve

quality of life, the efficiency of urban operation and services, and competitiveness while ensuring that it meets the needs of present and future generations concerning economic, social, environmental as well as cultural aspects.

8.2 URBAN DEVELOPMENT AND POPULATION GROWTH IN OMAN

The Gulf Cooperation Council (GCC) countries have seen an exceptional urban conversion ever since the finding and selling of oil and gas in the province in the early 1960s. Hasty development was imposed by the requirements of the recently established territories, but property speculation and race for the "top" were the very powerful instruments of this magnificent urban progress. Comparatively to the significance of this phenomenon that embossed all the GCC countries in general and Oman in particular, few studies have diagnosed the various aspects of urbanization in these countries and the drawbacks of this development. Urban development involves not merely the tangible variations in the municipalities (urbanization) but also their commercial, societal, and environmental conversions and the fundamental causes of these practices (Riad Kazem, 2019).

In the GCC countries, the center cities are the primary literary, financial, and governmental capitals in their countries. Due to their political, geographical, and historical values, they are intended to be the icons of the development and prosperity of the state. Hence, these cities were the core of the urbanization endeavor adopted by the governments. Consequently, the center cities attract most of the construction and advancement budget of these countries and most of the job opportunities were created there. Despite the success of the urban revolution in different facets, it has nonetheless left broad regions isolated and immature. Furthermore, the services, facilities, and government and companies' operations are radically centralized in the principal cities which have headed to their over-concretion and residents' migration from the towns, villages, and other occupied similar districts. Likewise, urbanization in Oman has been growing at a rapid level. From nearly 30% in 1970 and about 50% a decade after, it has increased to touch 80% these days and is projected to reach about 85% by 2040 as shown in Figure 8.2 (Naima Benkari, 2017).

Regrettably, the gap between governmental assertions and the actual application of the regulations on the land is still a true barrier in the GCC. Oman is not far distinct in this issue, although the worry about sustainability in the designing phase in Oman was stated much ahead than its nearby states (Law on Prevention of Pollution and Conservation of the Environment, Royal Decree 10/82 revised by 114/2001). Power and water expenditures are at their greatest (water utilization per head grew by 120%

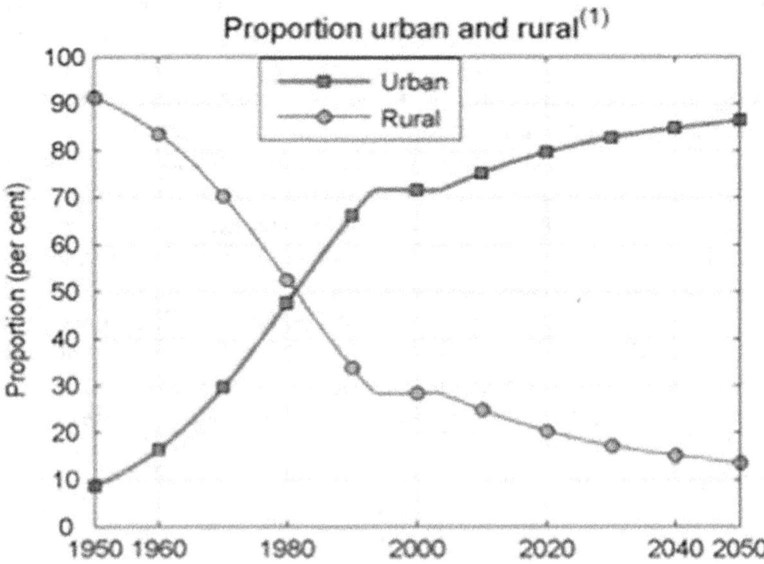

Figure 8.2 Proportion of urban vs. rural in Oman (UN, 2014) (Source: United Nations, Department of Economic and Social Affairs, Population Division, 2014. World Urbanization Prospects: The 2014 Revision.)

in Muscat between 2006 and 2009, and the same for garbage production level). However, the lack of coordination among different authorities like the Ministry of Housing and Urban Planning (formerly named as Ministry of Housing), the Supreme Committee for Town Planning (annulled later), and the Ministry of Regional Municipalities (combined later with the Ministry of Interior) and the absence of physical execution apparatuses have directed most of these strategies to the trash. Additionally, the carefully detailed suggestions and plans are vanished in the execution phase due to the intense bureaucracy (Xydis et al., 2021). As a result, sustainable urbanization in Oman is chaotic, unstable, and illogical that needs more thorough coordination and regulations (Naima Benkari, 2017).

8.3 EVOLUTION OF SUSTAINABLE SMART CITIES CONCEPT

The phrase sustainable mart cities (SSC) links sustainable cities and the smart cities notion. It is a recent phenomenon that turned out to be prevalent during the mid-2010s (Vishnivetskaya & Alexandrova, 2019). This notion has arisen because of urbanization, sustainability and environmental matters, and the revolution of ICTs and IoTs. The source of the notion of Smart Cities can be traced back to at least the Smart Growth Movement of the end of the 1990s or even the 1960s, however, it appeared more earnestly in 2010 (Vershinina et al., 2018). Though the phrase "smart city" is widening its application, it was primarily driven by the utilization of technology in a municipal condition that comprises numerous interconnected bodies. Converting the system to an intelligential condition means the permeation of radar technologies to collect data and recognize substances, communication aptitudes to unite and deliver data and information processing systems, and computational analytics to enhance city operations and protect natural assets to boost ecological performance. Thus, digitization bridges the gap and accelerates the transition between the city infrastructure and city services layers by inserting a new digital layer in between. The smart city goals can be classified into four groups based on urban challenges and business opportunities, namely: ecological sustainability, life quality and welfare, knowledge, and intelligent investment and contribution. After reviewing the academic, industrial, and governmental publications, for unraveling the smart sustainable city dimensions, the regularly quoted is the characterization of SCC issued by The Centre of Regional Science at the Vienna University of Technology, which divides the structure of the smart city into six axes: smart economy; smart mobility; smart environment; smart citizen; smart living; and, finally, smart governance as publicized in Figure 8.3 (Schipper & Silvius, 2018).

Overall, the SSC framework has several shared properties where all utilize innovative technology to gather, manage, and evaluate data; various

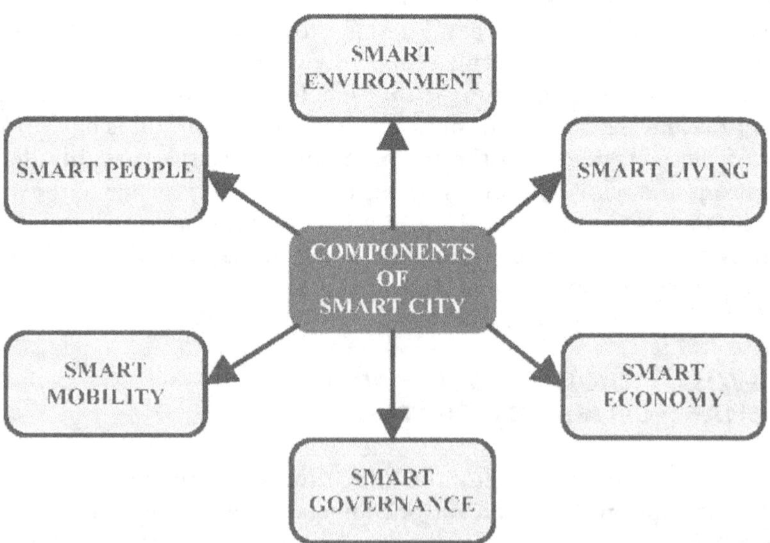

Figure 8.3 Component of sustainable smart city (Schipper & Silvius, 2018) (From smart to smarter cities: Bridging the dimensions of technology and urban planning | Semantic Scholar)

subsystems within the entire system recognize the instant information transfer without barriers; the life of urban citizen being more intellectual and more comfortable; consistently, the urban biological environment being healthier and more conservable (V. Verma, 2023, 2024). There are two strategies for establishing SSCs. The first one is to build new smart cities with integrated planning and the second aspect is the transformation of existing cities into cities with smart technologies (Huang, 2021).

8.4 AUTOMATION AND SUSTAINABILITY IN OMAN VISION 2040

Long-term planning is the best practice for successful countries to accomplish complete and sustainable development and to verify the nation's willingness toward the preferred future. Therefore, Oman Vision 2040 is the roadmap in the succeeding two decades to overtake challenges, bridge gaps, promote the economy, enrich welfare, and stimulate overall growth. It demonstrates incorporated structures for economic and community strategies with meticulous coordinates toward Oman's prosperity. It consists of three themes and 13 strategic directions, and each direction has its own goals, policies, and action plans with a precise timeframe. Additionally, the entire scheme implementation will be assessed and appraised over a set of national and international key performance indicators (KPIs). Strategic direction number 9 of the vision proposes a comprehensive governorate development to improve selective urban centers and develop lands in an idyllic and sustainable manner that delivers habitable situations and maintains environmental and social revitalization. The thorough and balanced growth of infrastructure, with all its segments, can encourage balanced development targeted to urban and rural areas, to stimulate socio-economic fortunes. Oman will embrace the concept of creating smart sustainable cities that practice advanced services. This orientation will utilize innovative technology and form three-dimensional municipalities involving the geographical, community, and financial parts. The implementation of this attitude relies on collaboration with private enterprises and employing it to energize urban societies to generate or convert their metropolises to be advanced and smart sustainable cities in terms of facilities, services, and operations. Hence, the balanced development that exploits the available resources will enhance the unified value chain among governorates and deliver better standards and quality of life. Tenets of decentralization will participate in the complete socio-economic development in the governorates so that ideal and sustainable handling of land and natural resources is in place, investing them markedly and efficiently. Figure 8.4 displays the assigned goals for this strategic direction (Oman Vision 2040, 2019).

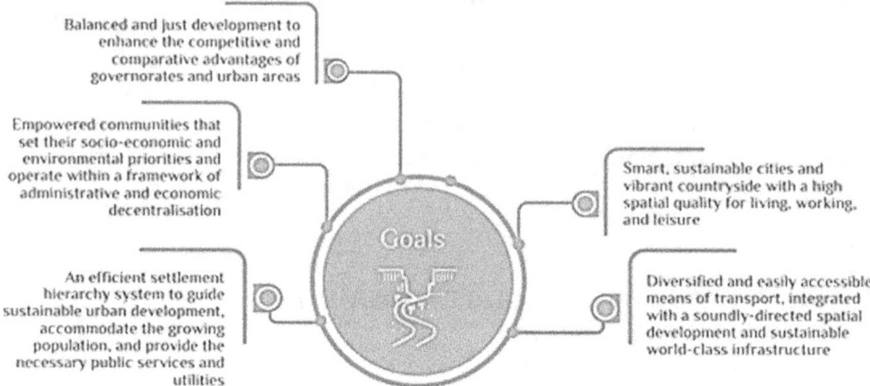

Figure 8.4 The assigned goals for Strategic# 9 direction (Oman Vision 2040, 2019) (Source: OmanVision2040-Preliminary-Vision-Document.pdf (national-day-of-oman.info))

8.5 METHODOLOGY

In 2017, smart cities platform was launched to highlight the significance of this scheme for the country. It has expanded to become a center for knowledge distribution and solution building. The Sohar free zone intends to be the first green hydrogen production hub in the Middle East (Greenport, 2020). In alliance with affording an agenda for innovation and professional enterprises to succeed, the Oman government has also been directing numerous smart city hackathons intended at nurturing the youth, entrepreneurs, and investors to create applications and utilize the digital revolution and social media for their commercial ideas. A new city named Madinat Al Irfan will accommodate 300,000 residents and is planned to be managed as part of a smart ecosystem. Innovations will involve utilizing smart meters to lessen resource demand, water recycling, urban farms, smart lighting, and buildings that employ thermal cooling architectural techniques (M. Ahmadizadeh, 2024 and M. Khashehchi, 2024). Oman is targeting to produce about 40% of renewable energy by 2040 through BBO projects.

A web-based survey assessing Oman residents' perception about the smart cites scheme was conducted. The objectives of that questionnaire were the following:

- To measure awareness, interest, and understanding toward sustainable smart cities and renewable energies among the citizens.
- To explore attitudes to a range of existing urbanization challenges and the propositions of using intelligent technologies in overcoming those challenges.

- To identify people's thoughts about the readiness of Oman IT infrastructure for the proposed conversion.
- To recognize and expose any barriers and risks to utilizing smart devices or digital services, e.g., data security, network weakness, competency, etc.
- To evaluate residents' willingness for accepting such technologies and examine the government's approach to this conversion.
- To discover the citizens' thoughts toward the significance of sustainability and recycling.

The questionnaire consisted of 17 questions related to sustainable solutions and intelligent technologies that were proposed to solve current urban challenges. It was widespread by using social media platforms like Twitter, Instagram, and WhatsApp (Facebook). 1000 responses were received out of 1000 participants.

The residents' answers are divided into different categories based on location, age, and level of education as illustrated in Tables 8.1, 8.2, and 8.3 and Figures 8.5, 8.6, and 8.7).

8.5.1 Data analysis

The analysis is divided into five main parts as shown below:

- Current urban challenges
- Residents' perception
- Transformation models
- Key performance indicators
- Risks and challenges

Table 8.1 Participant distribution location wise

Governorate	*Number of Participant*	*Percentage*
Muscat	331	33%
Musandam	7	1%
Al Dakhiliyah	319	32%
Al Dhahirah	68	9%
Al Buraimi	23	2%
Al Wusta	4	0%
Dhofar	15	1%
North Sharqia	57	6%
South Sharqia	61	6%
North Al Batinah	86	7%
South Al Batinah	29	3%
Total	**1,000**	**100.0%**

Table 8.2 Participant distribution based on age

Supplying Company	*Number of Participants*	*Percentage*
Between 15 and 25 years old	157	16%
Between 26 and 35 years old	307	31%
Between 36 and 45 years old	376	37%
Between 46 and 55 years old	137	14%
Between 56 and 65 years old	23	2%
Over 65 years old	0	0%
Total	**1,000**	**100%**

Table 8.3 Participant distribution education wise

Supplying Company	*Number of Participants*	*Percentage*
PhD	26	2%
Master's	137	14%
Bachelor's	414	41%
Higher Diploma	176	18%
General Diploma (High School)	206	21%
Less than General Diploma	41	4%
Total	**1,000**	**100%**

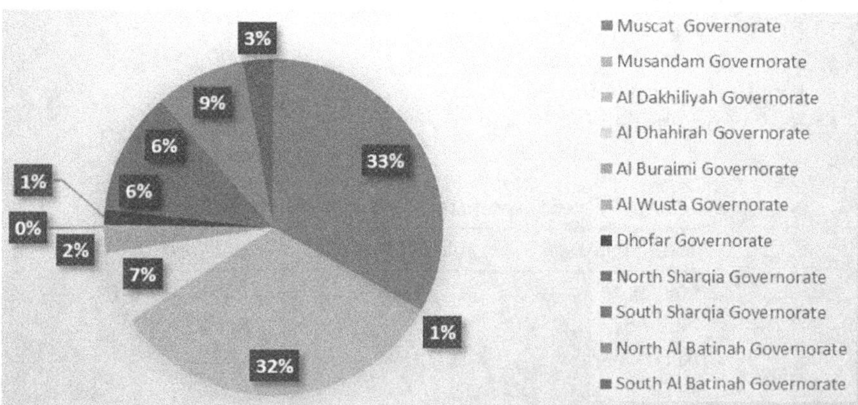

Figure 8.5 Sample distribution based on location

8.5.1.1 Current urban challenges

The common difficulties in cities facing urbanization are the increased demand for infrastructure and services, mobility restrictions, delayed access to housing, limited access to clean water, high levels of pollution, waste management, environmental sustainability, and other problems. In

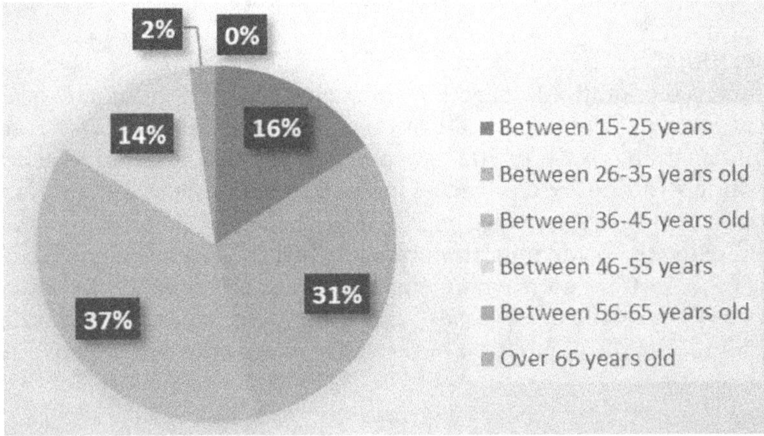

Figure 8.6 Sample distribution based on age

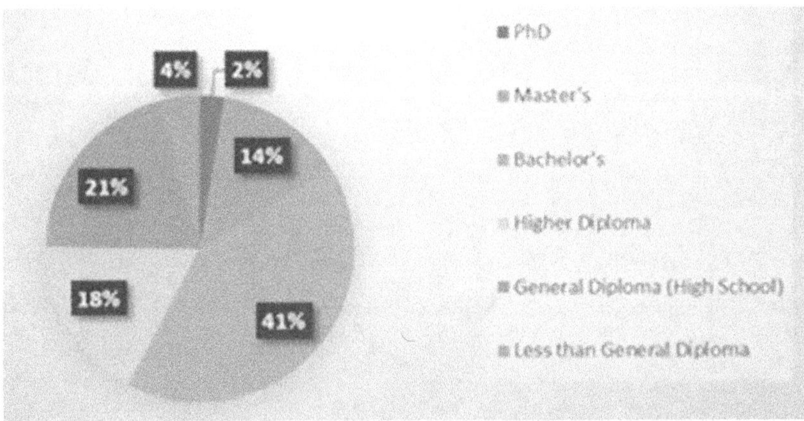

Figure 8.7 Sample distribution based on education

the presence of such challenges, Oman is facing a sharp rise in the urban population in addition to the demand for infrastructure, which resulted from the massive economic growth since the last 1970s era along with the wake of the oil economic boom in the region (Al Shezawi Mohammed, 2020). The exponential growth in the number of citizens and expatriates in Omani cities has led to the development of infrastructure, housing, and businesses on a large scale. The best example of the phenomenal growth in most of the major Omani cities, is the city of Muscat, as the city was pushed to grow outside its borders. With its low density and horizontal growth pattern, Muscat is currently suffering from urban scattering, traffic safety problems, high levels of energy emissions, heavy reliance on private cars as well as insufficient and inadequate infrastructure. For example, the population of Muscat has risen dramatically from 203, 509 in 1993 to more than 1,300,00 in 2020 (0.5 million persons is estimated to have resulted from internal migration), reflecting almost a third of the Sultanate of Oman population as per the latest publication of the National Centre for Statistics and Information (NCSI) in 2021 as shown in Figure 8.8. This has generated flourishing stress on the city and its territory where almost 30% of the overall population is overcrowded in 5.3% of the entire space of the country. Undoubtedly, the rising migration rate to Muscat causes numerous troubles such as the pressure on the municipal facilities, the demand for water and electricity, the traffic congestions, and other catering and civilization services. Similarly, the population in the other main cities in each government

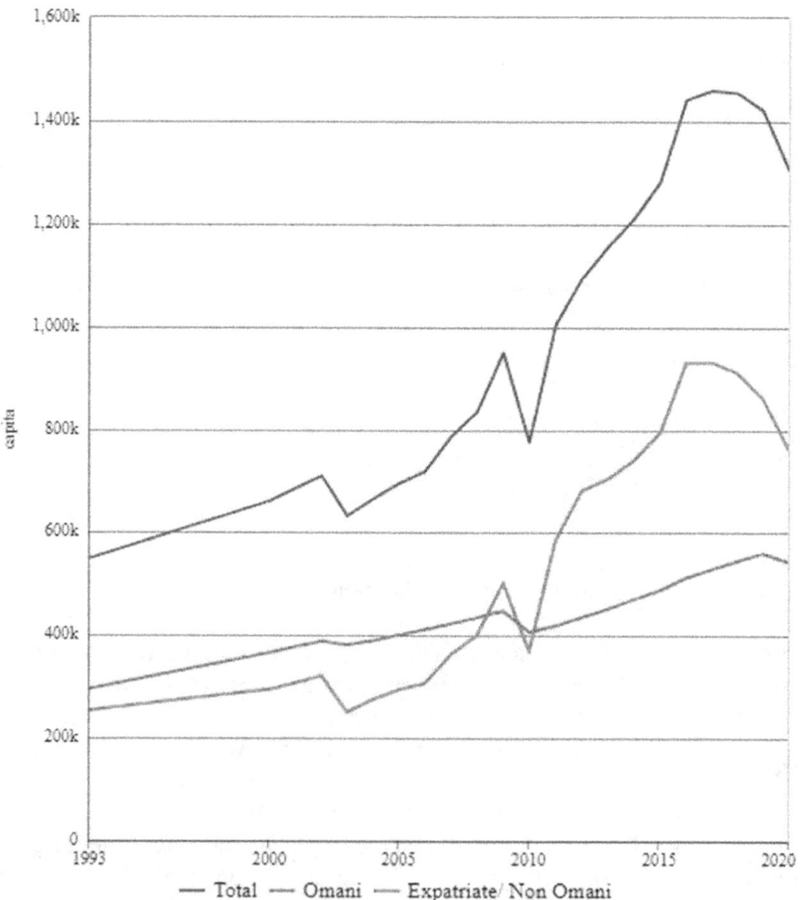

Figure 8.8 Population increases in Muscat governorate (NCSI, 2021) (Source: NCSI https://data.gov.om/OMPOP2016/population?regionId=OM-MA)

has boosted as can be seen in Figure 8.9. It can be seen easily that Salalah city has the largest increase in population from about 150,000 in 1993 to about 350,000 in 2020.

Similarly, the total registered vehicles with the Royal Oman Police (ROP) have increased from 1.3 million in January 2016 to 1.5 million in December 2020 according to NCSI. Water production has increased from more than 113.6 M^3 million in 2002 to more than 473.5 million in 2020 in Oman. For Muscat Governorate, the water production increased from 67.9 million in 2012 to about 276.5 million in 2020. Electricity production has increased from more than 10,300 GW/H in 2002 to more than 37,500 GW/H in 2020 in Oman. Natural Gas usage increased from more than 1.4 million standard cubic feet (MNSCF) in 2015 to more than 1.6 MNSCF in

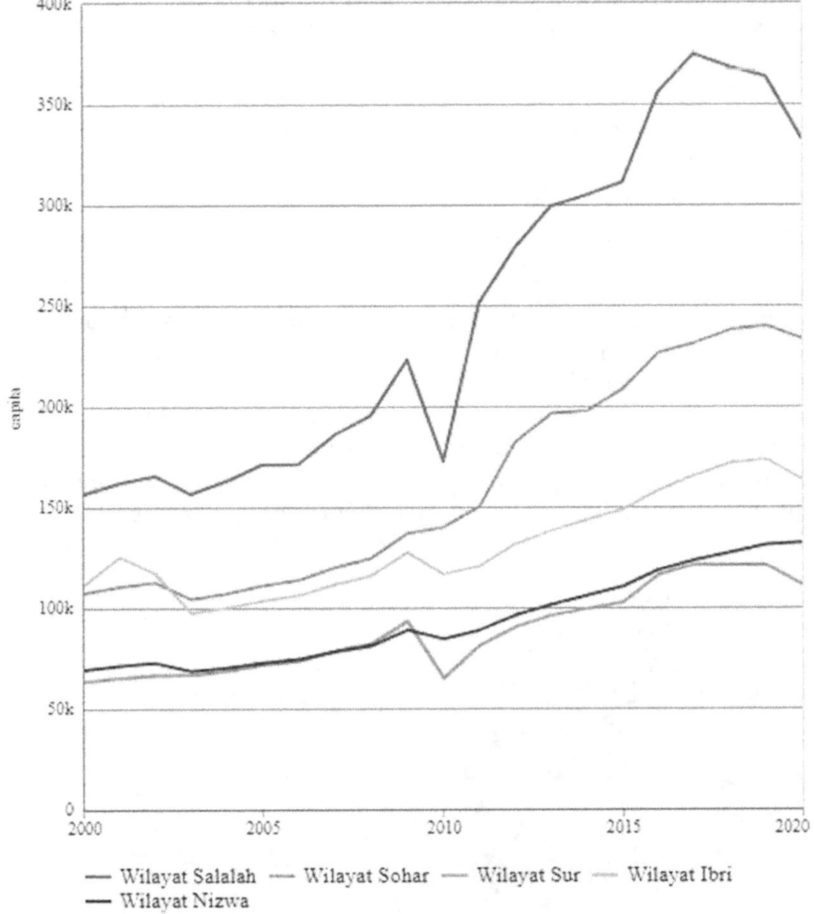

Figure 8.9 Population increases in Oman major cities (NCSI, 2021) (Source: https://data.gov.om/OMPOP2016/population?regionId=OM-MA)

2019. For greenhouse gas (GHG) emission, there was a dramatic drop from more than 1.4 million tons of CO2 in 2012 to 631,000 in 2018. In terms of safety and security, the fire cases have increased from 3,107 cases in 2010 to 3,409 cases in 2020 in all Oman, and in Muscat, it increased from 637 cases in 2010 to 1,046 cases in 2020. For the telephone/mobile and internet subscribers in Oman, the number increased from more than 464,000 and 48,000 in 2002, respectively, to about 6.4 million and about 474,000, respectively, in 2019. Therefore, the urban challenges in Oman can be summarized in the following points:

- Rapid increase in migration to the cities, particularly Muscat
- Centralization of government services, job opportunities, and prosperity in the capital
- Absence and shortage of public transportation
- Huge demand and expenses for water and electricity services
- Lack of recycling facilities for wastewater and garbage
- High cost of internet services and weakness of the signal coverage
- Inadequateness of planning for population expansion especially in the capital cities

8.5.1.2 Residents' perception

According to the result of the conducted public survey, the residents are confirming that the cities in Oman are having problems and challenges that need solutions as displayed the Figure 8.10. This reveals that there is a need to address the citizen views and to solve those issues by converting cities to be SSC and highlighting their benefits through intensive publicity and public relations. Additionally, data demonstrates that 78% of respondents believe that cities are having environmental planning and infrastructure problems. In the same context, 47% said that internet coverage is acceptable, and the

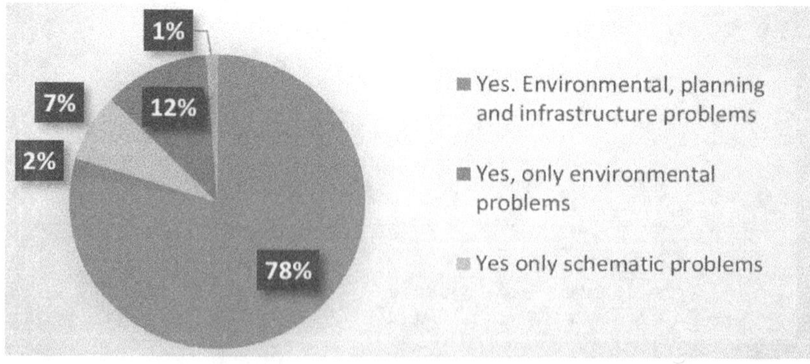

Figure 8.10 Residents' thought toward cities' problems

price is fair whereas 43% believed that network coverage is weak, and the price is high. Additionally, 36% stated they know about the concept of SSC and 33% declared they know a little as shown in Figure 8.11.

Furthermore, the obtained data reveals that 74% of the residents are using the internet in their daily activities to redeem some services and operations as shown in Figure 8.12, while 14% are using it for correspondence and following up on daily news. This is an excellent indication of residents' readiness to use the internet in executing their needs and requirements.

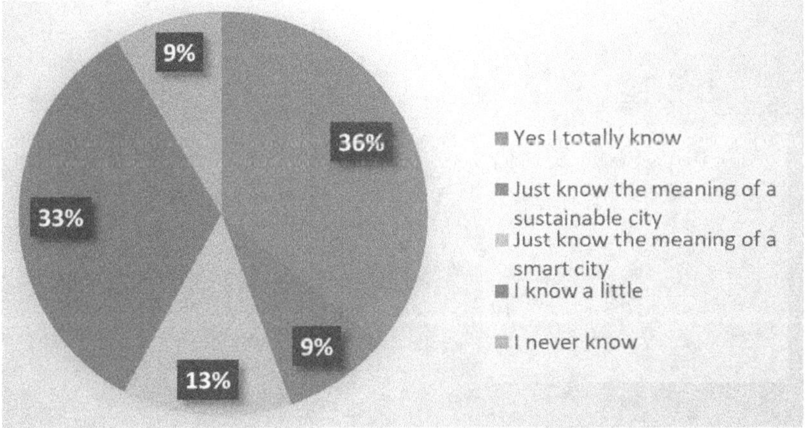

Figure 8.11 Residents' knowledge about SSC

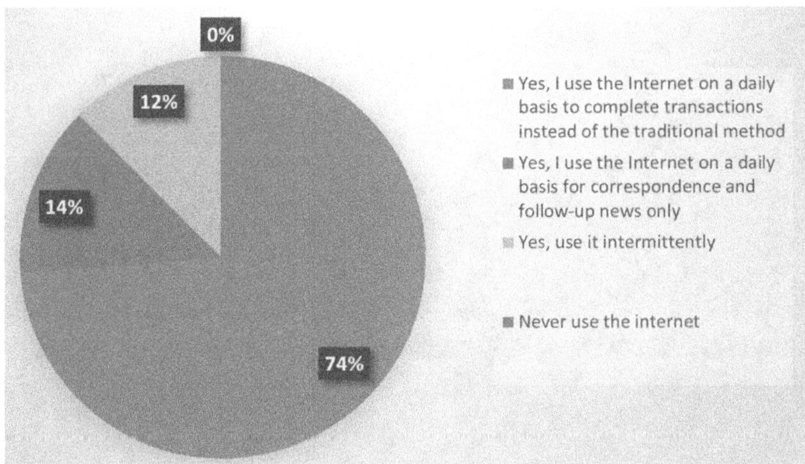

Figure 8.12 Purpose usage of internet by residents

The outcome of the survey appears that 36% of respondents have heard about the Internet of Things and smart home devices and they are using them in their daily life, while 49% stated that they have heard about them and are willing to own some due to their great benefits as illustrated in Figure 8.13.

Moreover, the questionnaire discovered massive support for converting governments and companies' services to be digitalized away from the traditional way of work using papers and personal attendance. However, as appears in Figure 8.14 only 19% think that this transformation should be applicable for certain institutions only.

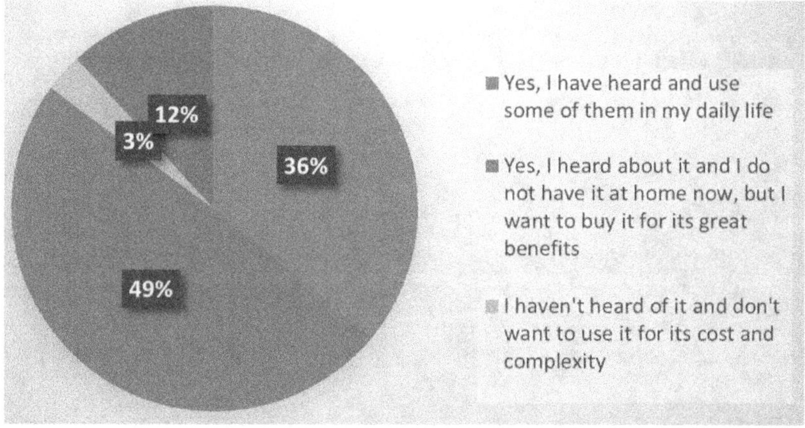

Figure 8.13 People awareness about IoT devices

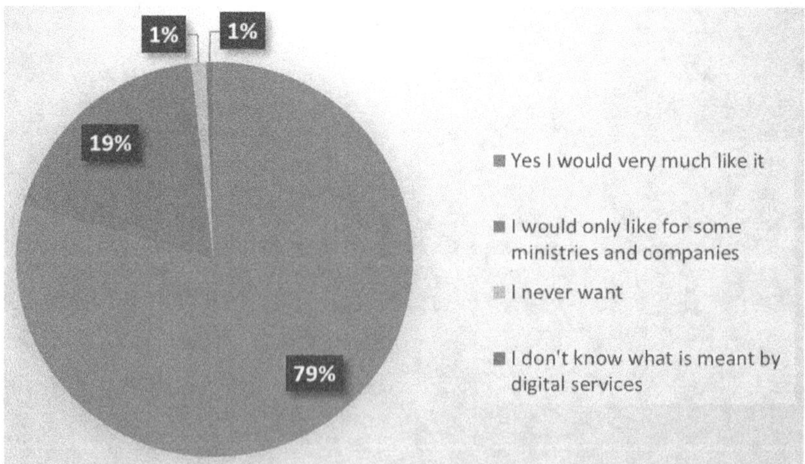

Figure 8.14 Citizen acceptance for digitalization

Besides that, 39% of the responses indicate that cities' infrastructure is ready to apply sustainable solutions and intelligent technologies in their provided services and facilities, but there is no effective and systematic approach from the government toward this transformation. Nevertheless, 32% think that IT infrastructure can be prepared within 5 years whereas 17% state that cities need 10 years of preparation as presented in Figure 8.15. These results reveal a vast gap between the government's willingness and people's vision toward this digital transformation. Social media and influencers nowadays are considered the best method for delivering messages to educate audiences, and the government should work on this orientation to clarify the digital facilities that are available for them.

In terms of sustainability and renewable energy solutions, the survey exposes that 40% of the audience have heard about the renewable energy projects in Oman but they believe those projects are simple and inadequate but 36% think those schemes are immense and will have great financial and environmental returns to the country. In the same context, 32% of the respondents are using devices that run on renewable energy and 64% of them are not using them at the moment but they are willing and encouraging others to utilize clean energy in their home operation as displayed in Figures 8.16 and 8.17, respectively. These findings are great signals for citizens' acceptance and knowledge about the significance of sustainability and its solutions.

Comparably, 38% of the replies exhibit unhappiness with the expense and efficiency of the "Sahem" program from the Electricity Regulatory Authority (ERA) which is about installing solar panels on the roofs of buildings to generate electric power. In contrast, 35% of the answers shows

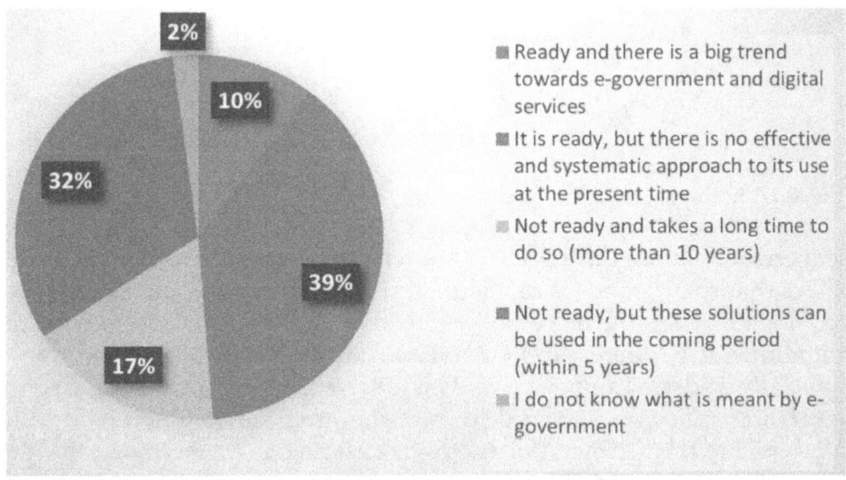

Figure 8.15 Citizens' attitude toward cities' readiness for electronic services

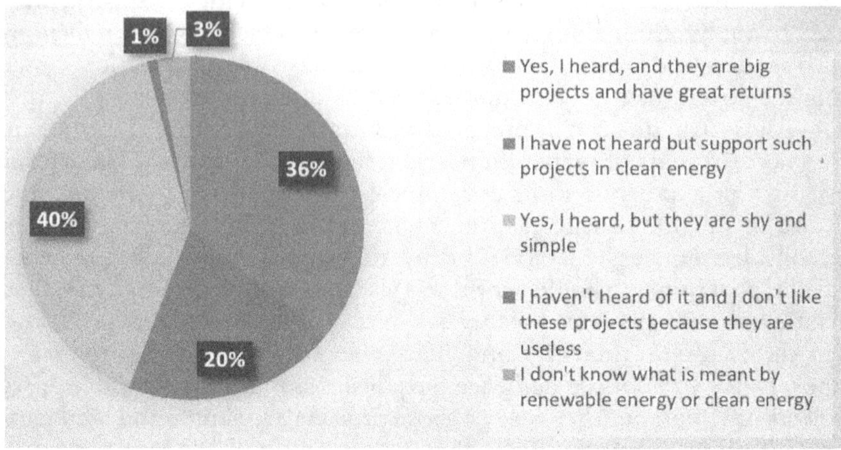

Figure 8.16 Residents' knowledge about Renewable energy projects in Oman

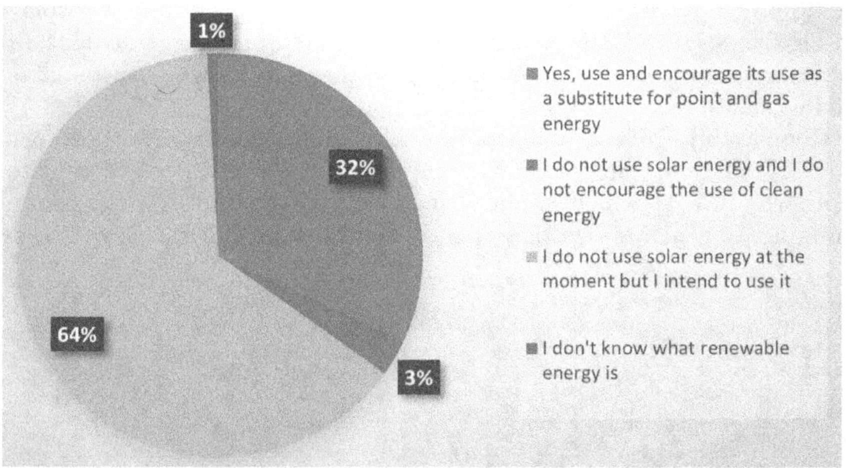

Figure 8.17 Respondents' usage of clean energy devices

inexperienced about this initiative but willing to learn more about it as they believe that solar energy is the best alternative option for energy in Oman. Additionally, 21% of the responses indicate a good satisfaction with this program as it is offered at a reasonable cost and delivers a good return, financially and environmentally. This difference in responses as clean in Figure 8.18 reflects a gap in delivering the program idea and its detailed structure and the benefits it offers to the customers. Furthermore, this program should be extended to all other governments in Oman with partnerships from the private sector to accommodate the implementation cost.

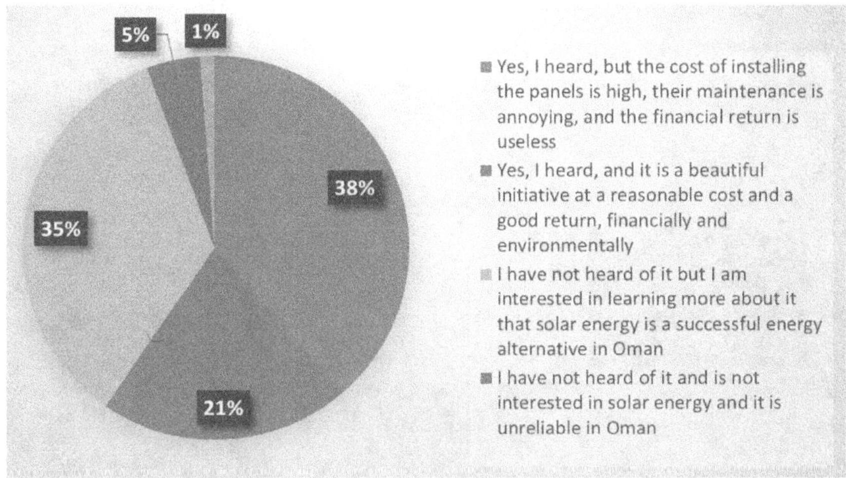

Figure 8.18 Customer experience with "Sahem" program from ERA

8.5.1.3 Transformation models

As mentioned earlier, and due to the swift urbanization, the shift toward SSCs is becoming a must. Hence, several organizations have designed structures to assist the conversion process. This comprises systems such as CISCO, IBM, Panasonic, and Huawei. Similarly, various governments worldwide including the European Union, the United Kingdom, China, and the USA have found their framework to deal with this matter. The SSC structures, as illuminated in Figure 8.19, are split into two models called the Greenfield model, where cities are built from zero, and the Brownfield model, when existing cities are transformed.

An instance of the first model is Masdar City in the United Arab Emirates. In contrast, Amsterdam signifies one of the famous examples of the Brownfield model. In the Arab province, a mixture of both models of SSC schemes exists as listed in Table 8.4. Undoubtedly, the Greenfield model involves substantial investments in ICT technologies for building new constructions, while the Brownfield model needs an advancement or renovation of the existing ICT facilities (M. Ibrahim, 2016).

Irrespective of the utilized model, the Greenfield, or Brownfield model, both, separate the conversion process into stages and sub-stages like planning, initiating, designing, delivering, consolidating, transforming, and operating. The sub-phases differ from one structure to another based on the requirements and properties of the intended city. For instance, the principal phases of a structure could be distributed into a list of sub-phases containing the change of the city strategy, development of residents' perception of SSC benefits, and improvement of IT infrastructure. Each current conversion structure trades the conversion process from its perspective depending

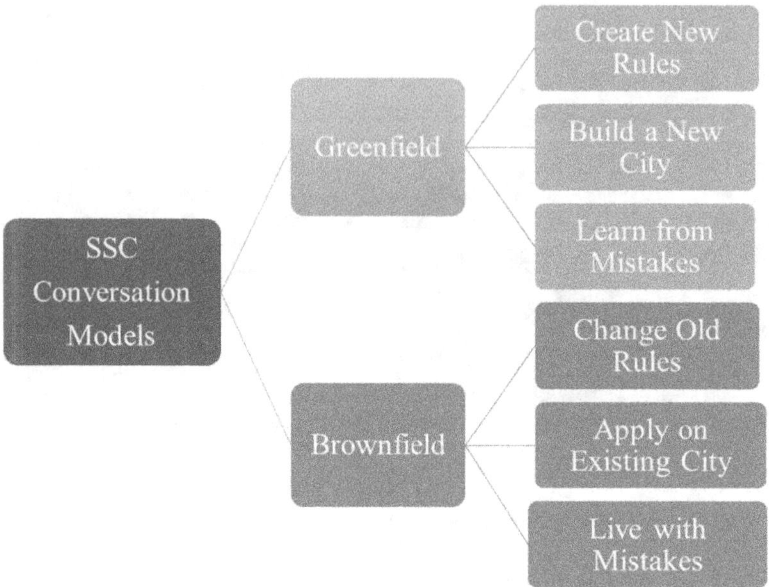

Figure 8.19 Greenfield versus Brownfield models

Table 8.4 Greenfield vs. Brownfield SSC initiatives in Arabic countries

Country	*Brownfield/Greenfield*	*Example of SSC Project*
Algeria	Greenfield	Cyberpark City of Sidi Abdullah
Bahrain	Brownfield	Manama
Egypt	Greenfield	Smart Village
Kuwait	Brownfield	Kuwait City
Morocco	Brownfield	Rabat
Oman	Brownfield	Muscat
Qatar	Brownfield/Greenfield	Lusail City – Greenfield Doha City – Brownfield
Saudi Arabia	Greenfield	King Abdullah Economic City
Tunisia	Greenfield	Tunisia Economic City
UAE	Brownfield/Greenfield	Masdar City – Greenfield Dubai City – Brownfield

on the requirements and features of the proposed city (CISCO, 2012 & EPIC, 2013). Subsequent are chosen instances of current conversion structures toward SSCs. The following frameworks are based on the Plan-Do-Check-Act (PDCA) model which can be described as:

Plan: Establish objectives and processes necessary to deliver results in accordance with community purposes.

Do: implement processes and achieve objectives.
Check: Monitor and measure processes against community policy, objectives, and commitments, and report the results.
Act: Take necessary actions to improve performance.

8.5.1.4 Key performance indicators

Although there is no unified official classification of indicators for measuring smart cities, there are a set of indicators that are based on measuring the extent of electronic readiness. Each group of indicators focuses on a specific breadth of the smart city dimensions such as e-government, citizen empowerment, and public-private partnership, smart applications, and national and regional considerations that measure the level of cooperation and homogeneity between cities in the same country. On the other hand, three indicators were recently presented that describe the competitiveness of smart cities in several geographical regions in the world, within the framework of an intensive study on smart cities and ways to measure their competitiveness (Jonathan Woetzel & others, 2018) including:

- The Smart Cities Technical Infrastructure Index measures the readiness of the technology infrastructure for the smart city.
- The Spread of Smart Applications Index measures the implementation of practical solutions to challenges and increases efficiency and comfortableness for users.
- Applications Usage Index measures users' satisfaction and interaction with the digital applications adopted by the launcher.

In parallel, for Oman, SSC proposal developing a practical system for quantifiable performance measurement of regional progress is essential for underlying the form of legalization, policy, figures, and framework. Subsequently, an apparent position of growth can be assessed, and methods can be put in to sustain growth and progress speed. If a governorate made poorer results, then it will benefit from capacity building team from the supervising authority. In contrast, a higher-scoring governorate will get distinctive, tailored interventions from the managing authority to increase their impact in releasing provincial progress and improvement opportunities. Based on the analysis and comparison of the collected data and the literature review, Oman SSC KPIs can be generated in line with a ranking criterion of an international organization such as IMD-SUTD Smart City Index and surely with customizing the index based on the local condition, culture, and people awareness. Five key fields that consist of mobility, activities, opportunities, governance, health, and safety can be covered by measuring the structures and technology levels of each field for each governorate. Other KPIs should be created for the government's authorities and companies to measure their seriousness in the digitalization transformation

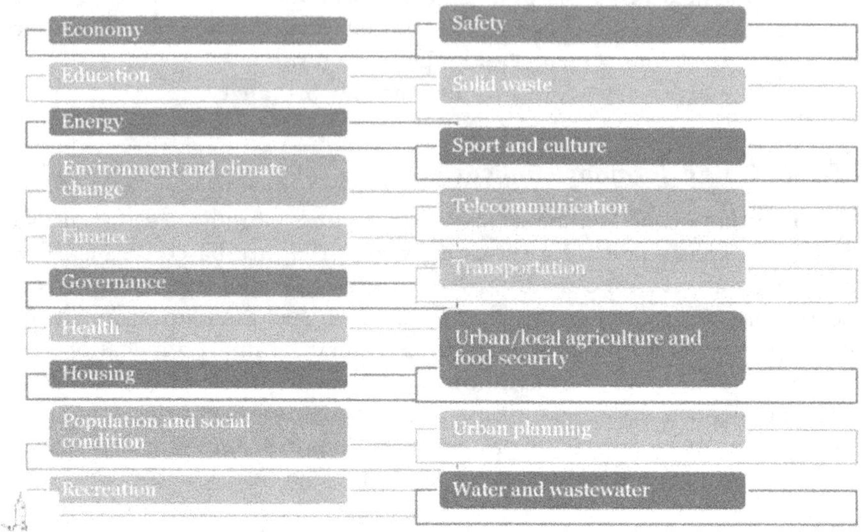

Figure 8.20 A list of SSC themes' indicators (Evelina, 2019)

journey. A list of the most common fields to be monitored and measured for evaluating the SSC success is illustrated in Figure 8.20.

8.6 CHALLENGES

Smart cities require substantial inputs of physical and human capital to produce and distribute the technology. The first task is to provide the appropriate infrastructure and technical cadres to run the projects, and then bring in the new technology in line with the needs of society, whether in the field of transportation, communications, safety, and others. Generally, challenges vary from economic and social to technological and regulatory where all must be overcome to transform Oman cities into SSCs. The most important of which is described as:

- Economic challenge: availability of sustainable financial investment to secure fundings
- Governance challenges: availability of the government willingness and systematic approach toward this process
- Technological challenge: availability of infrastructure and information systems that form the foundation of the transformation process
- Social challenge: availability of the human element and smart minds who are equipped with awareness and skills to utilize and positively interact with the technology

- Regulatory challenge: availability of regulations and legal institutional frameworks especially in terms of security and protection for both owners and users

8.7 FINDINGS AND RECOMMENDATIONS

The COVID-19 pandemic has acted as a booster for digital transformation across the board (public and private sectors) and led to a dramatic increase in technology adoption and upskilling of various parts of the economy and citizens. Hence, it creates further market opportunities for SSC solutions as user acceptance of new smart and innovative solutions increases. According to the received data, it is found that the sultanate of Oman has recently started a framework to implement smart cities concept and has commenced the digital transformation in most government services. By launching Sustainable Development Oman 2030 and Oman Vision 2040, a massive push is ongoing to achieve this conversion in terms of initiating clean energy projects and utilizing intelligent technologies. Additionally, it can be said that four types of governorates emerge where each type of governorate presents distinct opportunities for growth, development, and partnership. These governorates can be categorized as:

- The capital governorate and main hub of innovation and growth (Muscat)
- Governorates with cities that are engines of growth and have high levels of investment momentum (Dhofar, AL Wusta, and Al Batinah North)
- Low density or rural governorates with leverageable industry assets and opportunities (Al Dhahirah, Musandam, Al Buraimi, and AL Sharqiyah South)
- Governorates with some natural or historic assets with growth potential (Al Dakhilyah, AL Sharqiyah North, and Al Batinah South)

Oman has promised to become a leader in smart and clean mobility, underpinned by an ambitious cycle of transport and logistics investment. Without a question, transportation is a key area that needs investment: not only public transport but also transportation links (roads, rail) between the governorates and Muscat. Moreover, future focuses for port improvements should include digitalization and the creation of a customer portal to enable online transactions and payments. Besides that, Oman has already begun work on improving both its physical infrastructure, such as roads, and digital infrastructure, broadband penetration, and communication networks. Even though the internet penetration in Oman is about 95%, the fiber penetration is still under 20% nationally. Many smart city solutions and investors require high-speed internet as a key factor in locating outside of Muscat. This infrastructure must reach all governorates to support business, academia, and innovators, while also providing a key attraction to foreign investors looking to tap into the regional potential.

Moving forward in contemporary urban planning and management practices is critical and requires government authorities to be equipped with adequate power and resources. This includes local autonomy and clarity of roles and functions, along with the financial and technical capabilities to implement them smartly. Like many other countries, Oman faces difficulties in implementing the smart city agenda. Local institutions in Oman are constrained by the requirements. The smart city is a goal that everyone aspires to achieve. If cities want to reach this goal and succeed in entering the digital age, this requires, in addition to the national vision and efforts, the active participation of local authorities in the localization of the national agenda through the implementation of smart interventions on the scale of the residential neighborhood. Adequate capacity and autonomy of local authorities are extremely important for urban sustainability. Some recommendations are suggested to the policy-making level regarding the transformation of Oman cities into smart cities, including:

- Preparing an integrated strategy for sustainable smart cities based on a survey of stakeholders in various fields (urban planning, the business community, academia, civil society organizations, legislators, and others) to identify their views on the benefits of shifting to SSCs, and to face challenges in this regard.
- Leading a wide-ranging awareness campaign that includes various institutions and segments of society and aims to spread awareness and promote the SSC and demonstrate its importance for welfare, security, sustainability, and work to provide the prerequisites for the success of this proposal.
- Benefit from successful global experiences in the field of transition to SSC, especially by focusing on the targeted sectors to meet the needs of society such as transportation, health, energy, and education.
- Focusing on emerging technologies in data and information systems, especially through "open data," allowed a great deal of interaction between citizens, businesses, and relevant government institutions. E-government has contributed significantly to the ease of access of citizens to public services that are provided with high quality and at an acceptable cost.
- Benefiting from partnership contracts between the public and private sectors, including attracting foreign direct and specialized investments in the field of SSC, if this is preceded by the provision of the appropriate regulatory and legal framework.

- Study the feasibility of integrating sustainable smart infrastructure in new cities since the beginning of their establishment due to the ease of integration during the process of creating cities instead of turning cities into smart ones.

8.8 CONCLUSION

In the wake of the alarmingly growing urbanization challenges posed by cities around the world, SSC is becoming an increasingly popular target for modern urban planning. Efforts to transform cities into more livable, smarter, and resource-efficient places are feasible for governments and urban planners, as well as policy experts and the ICT industry. Experts believe that technology as it advances its innovations in recent decades are fully capable of providing the best solutions to traditional urban challenges, improving our lifestyles, and achieving sustainability.

SSC initiatives in Oman are still immature and limited to the use of technology and its implications for providing services. Certainly, there is no arguing about the importance of technology, but it represents only one component of smart cities. Therefore, the full concept must be achieved to become a large and comprehensive development agenda through which the smart city is viewed as a living environment that is integrated with effective management. The government should authorize governorates to laying the foundations for SSC transformation at the local level, benefiting from the governorate's economic strengths, and implementing the sustainable development agenda.

Governorates and municipality councils should participate effectively in defining urban developing policies to build a healthy environment that supports governments in implementing national priorities. These councils should deliver the correct information and clear pictures of local dynamics, along with culture and socio-economic characteristics, which are necessary to build a base for any successful interventions. A deep understanding of the local conditions and needs of each city helps to identify areas that need to be upgraded in the field of technology, and work to reach the best solutions tailored to solve current problems. By looking at the developed world where technology is present in all its areas of life, it is easily found that governments are increasingly dependent on information and communication technology solutions in the provision of daily urban services. This resulted not only in the efficient use of resources but also in the improvement of urban lifestyle and the strengthening of communication between citizens.

Undoubtedly, infrastructure and services are integrated largely determining the success of transformation toward SSC. Besides that, community awareness and social inclusion are the heart of building smart societies. This includes engaging civil society, academia, investors, and young entrepreneurs in the development process. Therefore, governorates and municipality councils should lead integration efforts to create coherence between local offices of ministries, as well as effectively manage the multiple stakeholders involved in the transformation process, to give feedback on the pilot programs to assist the government in implementing the SSC concept.

Effective management of services leads to efficiency in the use of resources, savings in overheads, and improvement in the overall quality of urban service delivery. The ICT network must therefore integrate information that allows data integration, improved coordination, and citizen access to government services, including real estate, industrial manufacturing, utilities, waste management, healthcare, education, mobility, safety, and security.

Sustainable financing for SSC projects is considered one of the most important global challenges. To develop long-term financing methods and ensure that the implementation of new projects is not hampered by financial constraints, the government should identify opportunities for public-private partnerships, facilitate investment procedures, and should also put in place strong control and monitoring systems to prevent corruption and misuse of fundings. Similarly, having an effective oversight and regulatory framework at the earlier level is important for such initiatives for young entrepreneurs. Also, an environment must be provided to nurture new start-ups, under the umbrella of government support, to help them take risks and raise their business ideas.

REFERENCES

Ahmadizadeh, M., Heidari, M., Thangavel, S., Naamani, E. A., Khashehchi, M., Verma, V., & Kumar, A. (2024). Technological advancements in sustainable and renewable solar energy systems. In *Highly Efficient Thermal Renewable Energy Systems: Design, Design, Optimization and Applications*; V. Verma, S. Thangavel, N. Dutt, A. Kumar, R. Weerasinghe Editors; CRC Press: Boca Raton, FL, Chapter 02. ISBN 9781032595641.

Al-Bahrani, A. (n.d.). Vision 2040 planning for the future: Market creating a thriving Omani Labor market. https://www.economy.com/oman/labor-force.

Al-Nasrawi, S., El-Zaart, A., & Adams, C. (2017). The anatomy of smartness of smart sustainable cities: An inclusive approach. 2017 International Conference on Computer and Applications, ICCA 2017, 348–353. https://doi.org/10.1109/COMAPP.2017.8079774.

Al Shezawi, M. (2020). *Duqm Smart City*. Duqm Special Economic Zones Magazine.

Al shueili, K. S. M. (2015). Towards a sustainable urban future in Oman: Problem & process analysis (Muscat as a case study). PhD diss., The Glasgow School of Art.

Anastasia, S. (2012). The concept of "smart cities". Towards community development? In *Networks and Communication Studies, NETCOM* (Vol. 26). http://www.digitalcommunities.com/survey/cities/.

Anthony, B. (2021). A case-based reasoning recommender system for sustainable smart city development. *AI and Society*, 36(1), 159–183. https://doi.org/10.1007/s00146-020-00984-2.

Assessment and Action on the Path to Maturity. International Data Corporation (IDC) Government Insights, Business Strategy #GI240620. Alexandria, VA.

Bednarska-Olejniczak, D., Olejniczak, J., & Svobodová, L. (2019). Towards a smart and sustainable city with the involvement of public participation-The case of Wroclaw. *Sustainability*, 11(2). https://doi.org/10.3390/su11020332.

Belli, L., Cilfone, A., Davoli, L., Ferrari, G., Adorni, P., di Nocera, F., Dall'Olio, A., Pellegrini, C., Mordacci, M., & Bertolotti, E. (2020). IoT-enabled smart sustainable cities: Challenges and approaches. *Smart Cities*, 3(3), 1039–1071. https://doi.org/10.3390/smartcities3030052.

Benkari, N. (2017). Urban development in Oman: An overview. *WIT Transactions on Ecology and the Environment*, 226(1), 143–156. https://doi.org/10.2495/SDP170131.

Bibri, S. E. (2019). On the sustainability of smart and smarter cities in the era of big data: An interdisciplinary and transdisciplinary literature review. In *Journal of Big Data* (Vol. 6, Issue 1). SpringerOpen. https://doi.org/10.1186/s40537-019-0182-7.

Bruno, L. (2020). Smart city observatory. https://www.imd.org/smart-city-observatory/home/.

BSI. (2014). *Smart Cities – Vocabulary*. British Standards Institute (BSI), BSI Standards Publication, PAS 180:2014. London.

Chaichan, M. T., Kazem, H. A., & Al-Waeli, A. H. (2020). Oman vision. *Renewable Energy Program*, 3(1). https://doi.org/10.13140/RG.2.2.20719.15524.

CISCO. (2012). Smart city framework – A systematic process for enabling smart+connected communities. Cisco Internet Business Solutions Group (IBSG): Falconer, G. & Mitchell, S.

Clarke, R. (2013). Business strategy: IDC government insights smart city maturity model.

Clemente, C. (n.d.). Technological solutions in residential building to improve quality of life and the environment social housing View project PED-Positive Energy District View project. https://www.researchgate.net/publication/38134227.

Doost Mohammadian, H., & Rezaie, F. (2020). I-sustainability plus theory as an innovative path towards sustainable world founded on blue-green ubiquitous cities case studies: Denmark and South Korea. *Inventions*, 5(2). https://doi.org/10.3390/inventions5020014.

EPIC. (2013). EPIC roadmap for smart cities. European Union, European Platform for Intelligent Cities (EPIC), Version 1.0, Project no. 270895.

Felipe, E., & Santana, Z. (n.d.). InterSCSimulator: A scalable, open source, smart city simulator. https://www.researchgate.net/publication/333340809.

GreenMatch. (2021). Sustainable and smart cities around the world. https://www.greenmatch.co.uk/blog/2017/04/sustainable-and-smart-cities-around-the-world.

Gonçalves, G. D. L., Filho, W. L., Neiva, S. da S., Deggau, A. B., Veras, M. de O., Ceci, F., de Lima, M. A., & Guerra, J. B. S. O. de A. (2021). The impacts of the fourth industrial revolution on smart and sustainable cities. *Sustainability*, 13(13). https://doi.org/10.3390/su13137165.

Huang, K., Luo, W., Zhang, W., & Li, J. (2021). Characteristics and problems of smart city development in China. *Smart Cities*, 4(4), 1403–1419. https://doi.org/10.3390/smartcities4040074.

Ibrahim, M., El-Zaart, A., & Adams, C. (2016). Paving the way to smart sustainable cities: Transformation models and challenges. *Journal of Information Systems and Technology Management*, 12(3). https://doi.org/10.4301/s1807-17752015000300004

Ibrahim, R., Asri, N. A. M., Jamel, S., & Wahab, J. A. (n.d.). Validation of requirements for transformation of an urban district to a smart city. In *IJACSA International Journal of Advanced Computer Science and Applications* (Vol. 12, Issue 7). www.ijacsa.thesai.org.

IEEE. (2015). IEEE smart cities. Institute of Electrical and Electronics Engineers (IEEE). Web. http://smartcities.ieee.org/about.html.

Irfan, A.-H., & Moneim, H. A. (2019). *Smart Cities in the Arab Countries: Lessons from Global Experiences.* Arab Monetary Fund.

Khashehchi, M., Thangavel, S., Rahmanivahid, P., Heidari, M., Moazzeni, T., Verma, V., & Kumar, A. (2024). Solar desalination techniques: Challenges and opportunities. In *Highly Efficient Thermal Renewable Energy Systems: Design, Design, Optimization and Applications*; V. Verma, S. Thangavel, N. Dutt, A. Kumar, R. Weerasinghe Editors; CRC Press: Boca Raton, FL; Chapter 19. ISBN 9781032595641.

Krause, A. (n.d.). The role of UNECE and the Key Performance Indicators for Smart and Sustainable Cities in improving smart and sustainable urban development. https://www.slideshare.net/OASC/the-role-of-unece-and-the-key-performance-indicators-for-smart-and-sustainable-cities-in-improving-smart-and-sustainable-urban-development

Kutty, S. (2020). Oman vision 2040 eyes increase on economic growth. https://www.omanobserver.om/article/16242/Local/vision-2040-eyes-increase-in-economic-growth.

Malmqvist Nilsson, M. (2020). Industrial revolution 4.0 as leverage to achieve the Sustainable Development Goals. https://iaf.news/2020/04/11/industrial-revolution-4-0-as-leverage-to-achieve-the-sustainable-development-goals/

Manville, C., Europe, R., Millard, J., Technological Institute, D., Liebe, A., & Massink, R. (2014). Directorate general for internal policies policy department a: Economic and scientific policy Mapping Smart Cities in the EU study.

Mersal, A. (2017). Eco city challenge and opportunities in transferring a city into green city. *Procedia Environmental Sciences*, 37, 22–33. https://doi.org/10.1016/j.proenv.2017.03.010.

Muhammad, A., & Khan, A. (n.d.). A stakeholder based assessment of developing country challenges and solutions in smart mobility within the smart city framework: A case of Lahore. https://doi.org/10.13140/RG.2.1.3516.9689.

Oman Observer. (December 2020). Oman vision to improve Oman's global standing. https://www.omanobserver.om/article/6459/Front%20Stories/vision-2040-to-improve-omans-global-standings.

Parthiban, P., Cyriac, M. G., Prabhu, T. S. B., Asok, G., Neeraja, P. G., & Athiappan, K. (2021). Sustainable solutions for effective housing environment. AIP Conference Proceedings, 2396. https://doi.org/10.1063/5.0066294.

PAS. (2014). Smart city framework – Guide to establishing strategies for smart cities and communities. The British Standards Institution (BSI), Department of

Business, Innovation & Skills (BIS), BSI Standards Limited Publication (PAS 181:2014).

Persaud, T., Amadi, U., Duane, A., Youhana, B., & Mehta, K. (2020, October 29). Smart city innovations to improve quality of life in urban settings. 2020 IEEE Global Humanitarian Technology Conference, GHTC 2020. https://doi.org/10.1109/GHTC46280.2020.9342905.

Riad Kazem. (2019). Promising Arab experiences in the field of smart city planning. The proceedings of the first international conference in Berlin entitled "Smart Cities in Light of the Current Changes: Reality and Prospects."

Roblek, V., Thorpe, O., Bach, M. P., Jerman, A., & Meško, M. (2020). The fourth industrial revolution and the sustainability practices: A comparative automated content analysis approach of theory and practice. *Sustainability*, 12(20), 1–28. MDPI. https://doi.org/10.3390/su12208497.

Schipper, R., & Silvius, A. (2018). Characteristics of smart sustainable city development: Implications for project management. *Smart Cities*, 1(1), 75–97. https://doi.org/10.3390/smartcities1010005.

Smart city and sustainable city: The case of Amsterdam by Jean Danielou. (2014). Smart city and sustainable city: The Case of Amsterdam (citego.org).

The Sultanate of Oman Public Authority for Electricity and Water Project for Energy Conservation Master Plan in the Power Sector in the Sultanate of Oman Final Report Tokyo Electric Power Company (TEPCO). (2013).

United Nations, Department of Economic and Social Affairs. (n.d.). World economic and social survey 2013: Sustainable development challenges.

Verma, V., Meena, C. S., Thangavel, S., Kumar, A., Choudhary, T., & Dwivedi, G. (2023). Ground and solar assisted heat pump systems for space heating and cooling applications in the northern region of India – A study on energy and CO2 saving potential. *Sustainable Energy Technologies and Assessments*, 59, 103405. ISSN 2213-1388. https://doi.org/10.1016/j.seta.2023.103405.

Verma, V., Thangavel, S., Dutt, N., Kumar, A., & Weerasinghe, R. (2024). Recent development of thermal energy storage: Solar, geothermal and hydrogen energy. In *Highly Efficient Thermal Renewable Energy Systems: Design, Design, Optimization and Applications*; V. Verma, S. Thangavel, N. Dutt, A. Kumar, R. Weerasinghe Editors; CRC Press: Boca Raton, FL, Chapter 01. ISBN 9781032595641.

Verma, V., Thangavel, S., Dutt, N., Kumar, A., & Weerasinghe, R., Editors. (2024). *Highly Efficient Thermal Renewable Energy Systems: Design, Design, Optimization and Applications*; CRC Press: Boca Raton, FL. ISBN 9781032595641.

Vershinina, I. A., & Volkova, L. V. (2018). Smart cities: Challenges and opportunities Ciudades inteligentes: Desafíos y oportunidades. In *ISSN* (Vol. 41).

Vishnivetskaya, A., & Alexandrova, E. (2019). "Smart city" concept. Implementation practice. IOP Conference Series: Materials Science and Engineering, 497(1). https://doi.org/10.1088/1757-899X/497/1/012019.

Whitepaper Narrowing the Gap between Industry and Academia to Establish Efficient Partnerships the Oman energy industry-academia R&D protocol. https://www.linkedin.com/pulse/whitepapers-bridging-gap-between-academia-industry-sairam-phd/

Woetzel, J., Remes, J., Boland, B., Lv, K., Sinha, S., Strube, G., Means, J., Law, J., Cadena, A., & von der Tann, V. (2018). *Smart Cities: Digital Solutions for a More Livable Future.* McKinsey Global Institute.

Xydis, G., Pagliaricci, L., Paužaitė, Ž., Grinis, V., Sallai, G., Bakonyi, P., & Vician, R. (2021). SMARTIES project: The survey of needs for municipalities and trainers for smart cities. *Challenges*, 12(1), 13. https://doi.org/10.3390/challe12010013.

Yates, N. (2020). *Towards Oman's First National Smart Cities Stack*. Ministry of Transport, Communications, and Information Technology.

Chapter 9

Data mining techniques in green building for evaluation of energy cost, grades, and adoption

Bhupendra, Varun Pratap Singh, and Ashwani Kumar

9.1 INTRODUCTION TO DATA MINING IN GREEN BUILDING

The introduction of data mining in green building commences by recognizing the historical origins of data mining technologies and their importance in many human endeavors. It focuses on the application of data mining techniques in green construction to assess energy costs, performance grades, and adoption rates. The historical backdrop of data mining, from early data interpretation to present sophisticated algorithms, highlights its continuing significance. The utilization of these strategies is imperative in the current global market landscape to promote sustainable practices in the building sector. This part offers a thorough examination and evaluation of the practical uses of data mining. It demonstrates how data mining is an effective technique for extracting meaningful information from intricate datasets in the context of green construction projects [1]. The topic encompasses the factors that drive and hinder the adoption of technology, specifically focusing on the necessity of overcoming obstacles to facilitate the wider use of data mining techniques. The analysis of predictions made by end-users combines insights from industry and history, with a particular focus on the transformative impact that data mining may have on the constructed environment. The introductory part explores the theoretical underpinnings and core principles of data mining, elucidating approaches to enhance performance and characteristics in green building operations. The primary topics being examined encompass the technological significance of data mining in assessing energy expenses, rating methodologies, and the wider implementation of sustainable construction methods. Throughout the chapter, it becomes clear that data mining is at the vanguard of innovation and has the potential to significantly transform the future of green construction [2].

DOI: 10.1201/9781003496656-9

9.1.1 Significance of data mining techniques in green building

Data mining techniques have a significant and complex role in the field of green construction, particularly in relation to sustainable development. As the world works toward a more environmentally friendly future, the use of data mining technology in green construction practices becomes an important tool for making well-informed decisions. Data mining enables the extraction of important insights from large and intricate information, providing a thorough knowledge of energy consumption trends, cost dynamics, and the overall effectiveness of green building technology. By acquiring this information, players in the construction sector are able to effectively use resources, improve energy efficiency, and reduce environmental impact. Data mining may enhance the achievement of sustainable development goals by enabling the design of buildings that not only save energy expenses but also comply with rigorous environmental regulations. Furthermore, the use of data-driven insights facilitates the detection of inventive solutions and the ongoing enhancement of environmentally friendly construction methods, promoting a flexible and adaptable approach to addressing sustainable development obstacles. The importance of data mining techniques rests in their capacity to match green construction activities with the principles of sustainable development, therefore facilitating the creation of a more resilient and environmentally conscious built environment [3].

9.2 OVERVIEW OF ENERGY COST EVALUATION IN GREEN BUILDING

Within the realm of green building practices, a comprehensive examination of energy cost assessment through the utilization of data mining tools reveals a revolutionary method for enhancing energy efficiency and reducing expenses. This procedure begins by carefully gathering comprehensive datasets that include various factors essential for comprehending the energy dynamics of a structure [4]. By utilizing sophisticated clustering methods like k-means, it becomes possible to categorize buildings according to their comparable energy usage patterns [5]. This allows for focused interventions and customized energy conservation plans. Regression analysis examines the complex connections between many factors and energy expenditures, offering a predictive framework for evaluating the influence of design components, occupancy patterns, and environmental conditions. By using machine learning techniques such as decision trees and neural networks, the accuracy of energy cost estimates is enhanced. These algorithms are able to learn from past data and adjust to changing trends, resulting in more precise predictions. The utilization of real-time data from intelligent building systems enhances this method, creating a dynamic and adaptable energy evaluation system that can make proactive modifications to maximize efficiency [6].

This data-centric approach not only enables a thorough comprehension of the present energy efficiency but also empowers individuals with a vested interest to actively influence the future of energy use in environmentally friendly structures. By utilizing data mining to extract actionable information, decision-makers may effectively adopt focused initiatives to decrease energy expenses and improve overall operational efficiency [7]. The comprehensive analysis emphasizes the crucial role of data mining techniques in aligning green building practices with the overall objectives of sustainability and environmental responsibility, whether by optimizing current building systems or incorporating renewable energy sources. Essentially, this technique serves as a fundamental basis for developing energy-efficient, cost-effective, and ecologically conscientious constructed environments [8].

9.3 DATA MINING TECHNIQUES FOR EXTRACTING INSIGHTS

The exploration of data mining techniques for extracting insights into green building is pivotal in unlocking the wealth of information embedded within complex datasets. These techniques offer sophisticated means to analyze, interpret, and harness valuable patterns and relationships, contributing significantly to the optimization of energy efficiency, cost-effectiveness, and overall sustainability in the built environment [9].

9.3.1 Clustering analysis

Clustering techniques, such as k-means, prove invaluable in categorizing similar data points within green building datasets. In the context of sustainable construction, this involves grouping buildings with analogous characteristics, enabling the identification of common energy consumption profiles or shared sustainable practices. By clustering, stakeholders gain a nuanced understanding of building types and can tailor strategies for improved energy performance based on these groupings. This not only facilitates targeted interventions but also ensures that energy optimization measures are customized to specific clusters, enhancing the overall effectiveness of sustainability initiatives [2].

9.3.2 Regression analysis

Regression analysis is employed to quantify the relationships between different variables within green building datasets. This technique helps in assessing the impact of various factors, such as building design, occupancy patterns, or environmental conditions, on energy costs and overall building performance. By quantifying these relationships, decision-makers can make informed choices to enhance energy efficiency and mitigate potential challenges. Additionally, regression analysis allows for the development of predictive models, enabling stakeholders to forecast future energy consumption trends based on historical data, further supporting proactive decision-making.

9.3.3 Machine learning algorithms

Machine learning algorithms, including decision trees and neural networks, bring a layer of adaptability and predictive power to green building data analysis. Decision trees create models that predict outcomes based on historical data, aiding in forecasting energy consumption patterns. Neural networks, with their ability to recognize complex, non-linear relationships, excel in uncovering intricate patterns that might be challenging for traditional analysis methods. These algorithms not only enhance the accuracy of predictions but also contribute to a deeper understanding of the underlying factors influencing green building performance, thereby informing more effective strategies for sustainability.

9.3.4 Real-time data integration

Real-time data integration from sensors and smart building systems amplifies the dynamic aspect of data mining in green construction. This enables the ongoing surveillance and fine-tuning of building systems, guaranteeing that energy-saving measures stay in sync with evolving environmental circumstances and occupancy trends. Real-time data integration offers a proactive strategy for managing buildings, allowing stakeholders to quickly address any deviations from planned performance and optimize systems to achieve immediate energy and cost savings. The capacity to react in real-time is essential for ensuring optimal sustainability performance in a continuously changing built environment [10].

9.3.5 Actionable insights for optimization

Stakeholders utilize data mining tools to get meaningful insights beyond basic data interpretation. These insights are used to make decisions that improve the efficiency of building systems, lower energy expenses, and promote overall sustainability. Data mining enables decision-makers to implement targeted strategies by changing HVAC schedules, optimizing lighting systems, and incorporating renewable energy sources. The practical observations derived from this analysis may be implemented as concrete strategies that enhance the ongoing enhancement of green building efficiency, in line with sustainability objectives and guaranteeing a lasting decrease in environmental effect. Data-driven optimization, which involves continuing analysis, allows green buildings to maintain efficiency and resilience in the face of shifting needs and obstacles due to its iterative nature [11].

9.3.6 Continuous improvement

Data mining in green construction is a perpetual process that promotes a culture of ongoing enhancement. The flexibility and variety of these methodologies enable stakeholders to modify their tactics over time, in response to developing patterns and trends in the data, in order to further refine and improve sustainability practices. Data mining techniques are a valuable set of tools for individuals and organizations involved in the green building industry. By utilizing these methods, they may effectively navigate the intricacies of sustainable building datasets, uncover significant observations, and facilitate well-informed decision-making to promote the development of energy-efficient and ecologically aware constructed environments [12].

Analyzing complicated links within green building information by utilizing data mining techniques is a vital component in unraveling complex patterns. Within the realm of sustainable construction, these patterns can be complex and diverse, encompassing factors like building design, occupant behavior, and environmental circumstances. Clustering analysis, regression models, and machine learning algorithms are crucial in identifying concealed patterns that may not be evident using conventional approaches. By unraveling these intricacies, stakeholders acquire the understanding of the subtle interplay of elements that affect energy usage and the overall efficiency of buildings [13]. This approach enables decision-makers to execute focused plans for maximizing energy efficiency and enhancing sustainability in green building practices. Data mining approaches in green construction address the fundamental challenge of identifying and measuring the elements that impact energy expenses. Regression analysis is an important technique that stakeholders use to evaluate the influence of several aspects, such as building materials, insulation, lighting systems, and HVAC efficiency, on total energy costs [14]. Machine learning algorithms contribute by discerning non-linear correlations and patterns that may defy conventional analytical methods [15]. This thorough analysis allows decision-makers to get a precise comprehension of the particular factors influencing energy expenses within the framework of sustainable construction. Equipped with this information, they may develop focused interventions and strategies to reduce factors that increase costs, maximize energy efficiency, and improve the financial viability of green building projects [16].

9.4 DATA MINING METHODS USED FOR ANALYSIS IN GREEN BUILDING

Several data mining methods are employed for analysis in the context of green building to extract valuable insights, optimize energy efficiency, and enhance sustainability. The application of these methods involves the exploration of intricate datasets to uncover patterns, correlations, and trends relevant to various aspects of green building practices. Here are some key data mining methods used for analysis in green building [2]:

- **Clustering techniques for pattern recognition:**

Clustering analysis is a fundamental data mining technique used in green construction to detect underlying patterns and groups within information. This approach commonly uses algorithms like k-means or hierarchical clustering to classify buildings into separate clusters based on similar attributes. Clustering, in the context of sustainable construction, aids in the identification of clusters of buildings that have similar energy consumption patterns or display common sustainable characteristics [17]. Through the process of grouping buildings, stakeholders are able to obtain a deeper understanding of shared characteristics that may not be immediately obvious. This enables them to develop focused strategies that are customized to specific clusters. This approach enables a more sophisticated comprehension of the many architectural structures present in the environmentally friendly construction field [18].

Clustering algorithms are crucial in pattern identification for green building since they provide an effective way to classify data points into groups based on similar properties. These approaches play a crucial role in deciphering intricate patterns and finding intrinsic structures within datasets [19]. Various clustering methods are frequently employed for the purpose of pattern identification in the analysis of green buildings:

- **K-means clustering:**

K-means clustering, a widely recognized algorithm in data science, has significant applications in green building, an emerging field focused on enhancing environmental sustainability and energy efficiency in building design and operation. The fundamental principle of k-means clustering involves partitioning a set of observations into k distinct subsets, minimizing the within-cluster sum of squares. Each observation is a d-dimensional real vector, and the goal is to assign each observation to a cluster such that the sum of the squared Euclidean distances between the observations and their respective cluster means (centroids) is minimized. In the context of green buildings, this algorithm can be employed to analyze and optimize various aspects, such as energy consumption patterns, resource utilization efficiency, and environmental impact. By clustering similar data points, such as buildings with comparable energy usage or areas with similar environmental conditions, stakeholders can identify patterns and inefficiencies, develop targeted strategies for energy conservation, and implement more effective building designs that align with sustainability goals [20].

The k-means algorithm operates through an iterative process that alternates between two primary steps: assignment and update. Initially, k means are arbitrarily chosen, and each observation is assigned to the cluster with the nearest mean based on the least squared Euclidean distance. The assignment step involves partitioning the observations according to the Voronoi diagram generated by the means. In the subsequent update step, the means of these clusters are recalculated based on the assigned observations. This process repeats until the cluster assignments stabilize, indicating convergence. However, k-means clustering is not guaranteed to find the global optimum and can converge to local minima. This characteristic necessitates multiple runs with different initial conditions to achieve more reliable results. In green building applications, this iterative refinement of clusters can be instrumental in categorizing buildings or areas based on specific performance metrics, facilitating the development of customized solutions for energy efficiency and sustainability. Moreover, the algorithm's versatility allows for its adaptation to various data types and scales, making it a valuable tool in the complex and multifaceted domain of green building [21].

K-means clustering is a method that seeks to divide a collection of observations ($x_1, x_2, ..., x_n$), where each observation is a d-dimensional real vector, into k ($\leq$ n) sets S = {$S_1, S_2, ..., S_k$}. The goal is to minimize the within-cluster sum of squares (WCSS), which represents the variance inside each cluster. The task at hand is to formally determine:

$$\underset{S}{\arg\min} \sum_{i=1}^{k} \sum_{x \in S_i} \left\| x - \mu_i \right\|^2 = \underset{S}{\arg\min} \sum_{i=1}^{k} \left| S_i \right| Var\, S_i$$

The objective of the argmin function is to find the set S that minimizes the sum of squared distances between each point x and its corresponding centroid μ_i. The set S is divided into k subsets, denoted as S_i. The centroid μ_i is the mean of the points in S_i.

$$\mu_i = \frac{1}{\left| S_i \right|} \sum_{x \in S_i} \mathbf{x},$$

The size of S_i is denoted as $|S_i|$. The L^2 norm is used to calculate the squared distance between each point and its centroid. This is tantamount to minimizing the squared variances between pairs of points inside the same cluster:

$$\underset{S}{\arg\min} \sum_{i=1}^{k} \frac{1}{\left| S_i \right|} \sum_{\mathbf{x},\mathbf{y} \in S_i} \left\| \mathbf{x} - \mathbf{y} \right\|^2$$

The expression represents the argument that minimizes the sum of squared distances between pairs of points in a set S. The set S is divided into k subsets, and the argument is found by minimizing the average squared distance within each subset.

The equivalence can be derived from the equation:

$$|S_i|\sum_{\mathbf{x}\in S_i}\|\mathbf{x}-\boldsymbol{\mu}_i\|^2=\frac{1}{2}\sum_{\mathbf{x},\mathbf{y}\in S_i}\|\mathbf{x}-y\|^2$$

Given that the overall variance remains constant, this is essentially the same as maximizing the sum of squared differences across points in distinct clusters, often known as the between-cluster sum of squares (BCSS). This deterministic is also associated with the law of total variance in probability theory.

- **Hierarchical clustering:**

Hierarchical clustering, a method distinct from partitioning algorithms like k-means, is highly advantageous in the context of green buildings. It involves creating a hierarchy of clusters without the need to predefine the number of clusters, which is especially useful in analyzing complex and multifaceted data typically found in green building applications. Green buildings, aiming for environmental sustainability and energy efficiency, generate diverse datasets encompassing various aspects such as energy consumption, building materials, and occupant behavior. Hierarchical clustering can be applied to group similar buildings or construction materials based on multiple attributes, enabling a more nuanced approach to improving building design, material selection, and energy policies. The use of dendrograms in hierarchical clustering provides a visual representation of these groupings, simplifying the interpretation of complex relationships between different green building parameters. This facilitates decision-making processes, allowing stakeholders to identify key areas for improvement and develop targeted strategies for sustainability [22].

The algorithm for hierarchical clustering, particularly the agglomerative type, starts with each observation as a distinct cluster and merges them step by step based on a chosen distance metric. The process involves calculating a proximity matrix to determine the similarity between individual observations. The algorithm then iteratively merges the closest clusters, updating the distance matrix at each step. The choice of the distance metric (e.g., Min, Max, Centroid, or Ward linkage) can significantly affect the clustering outcome, and the selection depends on the specific characteristics of the data, such as noise level or cluster shape. In green building research, where data might range from highly structured to irregular and noisy, different linkage methods can be applied to best suit the data characteristics. The Min linkage is suitable for non-elliptical clusters but is sensitive to noise, while the Max linkage method is less sensitive to noise and outliers. The Centroid and Average linkage methods provide a balance, considering all distances within the clusters, and the Ward method minimizes the within-cluster variance, offering robustness against noise and outliers. Employing hierarchical clustering in green building analysis can reveal insightful patterns and relationships, aiding in the development of more efficient and sustainable building practices. This method's flexibility and adaptability make it a powerful tool for handling the diverse and complex datasets encountered in the field of green buildings [23–26].

- **DBSCAN (density-based spatial clustering of applications with noise):**

DBSCAN (density-based spatial clustering of applications with noise) stands out as an effective algorithm for clustering in the field of green building analysis, particularly due to its proficiency in handling data with noise and outliers, common in environmental and sustainability datasets. Unlike partitioning methods like k-means, which are ideal for spherical clusters, DBSCAN excels in identifying clusters of arbitrary shapes, making it well-suited for complex, real-world data sets encountered in green buildings. Such datasets often involve diverse parameters like energy usage patterns, material properties, or spatial distributions of buildings, which may not conform to standard cluster shapes. DBSCAN's ability to detect these non-linear patterns and its insensitivity to outliers enables the identification of meaningful clusters that can inform sustainable design, energy efficiency strategies, and resource allocation [27].

The DBSCAN algorithm operates on two key parameters: the neighborhood size (eps) and the minimum number of points required to form a dense region (MinPts). The eps parameter determines the radius around each data point to search for neighboring points. If the distance between two points is less than or equal to eps, they are considered neighbors. The selection of eps is crucial; too small a value may lead to many data points being classified as outliers, while too large a value may cause distinct clusters to merge. The MinPts parameter defines the minimum number of points required within an eps radius to classify a point as a core point. The general rule of thumb is that MinPts should be higher for larger datasets and is typically set to a value greater than or equal to the number of dimensions in the dataset plus one [28].

DBSCAN's algorithmic steps involve marking each point as either a core point, border point, or noise. Core points have at least MinPts within their eps neighborhood, border points have fewer than MinPts within eps but are in the neighborhood of a core point, and noise points are neither core nor border points. The process starts by identifying all neighbor points within eps of each point and marking the core points. For each core point, if it is not already part of a cluster, a new cluster is created, and all points density-connected to it are added to this cluster. Points are considered density-connected if there is a chain of points, each within eps of the next, connecting them. The algorithm iterates through all points, assigning them to clusters or marking them as noise [29].

In the context of green building analysis, DBSCAN can be particularly useful for categorizing buildings based on energy consumption patterns, identifying groups of materials with similar properties, or clustering urban areas with similar environmental characteristics. Its ability to handle irregular data and outliers enables more accurate and meaningful analysis, leading to insightful conclusions that can drive sustainable practices in the field of green buildings.

- **Mean shift clustering:**

Mean shift clustering is a robust, non-parametric clustering algorithm that is highly effective in the analysis of green buildings, a domain characterized by complex, multi-dimensional data. This algorithm is especially useful when dealing with datasets where the clusters are of arbitrary shape and the number of clusters is not known beforehand. In green building analysis, Mean shift can be applied to segment buildings based on energy consumption patterns, identify regions with similar environmental conditions, or cluster materials based on their properties. The algorithm's ability to discover the number of clusters inherent in the data makes it suitable for exploratory data analysis in this field [30].

The working principle of mean shift involves shifting each data point toward the mode (the highest density of data points) within a certain radius. This process iteratively updates the location of each point, moving it toward the densest area in its vicinity, based on a kernel function. The kernel function, often a Gaussian kernel, determines the weight of neighboring points based on their distance, thereby influencing the shift direction and magnitude. The choice of the kernel's bandwidth, a critical parameter in mean shift, affects the scale of clustering: a smaller bandwidth can lead to many small clusters, while a larger one might merge distinct clusters [31].

In practice, mean shift starts with each data point as a cluster centroid and iteratively updates these centroids by calculating the mean of all points within a specified radius (the kernel). Points shift toward the mean until convergence, typically when they reach a local maximum of the density function. At the end of the process, points that converge to the same maxima are grouped into the same cluster. The algorithm's non-reliance on prespecified cluster numbers and its adaptability to arbitrary cluster shapes make it suitable for the heterogeneous and often unpredictable data patterns encountered in green building datasets [32].

However, mean shift's computational complexity, typically $O(n^2)$, and its sensitivity to the kernel's bandwidth are notable downsides. The bandwidth selection, not always straightforward, significantly influences the clustering outcome and requires careful tuning, especially in green building applications where data can be highly varied. Despite these challenges, mean shift's model-free nature and its ability to find variable numbers of modes robustly make it a valuable tool in the ever-evolving field of green building analysis. Its application extends beyond clustering, including image processing and computer vision, demonstrating its versatility in handling complex datasets [20].

- **Fuzzy C-means clustering:**

Fuzzy C-means clustering, a variant of the traditional clustering methods like k-means, is particularly beneficial in the field of green building analysis. This clustering technique is unique as it allows each data point to belong to multiple clusters with varying degrees of membership, a concept stemming from fuzzy logic. In green building analysis, where data can often be complex and not distinctly separable, Fuzzy C-means provides a more nuanced approach to clustering. It is adept at handling ambiguities and overlapping data points, making it suitable for categorizing buildings based on energy usage, environmental impact, or other multifaceted criteria.

The algorithm of Fuzzy C-means includes several key steps:

Initialization: The process begins by assigning each data point a membership value for each cluster. For instance, in a scenario with two clusters, each data point will have two membership values indicating the degree to which it belongs to each cluster.

Centroid calculation: The algorithm then computes the centroid for each cluster. The centroid calculation takes into account the degree of membership of each data point and its respective value. The formula for calculating the centroid involves a weighted average of all data points, where the weights are the membership values raised to the power of the fuzziness parameter, usually denoted by "m."

Updating membership values: The next step involves updating the membership values for each data point based on its distance from the centroids of the clusters. The new membership values are calculated using a formula that accounts for the inverse distance of the data point from each centroid. This step ensures that the membership values are updated to reflect the current positioning of the centroids.

Iteration: Steps 2 and 3 are repeated iteratively. With each iteration, the centroids shift according to the updated membership values, and the membership values change according to the new centroid positions. This iterative process continues until the membership values stabilize, indicating the formation of clusters.

Defuzzification: Finally, the fuzzy membership values are defuzzified to assign each data point to one or more clusters. This can be done by setting a threshold for membership values or by assigning each data point to the cluster for which it has the highest membership value.

In the context of green building analysis, Fuzzy C-means clustering can be instrumental in identifying patterns and categories that are not immediately apparent. For instance, it can help in segmenting buildings into various efficiency categories based on a range of overlapping characteristics like energy consumption, building materials, and occupancy patterns. The flexibility of Fuzzy C-means clustering in handling overlapping and ambiguous data makes it a powerful tool in the nuanced field of green building analysis. However, it is important to note that the method can be computationally intensive and requires careful selection of the number of clusters and the fuzziness parameter. Despite these challenges, its robustness against noise and outliers and its ability to reveal the subtle structure in complex datasets make it a valuable technique in this domain [33].

- **Agglomerative clustering:**

Agglomerative clustering, a type of hierarchical clustering, plays a significant role in the analysis of green building data. This method is particularly useful for organizing complex and multi-dimensional data into clusters based on similarity. In the context of green building, agglomerative clustering can be employed to categorize buildings, materials, or environmental conditions into distinct groups, facilitating better decision-making for sustainable practices and energy efficiency. The algorithm for agglomerative clustering operates in a bottom-up approach. It begins by treating each data point as a single cluster and then iteratively merges the closest pairs of clusters. This merging process continues until all the data points are clustered into a single group, or until a desired number of clusters is reached. The result of this process is a tree-like structure called a dendrogram, which visually represents the arrangement and the hierarchical relationship between the clusters [34].

Agglomerative clustering is a hierarchical clustering technique that follows a series of key steps to group data points based on their similarity. The initial step in this process is preparing the data, which involves organizing it into a matrix format where rows are observations and columns are variables. Standardizing the data is a critical part of this step, ensuring consistent scaling across all variables.

Once the data is prepared, the next step is computing similarity. This involves calculating the similarity between every pair of observations, typically using distance measures such as Euclidean or Manhattan distance. The outcome of these calculations is a distance matrix, which serves as the foundation for the clustering process. The third step involves determining the linkage criteria, a method used to measure the distance between clusters. This choice significantly impacts the shape and composition of the resulting clusters. The common methods of linkage include complete linkage, which maximizes the distance between elements in different clusters; single linkage, which minimizes this distance; average linkage, which considers the average distance between all pairs of elements in any two clusters; and Ward's method, which aims to minimize the variance within each cluster [35].

Building the dendrogram is the subsequent step. This process starts with considering each data point as an individual cluster. The algorithm then iteratively merges the closest pairs of clusters. This merging continues until all points are grouped into a single cluster. Each merge is visually represented in the dendrogram, providing a tree-based representation of the hierarchical relationships among the clusters. The final step is cutting the dendrogram, which is done to determine the number of clusters. This decision is based on the desired level of granularity in the clustering and the specific requirements of the analysis. The point at which the dendrogram is cut will dictate the final number of clusters, with each branch representing a separate cluster. This step is crucial as it determines the final output of the clustering process [36].

In green building analysis, agglomerative clustering can be used to find patterns in building energy usage, group similar types of buildings for benchmarking, or cluster regions based on environmental impact. Its flexibility in handling different types of data and its intuitive hierarchical structure make it a valuable tool for uncovering insights from complex green building datasets. However, it is important to choose an appropriate linkage method and to interpret the dendrogram carefully to ensure meaningful clustering results [37].

- **Self-organizing maps (SOM):**

Teuvo Kohonen's unsupervised neural network self-organizing maps (SOMs) are ideal for green building analysis and visualization of complicated, high-dimensional data. Competitive learning in SOMs depicts multi-dimensional data onto a two-dimensional grid by competing neurons to represent the input data. This unique trait makes SOMs ideal for discovering patterns and linkages in green building data, which generally includes many interrelated factors. SOMs can visualize energy usage, building materials' environmental effect, and occupant behavior in green building analysis. By entering energy use, material kinds, and other environmental elements into a SOM, one may see clusters or patterns on the map. Each SOM grid neuron represents a data pattern or combination of characteristics. This visualization helps identify buildings or activities with comparable energy efficiency or environmental effect, supporting sustainable practice decision-making [38].

Green building analysis uses SOMs by initializing them with random weights, giving input data to them, and modifying their neuron weights through competitive learning. As it trains, the SOM organizes itself, with distinct map sections representing input data patterns. This topological map places comparable data points near together and different points far apart. Complex multi-dimensional data is simpler to grasp with this map layout. SOMs minimize dimensionality while keeping topological and metric linkages, making them useful for green building studies. The complicated and non-linear interactions between components in green building data make this feature important. For instance, SOMs can split buildings into energy efficiency groups, identify outlier buildings with odd energy consumption trends, or show environmental factor correlations.

SOMs are useful in green building analysis when the map size, learning rate, and neighborhood function are selected properly. Understanding SOM findings takes knowledge since the map's layout must be analyzed to reveal data patterns. Although SOMs have problems, their capacity to give visual insights and simplify complicated data makes them a strong tool for green building energy efficiency and sustainability [39].

- **Spectral clustering:**

Spectral clustering has become a valuable tool in green building analysis, offering a unique approach to dissecting complex data sets. By leveraging the eigenvalues of the similarity matrix of data, it excels in identifying clusters that traditional methods might miss, particularly in datasets where relationships are non-linear or clusters are non-globular. This technique is particularly useful in segmenting buildings based on energy consumption patterns or environmental impact, even when these buildings vary significantly in other attributes like size or location. In the realm of sustainable urban planning and development, spectral clustering reveals intricate geospatial patterns and subgroup characteristics, guiding targeted sustainability initiatives and energy-efficient strategies. However, its effective application requires careful selection of parameters and an adept interpretation of results, as the underlying eigenvalues and eigenvectors can present complex insights into the data's structure. Despite these challenges, spectral clustering's ability to unravel complex relationships in data makes it an invaluable asset in advancing green building practices and sustainable development [40].

- **Affinity propagation:**

Affinity propagation is a unique clustering algorithm, distinct in its approach as it does not require pre-specifying the number of clusters. It operates by identifying exemplars among data points and forming clusters around these key points. Using a similarity matrix, typically based on the negative squared Euclidean distance, the algorithm iteratively exchanges two types of messages – "responsibility" and "availability" – between data points. These messages reflect the suitability of points to serve as exemplars and the appropriateness of choosing certain points as exemplars, respectively. The process converges when these messages stabilize, resulting in a set of clusters, each represented by an exemplar. In green building analysis, affinity propagation can effectively categorize buildings or materials by their energy efficiency or environmental impact, offering insights where the optimal number of clusters is not initially clear. However, the algorithm requires careful parameter tuning and interpretation to derive meaningful results [41].

Each clustering technique has its strengths and is chosen based on the characteristics of the green building dataset and the specific goals of the analysis. Through the application of these techniques, pattern recognition becomes a potent tool for understanding the diversity and complexity inherent in sustainable construction practices.

9.5 REGRESSION ANALYSIS FOR UNDERSTANDING ENERGY COST PATTERNS

Regression analysis is a quantitative data mining method applied to examine relationships between different variables within green building datasets. Utilizing techniques like linear regression or multiple regression, this method quantifies how changes in one variable influence another. In green building, regression analysis proves invaluable for assessing the impact of various factors, such as building design, occupancy patterns, or environmental conditions, on energy costs and overall building performance. The results provide stakeholders with a numerical understanding of these relationships, enabling them to make informed decisions aimed at improving energy efficiency and mitigating potential challenges [42].

Regression analysis serves as a powerful statistical method for understanding energy cost patterns within the realm of green building analysis. This technique allows stakeholders to quantitatively examine the relationships between various factors and energy costs, providing valuable insights into the determinants of energy consumption. The application of regression analysis involves several key components:

1. **Methodology:** Regression analysis employs statistical models to examine the relationship between a dependent variable (in this case, energy costs) and one or more independent variables (such as building characteristics, occupancy patterns, or environmental conditions). Common regression models include linear regression, multiple regression, and polynomial regression, each suited for different complexities in the data.
2. **Energy cost determinants:** Regression analysis helps identify the factors that significantly influence energy costs within green building datasets. These factors may include building size, insulation efficiency, the presence of renewable energy sources, or the implementation of energy-efficient technologies. By quantifying the impact of each variable, stakeholders gain a comprehensive understanding of the key determinants shaping energy cost patterns.
3. **Predictive modeling:** One of the primary advantages of regression analysis is its ability to create predictive models. These models can

forecast energy costs based on specified input variables. Stakeholders can use regression equations to estimate how changes in building features or occupancy behaviors may impact future energy expenditures. This predictive capability is invaluable for planning and implementing proactive energy efficiency measures.

4. **Data interpretation:** Regression analysis provides coefficients that quantify the strength and direction of the relationships between variables. Positive coefficients indicate a positive correlation, meaning an increase in the independent variable leads to an increase in energy costs. Negative coefficients signify an inverse relationship. By interpreting these coefficients, stakeholders can discern the magnitude and direction of the impact each variable has on energy costs.
5. **Variable selection:** Variable selection is a critical aspect of regression analysis. Through techniques like backward elimination or stepwise regression, stakeholders can refine the model by identifying the most influential variables. This process helps streamline the analysis, focusing on key factors that significantly contribute to the variability in energy costs.
6. **Model validation:** To ensure the reliability of the regression model, stakeholders engage in model validation. This involves assessing the model's performance on independent datasets or using techniques like cross-validation. Validated models provide confidence in their predictive capabilities, allowing stakeholders to make informed decisions based on the identified energy cost patterns.
7. **Continuous improvement:** Regression analysis fosters a culture of continuous improvement within green building practices. As new data becomes available or as building features evolve, stakeholders can adapt the regression model to reflect changing circumstances. This iterative approach ensures that energy cost patterns are continually analyzed and understood in light of evolving variables and technologies.

Regression analysis is a versatile and indispensable tool for understanding energy cost patterns in green building. By quantifying relationships, identifying key determinants, and creating predictive models, stakeholders can make data-driven decisions to optimize energy efficiency, reduce costs, and enhance the overall sustainability of built environments [43].

9.6 MACHINE LEARNING ALGORITHMS IN GREEN BUILDING ASSESSMENT

Machine learning algorithms, including decision trees, random forest, support vector machines (SVMs), and neural networks, enhance the sophistication of data mining in green building analysis. These algorithms contribute by developing predictive models for energy consumption patterns and recognizing complex relationships within datasets. Decision trees, for example, create models that predict outcomes based on historical data, while neural networks excel in identifying intricate, non-linear patterns. By leveraging machine learning, stakeholders can gain insights into future energy usage based on historical trends, providing a more adaptive and precise approach to green building analysis [44].

Machine learning algorithms have emerged as powerful tools in the assessment and optimization of green building practices. These algorithms leverage computational models to analyze vast and complex datasets, extracting valuable insights to enhance energy efficiency, sustainability, and overall performance in the built environment. Here's an exploration of the role of machine learning algorithms in green building assessment [45, 46]:

1. **Clustering algorithms:** Clustering algorithms play a pivotal role in green building assessment by categorizing buildings with similar characteristics into distinct groups. Methods such as k-means clustering and hierarchical clustering are applied to identify patterns in energy consumption, sustainability features, or other relevant factors. The benefit lies in the ability to tailor interventions and strategies based on specific building clusters, allowing for more targeted and efficient assessments. Clustering algorithms contribute to a nuanced understanding of the diverse building types within the green building landscape.
2. **Regression analysis:** Regression analysis is a fundamental machine learning technique employed to quantify relationships between various factors and energy performance in green building assessment. Techniques like linear regression and multiple regression help stakeholders understand the impact of specific variables on energy efficiency. This method provides a quantitative understanding of how changes in building design, occupancy patterns, or other factors influence energy consumption. Regression analysis aids in predictive modeling, allowing for informed decision-making and targeted improvements in building performance.
3. **Decision trees:** Decision trees are employed in green building assessment to create models that predict outcomes based on various criteria. These tree-like structures visually represent decision-making processes, making them accessible for stakeholders. Decision trees are beneficial for developing predictive models that guide decision-making on energy efficiency measures. Their intuitive nature allows stakeholders to understand and interpret the factors influencing green building performance effectively.
4. **Random forest:** Random forest is an ensemble learning method that leverages multiple decision trees to enhance the accuracy and reliability of predictions in green building assessment. By aggregating

outputs from various decision trees, random forest reduces overfitting and provides a more comprehensive understanding of complex patterns. This method is particularly useful when dealing with diverse and intricate relationships within green building datasets.

5. **Support vector machines (SVMs):** Support vector machines (SVMs) classify data points into categories based on a defined boundary. In green building assessment, SVM can be employed to categorize buildings into energy efficiency classes or predict the adoption of sustainable practices. SVM is effective in handling high-dimensional data, providing accurate classifications even in complex scenarios, making it a valuable tool for categorization tasks in green building assessment.
6. **Neural networks:** Neural networks mimic the structure and functioning of the human brain, making them powerful tools for analyzing complex relationships in green building datasets. These networks, composed of interconnected layers of nodes, excel in recognizing nonlinear patterns and relationships. In green building assessment, neural networks contribute sophisticated insights into the intricate dynamics of factors influencing building performance and sustainability.
7. **Ensemble methods:** Ensemble methods combine predictions from multiple machine learning models to improve the overall accuracy and reliability of green building assessments. By leveraging diverse algorithms, ensemble methods mitigate the weaknesses of individual models, resulting in more robust and trustworthy assessments. This approach enhances predictive performance, providing stakeholders with a more comprehensive and accurate understanding of green building dynamics.
8. **Reinforcement learning:** Reinforcement learning, characterized by learning through trial and error, is applied in green building assessment to optimize building automation systems for energy efficiency. The algorithm learns and adapts in real-time, making adjustments to building systems to improve efficiency over time. Reinforcement learning contributes to the development of adaptive systems that respond dynamically to changing conditions, enhancing energy performance in the long run.
9. **Time series analysis with LSTM (long short-term memory):** Time series analysis with long short-term memory (LSTM) is a specialized technique for analyzing sequential data. In green building assessment, LSTM is particularly useful for understanding temporal patterns in energy consumption. It captures long-term dependencies in time-series data, making it effective for predicting future energy usage or optimizing building operations based on historical patterns. LSTM contributes to a more sophisticated understanding of how energy consumption evolves over extended periods, aiding in strategic decision-making for sustainable building practices.

Machine learning algorithms in green building assessment offer a data-driven approach, enabling stakeholders to make informed decisions for optimizing energy efficiency, predicting performance, and advancing sustainability goals. These algorithms adapt and learn from data, providing valuable insights that contribute to the ongoing improvement of green building practices [47–51]

9.7 TEXT MINING FOR BUILDING DOCUMENTATION

Text mining for building documentation involves the extraction of valuable insights and knowledge from textual information associated with construction projects, building designs, and performance documentation. This process employs natural language processing (NLP) techniques to analyze and interpret unstructured text data. NLP, as a fundamental component of text mining, allows computers to understand, interpret, and generate human like language. In the realm of building documentation, NLP processes text data to extract pertinent information, identify patterns, and derive meaningful insights from unstructured textual content. This includes information extraction techniques that identify key entities, relationships, and attributes within the text, aiding in the creation of structured datasets from unstructured information. Sentiment analysis assesses the emotional tone expressed in building documentation, offering insights into stakeholder satisfaction and areas for improvement. Document summarization condenses lengthy documentation, facilitating efficient review for stakeholders and decision-makers. Text mining also enables trend analysis, keyword extraction, pattern recognition, and risk assessment within building documentation. By ensuring regulatory compliance and facilitating knowledge discovery, text mining proves to be a versatile and powerful approach that transforms unstructured text into actionable insights. Leveraging NLP techniques, stakeholders can extract, analyze, and apply valuable information embedded in textual data, thereby contributing to more informed decision-making and enhancing overall efficiency in the construction and building [52–55].

9.8 DATA-DRIVEN ENHANCEMENT OF GRADING SYSTEMS ASSESSMENTS FOR EVALUATING GREEN BUILDING PERFORMANCE

The data-driven enhancement of grading systems assessments for evaluating green building performance signifies a monumental shift toward precision and efficacy in sustainable construction. This strategic approach integrates advanced data analysis techniques to revolutionize traditional grading systems, offering a more nuanced and comprehensive evaluation of green building initiatives. By harnessing the power of data-driven insights, grading systems transition from subjective assessments to incorporate quantifiable metrics, enabling an objective and standardized evaluation of environmental, energy, and resource efficiency. The parameters considered in grading encompass a broad spectrum, including energy consumption, water usage, material sourcing, indoor air quality, and overall ecological impact. Informed by data-driven insights, these parameters allow for a holistic assessment that captures the multifaceted aspects of a building's sustainability. The integration of machine learning algorithms enhances the accuracy of assessments by identifying patterns, trends, and correlations within large datasets. This not only refines the grading process but also facilitates targeted improvements in areas demonstrating suboptimal performance. The continuous monitoring and feedback loop, benchmarking against industry standards, incorporation of occupant feedback, dynamic adaptation to evolving standards, and transparent reporting through visualization tools foster collaboration among stakeholders. In conclusion, the data-driven enhancement of grading systems for green building assessments signifies a paradigm shift toward more informed, precise, and dynamic evaluations, ultimately driving advancements in sustainable construction practices [56].

9.9 TRANSFORMATIVE IMPACT OF DATA MINING AND DATA ANALYTICS IN GREEN BUILDING AND SUSTAINABLE CONSTRUCTION FOR A RESILIENT BUILT ENVIRONMENT

The transformative impact of data mining and data analytics in green building and sustainable construction is instrumental in creating a resilient built environment. These technologies empower stakeholders to harness valuable insights from vast datasets, leading to informed decision-making, optimized resource utilization, and the advancement of sustainable practices [57]. A detailed exploration of the transformative impact of data mining and data analytics in this context:

1. **Data-driven decision-making in green building: Leveraging insights for strategic choices**

   In the realm of green building, data-driven decision-making has emerged as a transformative force, revolutionizing how stakeholders approach strategic choices. Through the application of data mining and analytics, valuable insights are extracted from diverse datasets related to energy consumption, materials usage, and environmental impact. This wealth of information empowers decision-makers to base their choices on evidence, fostering a more informed and strategic approach to green building design, construction, and operation.

2. **Resource optimization in sustainable construction: Harnessing data analytics for efficiency**
   Data analytics plays a pivotal role in optimizing resource utilization in sustainable construction practices. By analyzing historical and real-time data, stakeholders can identify patterns and trends that inform more efficient resource allocation. This includes the judicious use of energy, water, and materials. The result is a more streamlined and eco-friendly construction process that reduces environmental impact while maximizing efficiency and sustainability.
3. **Predictive modeling for resilient building performance: Anticipating challenges through data insights**
   Predictive modeling, fueled by data insights, stands as a cornerstone for achieving resilient building performance. By examining historical data, stakeholders can develop models that forecast future challenges and opportunities. This proactive approach allows for the anticipation of potential issues, enabling timely interventions and the optimization of building operations. The result is a built environment that is not only sustainable but also resilient in the face of evolving conditions.
4. **Adaptive building strategies: Real-time adjustments for dynamic sustainability**
   In the context of sustainable construction, the concept of adaptive building strategies is empowered by real-time adjustments facilitated through data analytics. Continuous monitoring of performance metrics, coupled with dynamic data feeds from sensors and building systems, enables buildings to adapt to changing conditions. This adaptability ensures that sustainable strategies remain effective over time, aligning with evolving environmental considerations, occupant needs, and operational requirements.
5. **Transformative impact on green building certification: Streamlining compliance through data analytics**
   The transformative impact of data analytics extends to the realm of green building certification, streamlining the often-complex process of documentation and verification. By systematically collecting and analyzing data relevant to sustainability criteria, stakeholders can navigate the certification landscape more efficiently. This not only ensures compliance with established standards but also expedites the certification process, providing a tangible demonstration of a building's commitment to sustainable practices.

The transformative impact of data mining and data analytics in green building and sustainable construction is multifaceted. It extends from improving decision-making and resource efficiency to enabling adaptive design and operation, fostering resilience, and promoting collaboration among stakeholders. These technologies play a pivotal role in shaping a resilient and sustainable built environment for the present and future [58–61].

9.10 CONCLUSION AND FUTURE DIRECTIONS

In conclusion, the integration of data-driven decision-making in green building practices marks a transformative shift in the industry. By leveraging insights from data mining and analytics, stakeholders can make informed choices, fostering a more sustainable and efficient built environment. This approach not only enhances the decision-making process but also contributes to the overall success of green building initiatives.

- The journey toward sustainable construction is significantly enriched by the optimization of resource efficiency through data analytics. By harnessing the power of data insights, stakeholders can achieve a more judicious use of resources such as energy, water, and materials. This not only reduces environmental impact but also sets the stage for a more resilient and resource-efficient future in the realm of sustainable construction.
- The application of predictive modeling, driven by data insights, emerges as a key factor in achieving resilient building performance. Anticipating challenges through historical data analysis allows stakeholders to proactively address issues and optimize operations. As a result, buildings become more adaptable, capable of withstanding evolving conditions, and ensuring sustained sustainability over the long term.
- The concept of adaptive building strategies, facilitated by real-time adjustments through data analytics, concludes as a cornerstone of dynamic sustainability. The continuous monitoring of performance metrics and dynamic data feeds empower buildings to adapt to changing conditions seamlessly. This adaptability ensures that sustainable strategies remain effective, aligning with evolving environmental considerations and operational needs.
- The transformative impact of data analytics extends to the certification landscape, streamlining compliance and verification processes. Through systematic data collection and analysis, green building stakeholders can navigate certification requirements more efficiently. This not only expedites the certification process but also serves as a testament to a building's commitment to sustainable practices.

9.10.1 Future directions of data-driven green building research

Looking ahead, the future directions of data-driven green building research hold immense promise. The ongoing evolution of data analytics technologies, machine learning, and artificial intelligence will further refine our understanding of sustainable construction. Future research should focus on developing more advanced predictive models, expanding the scope of adaptive strategies, and exploring novel applications of data analytics in addressing emerging challenges. By embracing these future directions, the green building industry can continue to lead in creating a sustainable, resilient, and environmentally conscious built environment.

REFERENCES

1. H. Son and C. Kim, "Early prediction of the performance of green building projects using pre-project planning variables: Data mining approaches," *J. Clean. Prod.*, vol. 109, pp. 144–151, 2015, doi: 10.1016/j.jclepro.2014.08.071.
2. Y.-K. Juan and P.-H. Lee, "Applying data mining techniques to explore technology adoptions, grades and costs of green building projects," *J. Build. Eng.*, vol. 45, p. 103669, 2022, doi: 10.1016/j.jobe.2021.103669.
3. A. Motaghifard, M. Omidvari, and A. Kazemi, "Forecasting of safe-green buildings using decision tree algorithm: Data mining approach," *Environ. Dev. Sustain.*, vol. 25, no. 9, pp. 10323–10350, 2023, doi: 10.1007/s10668-022-02491-4.
4. A. Saxena, A. N. Prajapati, G. Pant, C. S. Meena, A. Kumar, and V. P. Singh, "Water consumption optimization of hybrid heat pump water heating system," in *Lecture Notes in Mechanical Engineering*. Springer, Singapore, 2023, pp. 721–732, doi: 10.1007/978-981-99-1894-2_61.
5. V. P. Singh, S. Jain, and A. Kumar, "Establishment of correlations for the Thermo-Hydraulic parameters due to perforation in a multi-V rib roughened single pass solar air heater," *Exp. Heat Transf.*, vol. 35, no. 5, pp. 1–20, 2022, doi: 10.1080/08916152.2022.2064940.
6. X. Liu, M. Wang, and H. Fu, "Visualized analysis of knowledge development in green building based on bibliographic data mining," *J. Supercomput.*, vol. 76, no. 5, pp. 3266–3282, 2020, doi: 10.1007/s11227-018-2543-y.
7. Y. G. Ashwani Kumar, A. K. S. Gangwar, A. Kumar, C. S. Meena, V. P. Singh, N. Dutt, and A. Prasad, "Biomedical study of femur bone fracture and healing," in *Advanced Materials for Biomedical Applications*, 2022, 1st ed., A. Kumar, Y. Gori, A. Kumar, C. S. Meena, N. Dutt, Eds., Boca Raton, CRC Press, pp. 212–235.
8. A. Jeerage, B. Erwine, S. Mallory, and V. Agarwal, Predicting green building performance over time: Data mining untapped information in LEED, vol. 128, pp. 355–365, 2010, doi: 10.2495/ARC100301.
9. M. Jun and J. C. P. Cheng, "Analysis of the related credits in LEED green building rating system using data mining techniques," *Computing in Civil and Building Engineering*, pp. 1917–1924, 2014, doi: 10.1061/9780784413616.238.
10. O. B. Abounia, C. Qian, and J. Ruoyu, "Comparison of data mining techniques for predicting compressive strength of environmentally friendly concrete," *J. Comput. Civ. Eng.*, vol. 30, no. 6, p. 4016029, Nov. 2016, doi: 10.1061/(ASCE)CP.1943-5487.0000596.
11. J. Zhao and K. P. Lam, "Influential factors analysis on LEED building markets in U.S. East Coast cities by using Support Vector Regression," *Sustain. Cities Soc.*, vol. 5, pp. 37–43, 2012, doi: 10.1016/j.scs.2012.05.009.

12. I. Khan, A. Capozzoli, S. P. Corgnati, and T. Cerquitelli, "Fault detection analysis of building energy consumption using data mining techniques," *Energy Procedia*, vol. 42, pp. 557–566, 2013, doi: 10.1016/j.egypro.2013.11.057.
13. R. K. Shubham Srivastava, D. Verma, S. Thusoo, A. Kumar, and V. P. Singh, "Nanomanufacturing for energy conversion and storage devices," in *Nanomanufacturing and Nanomaterials Design: Principles and Applications*, 1st ed., S. Singh, S. K. Behura, A Kumar, K. Verma, Eds., Boca Raton, CRC Press, 2022, pp. 165–174, doi: 10.1201/9781003220602.
14. A. Datta, A. Kumar, A. Kumar, A. Kumar, and V. P. Singh, "Advanced materials in biological implants and surgical tools," in *Advanced Materials for Biomedical Applications*, 1st ed., N. D. Ashwani Kumar, Y. Gori, A. Kumar, C. S. Meena, Eds., Boca Raton, CRC Press, 2022, pp. 283–298.
15. A. K. Sharma, M. Nayal, S. Jain, and V. P. Singh, "Chapter 07 optimization techniques of solar thermal and hybrid energy systems," in *Highly Efficient Thermal Renewable Energy Systems Design, Optimization and Applications*, 1st ed., R. W. V. Verma, S. Thangavel, N. Dutt, A. Kumar, Eds. CRC Press, 2024, pp. 1–14.
16. K. Li, Y. Sun, D. Robinson, J. Ma, and Z. Ma, "A new strategy to benchmark and evaluate building electricity usage using multiple data mining technologies," *Sustain. Energy Technol. Assessments*, vol. 40, p. 100770, 2020, doi: 10.1016/j.seta.2020.100770.
17. V. P. Singh *et al.*, "Nanomanufacturing and design of high-performance piezoelectric nanogenerator for energy harvesting," in *Nanomanufacturing and Nanomaterials Design: Principles and Applications*, 1st ed., S. Singh, S. K. Behura, A Kumar, K. Verma, Eds., Boca Raton, CRC Press, 2022, pp. 241–272, doi: 10.1201/9781003220602.
18. M. Alam and M. R. Devjani, "Analyzing energy consumption patterns of an educational building through data mining," *J. Build. Eng.*, vol. 44, p. 103385, 2021, doi: 10.1016/j.jobe.2021.103385.
19. T. Sivasakthivel, V. Verma, R. Tarodiya, C. S. Meena, V. P. Singh, and R. Kumar, "Chapter 11: Analysis of optimum operating parameters for ground source heat pump system for different cases of building heating and cooling mode operations," in *Thermal Energy Systems: Design, Computational Techniques, and Applications*, 1st ed., A. Kumar, N. Dutt, V. Singh, C. Meena, Eds. CRC Press, 2023, pp. 183–207.
20. M. Sadeghi, R. Naghedi, K. Behzadian, A. Shamshirgaran, M. R. Tabrizi, and R. Maknoon, "Customisation of green buildings assessment tools based on climatic zoning and experts judgement using K-means clustering and fuzzy AHP," *Build. Environ.*, vol. 223, p. 109473, 2022, doi: 10.1016/j.buildenv.2022.109473.
21. J. Yang *et al.*, "k-Shape clustering algorithm for building energy usage patterns analysis and forecasting model accuracy improvement," *Energy Build.*, vol. 146, pp. 27–37, 2017, doi: 10.1016/j.enbuild.2017.03.071.
22. H. Plaisance, A. Blondel, V. Desauziers, and P. Mocho, "Hierarchical cluster analysis of carbonyl compounds emission profiles from building and furniture materials," *Build. Environ.*, vol. 75, pp. 40–45, 2014, doi: 10.1016/j.buildenv.2014.01.014.
23. A. R. Singh, S. K. Singh, and V. P. Singh, "Process parameters optimization of carbon nano tube based catalytic transesterification of algal oil," *Mater. Today Proc.*, no. xxxx, 2023, doi: 10.1016/j.matpr.2023.01.418.

24. Z. Li, Y. Jin, J. Zhao, X. Tan, D. Zhang, and J. Wei, "Differential clustering analysis of power grid green development based on hierarchical cluster method," *IOP Conf. Ser. Earth Environ. Sci.*, vol. 634, no. 1, 2021, doi: 10.1088/1755-1315/634/1/012077.
25. S. Jain, N. Kumar, V. P. Singh, S. Mishra, and N. K. Sharma, "Transesterification of algae oil and little amount of waste cooking oil blend at low temperature in the presence of NaOH," *Energies*, vol. 16, no. 1, pp. 1–12, 2023, doi: 10.3390/ en16031293.
26. S. Bawankar *et al.*, "Environmental impact assessment of lithium ion battery employing cradle to grave," *Sustain. Energy Technol. Assessments*, vol. 60, no. May, 2023, doi: 10.1016/j.seta.2023.103530.
27. X. Tu, C. Fu, A. Huang, H. Chen, and X. Ding, "DBSCAN spatial clustering analysis of urban "Production–Living–Ecological" space based on POI data: A case study of central urban Wuhan, China," *Int. J. Environ. Res. Public Health*, vol. 19, no. 9, 2022, doi: 10.3390/ijerph19095153.
28. V. P. Singh *et al.*, "Heat transfer and friction factor correlations development for double pass solar air heater artificially roughened with perforated multi-V ribs," *Case Stud. Therm. Eng.*, vol. 39, no. September, p. 102461, 2022, doi: 10.1016/j.csite.2022.102461.
29. G. Yao, C. Guo, Q. Ge, and M. Ait-Ahmed, "A practical building energy consumption anomaly detection method based on parameter adaptive setting DBSCAN," *Cogn. Comput. Syst.*, vol. 3, no. 2, pp. 154–168, Jun. 2021, doi: 10.1049/ccs2.12015.
30. V. P. Singh, A. Karn, G. Dwivedi, T. Alam, and A. Kumar, "Experimental assessment of variation in open area ratio on thermohydraulic performance of parallel flow solar air heater," *Arab. J. Sci. Eng.*, vol. 41, no. 12, pp. 1–17, 2022, doi: 10.1007/s13369-022-07525-7.
31. B. Yu, Z. Niu, and L. Wang, "Mean shift based clustering of neutrosophic domain for unsupervised constructions detection," *Optik (Stuttg).*, vol. 124, no. 21, pp. 4697–4706, 2013, doi: 10.1016/j.ijleo.2013.01.117.
32. V. P. Singh, S. Jain, A. Karn, A. Kumar, and G. Dwivedi, "Mathematical modeling of efficiency evaluation of double pass parallel flow solar air heater," *Sustainability*, vol. 14, no. 17, pp. 1–22, 2022, doi: 10.3390/su141710535.
33. A. Alghamdi, G. Hu, H. Haider, K. Hewage, and R. Sadiq, "Benchmarking of water, energy, and carbon flows in academic buildings: A fuzzy clustering approach," *Sustainability*, vol. 12, no. 11, 2020, doi: 10.3390/ su12114422.
34. N. Kaza, T. W. Lester, and D. A. Rodriguez, "The spatio-temporal clustering of green buildings in the United States," *Urban Stud.*, vol. 50, no. 16, pp. 3262–3282, May 2013, doi: 10.1177/0042098013484540.
35. G. S. Rathore, S. Rathor, A. Nazeer, A. Karn, and V. P. Singh, "Recent progress in solar dryers using phase changing material: A review," *Int. J. Energy Resour. Appl.*, vol. 2, no. 1, pp. 57–77, 2023, doi: 10.56896/ IJERA.2023.2.1.005.
36. M. Saini, A. Sharma, V. P. Singh, S. Jain, and G. Dwivedi, "Solar thermal receivers A review," *Lect. Notes Mech. Eng.*, vol. II, pp. 1–25, 2022, doi: 10.1007/978-981-16-8341-1.
37. Y. Hong, C. I. Ezeh, H. Zhao, W. Deng, S.-H. Hong, and Y. Tang, "A target-driven decision-making multi-layered approach for optimal building retrofits

via agglomerative hierarchical clustering: A case study in China," *Build. Environ.*, vol. 197, p. 107849, 2021, doi: 10.1016/j.buildenv.2021.107849.
38. X. Zheng *et al.*, "A factor analysis and self-organizing map based evaluation approach for the renewable energy heating potentials at county level: A case study in China," *Renew. Sustain. Energy Rev.*, vol. 165, p. 112597, 2022, doi: 10.1016/j.rser.2022.112597.
39. A. Abdelaziz, V. Santos, M. S. Dias, and A. N. Mahmoud, "A hybrid model of self-organizing map and deep learning with genetic algorithm for managing energy consumption in public buildings," *J. Clean. Prod.*, vol. 434, p. 140040, 2024, doi: 10.1016/j.jclepro.2023.140040.
40. K. Jonghoon, N. Hariharan, M. Soo-Young, W. K. O. Chong, and S. T. Ariaratnam, "Applications of clustering and isolation forest techniques in real-time building energy-consumption data: Application to LEED certified buildings," *J. Energy Eng.*, vol. 143, no. 5, p. 4017052, Oct. 2017, doi: 10.1061/(ASCE)EY.1943-7897.0000479.
41. H. Xia, X. Chen, and P. Guo, "A shadow detection method for remote sensing images using Affinity Propagation algorithm," in *2009 IEEE International Conference on Systems, Man and Cybernetics*, 2009, pp. 3116–3121, doi: 10.1109/ICSMC.2009.5346147.
42. X. Gui and Z. Gou, "Understanding green building energy performance in the context of commercial estates: A multi-year and cross-region analysis using the Australian commercial building disclosure database," *Energy*, vol. 222, p. 119988, 2021, doi: 10.1016/j.energy.2021.119988.
43. L. Zhang and R. Li, "Impacts of green certification programs on energy consumption and GHG emissions in buildings: A spatial regression approach," *Energy Build.*, vol. 256, p. 111677, 2022, doi: 10.1016/j.enbuild.2021.111677.
44. T. Mohd, S. Jamil, and S. Masrom, "Machine learning building price prediction with green building determinant," *IAES Int. J. Artif. Intell.*, vol. 9, no. 3, pp. 379–386, 2020, doi: 10.11591/ijai.v9.i3.pp379-386.
45. O. Alshboul, A. Shehadeh, G. Almasabha, R. E. Mamlook, and A. S. Almuflih, "Evaluating the impact of external support on green building construction cost: A hybrid mathematical and machine learning prediction approach," *Buildings*, vol. 12, no. 8, 2022, doi: 10.3390/buildings12081256.
46. M. Rajabi, J. M. Sardroud, and A. Kheyroddin, "Green standard model using machine learning: Identifying threats and opportunities facing the implementation of green building in Iran," *Environ. Sci. Pollut. Res.*, vol. 28, no. 44, pp. 62796–62808, 2021, doi: 10.1007/s11356-021-14991-3.
47. N. N. Barros and R. C. Ruschel, "Machine learning for whole-building life cycle assessment: A systematic literature review BT – Proceedings of the 18th international conference on computing in civil and building engineering," ICCCBE 2020. Lecture Notes in Civil Engineering, Springer, Cham., 2021, vol. 98, pp. 109–122, doi: 10.1007/978-3-030-51295-8_10.
48. V. P. Singh and S. Jain, "Economic analysis of a large scale solar updraft tower power plant," *Sustain. Energy Technol. Assessments*, vol. 58, p. 103325, 2023, doi: 10.1016/j.seta.2023.103325.
49. V. P. Singh, C. S. Meena, A. Kumar, and N. Dutt, "Double pass solar air heater: A review," *Int. J. Energy Resour. Appl.*, vol. 1, no. 2, pp. 22–43, 2022, doi: 10.56896/IJERA.2022.1.2.009.

50. V. P. Singh, S. Jain, and J. M. L. Gupta, "Analysis of the effect of perforation in multi-v rib artificial roughened single pass solar air heater: Part A," *Exp. Heat Transf.*, pp. 1–20, Oct. 2021, doi: 10.1080/08916152.2021.1988761.
51. M. Nayal, A. K. Sharma, S. Jain, and V. P. Singh, "Chapter 11 green hydrogen production methods, designs and applications," in *Highly Efficient Thermal Renewable Energy Systems Design, Optimization and Applications*, 1st ed., R. W. V. Verma, S. Thangavel, N. Dutt, A. Kumar, Eds. CRC Press, 2024, pp. 178–193.
52. X. Xiao, M. Skitmore, and X. Hu, "Case-based reasoning and text mining for green building decision making," *Energy Procedia*, vol. 111, pp. 417–425, 2017, doi: 10.1016/j.egypro.2017.03.203.
53. M. Nayal, A. K. Sharma, S. Jain, and V. P. Singh, "Chapter 06 design and modelling of solar, geothermal and hybrid energy systems.pdf," in *Highly Efficient Thermal Renewable Energy Systems Design, Optimization and Applications*, 1st ed., R. W. V. Verma, S. Thangavel, N. Dutt, A. Kumar, Eds. CRC Press, 2024, pp. 1–15.
54. V. P. Singh and G. Dwivedi, "Technical analysis of a large-scale solar updraft tower power plant," *Energies*, vol. 16, no. 1, p. 103325, 2023, doi: 10.3390/en16010494.
55. A. Kundu, A. Kumar, V. P. Singh, C. S. Meena, and N. Dutt, "Chapter 1: Introduction to thermal energy resources and their smart applications," in *Thermal Energy Systems: Design, Computational Techniques, and Applications*, 1st ed., A. Kumar, V. P. Singh, C. S. Meena, N. Dutt, Eds., Boca Raton, CRC Press, 2023, pp. 1–15.
56. Z. Tian, X. Zhang, S. Wei, S. Du, and X. Shi, "A review of data-driven building performance analysis and design on big on-site building performance data," *J. Build. Eng.*, vol. 41, p. 102706, 2021, doi: 10.1016/j.jobe.2021.102706.
57. C. Debrah, A. P. C. Chan, and A. Darko, "Artificial intelligence in green building," *Autom. Constr.*, vol. 137, p. 104192, 2022, doi: 10.1016/j.autcon.2022.104192.
58. M. Ahmadizadeh, M. Heidari, S. Thangavel, E. A. Naamani, M. Khashehchi, V. Verma, and A. Kumar, "Technological advancements in sustainable and renewable solar energy systems," In *Highly Efficient Thermal Renewable Energy Systems: Design, Design, Optimization and Applications*; V. Verma, S. Thangavel, N. Dutt, A. Kumar, R. Weerasinghe, Eds. CRC Press, 2024. Chapter 02. ISBN 9781032595641.
59. V. Verma, S. Thangavel, N. Dutt, A. Kumar, and R. Weerasinghe, Eds., *Highly Efficient Thermal Renewable Energy Systems: Design, Design, Optimization and Applications*. CRC Press, 2024. ISBN 9781032595641.
60. N. Dutt, A. Hedau, A. Kumar, M. K. Awasthi, S. Hedau, and C. S. Meena, "Thermo-hydraulic performance investigation of solar air heater duct having staggered D-shaped ribs: Numerical approach," *Heat Transfer*, pp. 1–31, 2024, doi: 10.1002/htj.22998.
61. M. K. Awasthi, A. Kumar, N. Dutt, and S. Singh, Eds., *Computational Fluid Flow and Heat Transfer: Advances, Design, Control and Applications*. CRC Press, 2024. ISBN 9781032603186.

Chapter 10

Energy policies and infrastructure requirements for efficient green buildings

Wasswa Shafik

10.1 INTRODUCTION

The emergence of sustainable technology has become crucial in addressing urgent environmental concerns and the imperative to save resources. Amidst growing apprehensions regarding climate change and the depletion of natural resources, there has been a heightened emphasis on sustainability across several sectors. Consequently, this has spurred the advancement and acceptance of novel technology (Ashwathi et al., 2023). The primary objective of these solutions is to mitigate ecological footprints, decrease carbon emissions, and foster the effective exploitation of resources. The expeditious advancement of renewable energy technologies, including solar, wind, hydroelectric, and geothermal sources, highlights a fundamental transformation in the production of energy (Baronin et al., 2023). Many advancements, including increased efficiency, cost reductions, and improved scalability, have driven the broad adoption of these systems. The knowledge obtained from these technologies emphasizes their crucial contribution to the worldwide shift toward sustainable and clean energy sources (Meddah et al., 2023). These technologies have the potential to significantly modify traditional energy systems while providing scalable and environmentally friendly alternatives.

The exemplification of sustainable construction practices can be observed through the exploration and exploitation of energy-efficient materials, including recycled steel, bamboo, cross-laminated timber, and aerogel-based insulation (Meddah et al., 2023). These materials demonstrate exceptional attributes in terms of durability, thermal efficiency, and minimizing environmental effects, hence bringing about a significant transformation in architectural and construction paradigms. The observations obtained from their implementation highlight the significant influence of material innovation in promoting environmentally friendly constructed environments shaping design strategies and construction practices toward the development of more sustainable and ecologically conscious structures (Binega Yemesegen & Memari, 2023). The incorporation of intelligent technologies into building infrastructure, including Internet of Things (IoT) sensors, artificial intelligence (AI)-powered systems, and automated controls,

 DOI: 10.1201/9781003496656-10

represents a significant transition toward more energy-efficient operations. These technological advancements provide the continuous monitoring of systems, the ability to foresee maintenance needs, and the optimization of energy usage. As a result, operational efficiency is improved, while waste is minimized (Assef & Mangold, 2022). The knowledge obtained from these breakthroughs demonstrates their significance in redefining the management of buildings, providing tools for decision-making based on data and systems that adapt to better energy efficiency and the comfort of occupants.

The increase in the adoption of electric mobility solutions signifies a notable advancement in the realm of environmentally friendly transportation. The progression of electric vehicle (EV) technology, in conjunction with improvements in battery capacity, charging infrastructure, and range, signifies a significant transition away from cars reliant on fossil fuels toward zero-emission alternatives (Bapat et al., 2022). The advancements highlight the significant impact that electric vehicles can have on mitigating greenhouse gas emissions, improving air quality, and restructuring the worldwide transportation sector. The implementation of circular economy principles has catalyzed technological advancements in the field of waste management. The technologies, including recycling, upcycling, and waste-to-energy conversion, prioritize the minimization of waste disposed in landfills and the implementation of closed-loop systems (Cooper et al., 2023). The observations obtained from these activities emphasize their crucial significance in facilitating resource efficiency, minimizing waste, and establishing sustainable material cycles. Figure 10.1 presents some identified differences between standard and energy-smart homes.

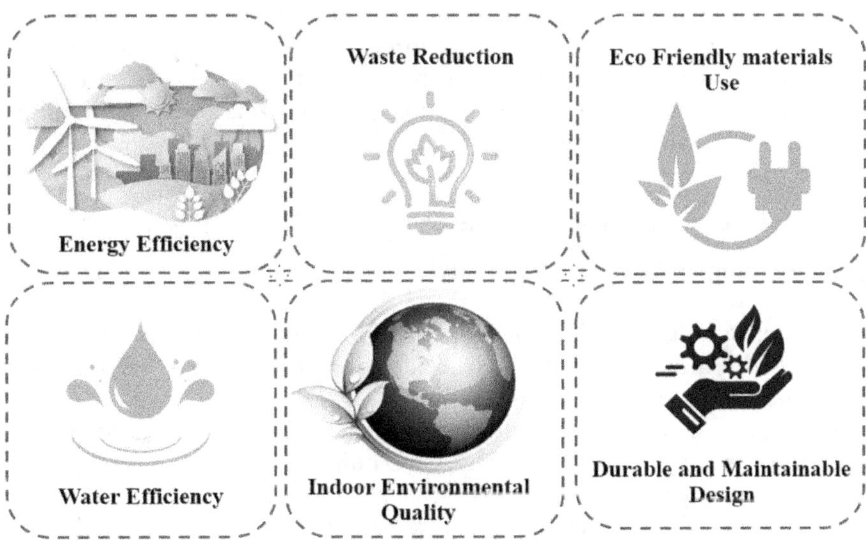

Figure 10.1 Standard home and energy smart home comparison

Water-saving technologies have seen significant advancements, encompassing several innovative solutions such as efficient irrigation systems, graywater recycling, and low-flow fixtures. These technologies aim to mitigate the urgent problem of water scarcity by advocating for responsible water management in diverse sectors (Kanagaraj et al., 2023). The insights derived from these solutions highlight their importance in the conservation of freshwater resources, reduction of water wastage, and promotion of sustainable water usage practices. Technological interventions in the agricultural sector, such as precision farming techniques, IoT-enabled monitoring, and data analytics, play a crucial role in promoting sustainable farming practices (Kayaci & Kanbur, 2023). The primary objective of these innovations is to enhance the efficiency of resource usage, reduce the use of chemical inputs, and address the environmental consequences. The observations derived from precision agriculture highlight its capacity to improve crop productivity, minimize resource inefficiencies, and promote ecologically sustainable agricultural practices (Indunil et al., 2023).

The development of carbon capture and storage (CCS) technology signifies a crucial frontier in addressing the challenges posed by climate change. The objective of these technologies is to effectively capture carbon dioxide emissions originating from industrial operations and power generation, subsequently ensuring their secure storage or usage (Kovalchuk & Stepnov, 2023). The analysis of CCS advancements reveals its capacity to effectively reduce greenhouse gas emissions, serving as a transitional mechanism toward a future characterized by reduced carbon emissions. Furthermore, CCS technologies can supplement endeavors aimed at attaining climate neutrality (Rathnasiri et al., 2021). These observations encompass a wide range of sustainable technologies that cover several sectors, demonstrating the complex and diversified nature of innovation in addressing global environmental issues. The combined effects of these technologies offer a promising trajectory toward a future characterized by enhanced sustainability, resilience, and environmental awareness (Harmathy, 2021).

10.1.1 The chapter contribution

The following insights are heightened and contributed to as summarized:

- Energy policies include a range of incentives, such as tax credits and rebates, aimed at promoting the adoption of renewable energy sources and energy-efficient technologies.
- Policies are implemented to enforce rigorous building codes that require the adoption of sustainable building methods, thereby guaranteeing adherence to energy efficiency standards.
- These projects effectively mitigate greenhouse gas emissions through the promotion and adoption of renewable energy sources and environmentally sustainable products.

- The implementation of infrastructure prerequisites, like the integration of sophisticated insulation and smart technologies, has been found to result in a decrease in operational expenses and an augmentation in property valuations.
- Policies play a crucial role in promoting innovation within the realm of green building technologies, hence stimulating research and development efforts aimed at advancing more efficient infrastructure solutions.
- The incorporation of various infrastructure components enhances the resilience of buildings, enabling them to survive environmental adversities and reduce their reliance on resources.
- Case studies serve as valuable illustrations of the pragmatic feasibility and efficacy of incorporating policies and infrastructure to attain energy-efficient buildings.
- These efforts exemplify a worldwide trend toward sustainable urban development, establishing standards for construction techniques that prioritize environmental responsibility on a global scale.

10.1.2 The chapter organization

The rest of this study is arranged as follows. Section 10.2 demonstrates an overview of energy policies and their impact on green buildings and infrastructure requirements for energy-efficient green buildings. Section 10.3 presents energy policies for green buildings, a description of relevant energy policies at national and local levels, their goals, and their impact on the construction and operation of green buildings. Section 10.4 details the infrastructure requirements. It further details a deep discussion of indispensable infrastructure elements for energy-efficient green buildings, including smart technologies, renewable energy systems, insulation, and water conservation features, among others. Section 10.5 presents some selected case studies specific to green building projects that showcase the successful integration of energy policies and infrastructure requirements. Section 10.6 illustrates the future directions of energy and infrastructure policies and discusses the consequences of the findings for energy policies and infrastructure development and future trends of this approach to sustainable green building. Finally, Section 10.7 presents some lessons learned from the chapter and the conclusion.

10.2 AN OVERVIEW OF ENERGY POLICIES

Energy policies function as overarching frameworks that governments enforce to regulate and provide incentives for the adoption of sustainable energy practices within the building industry (Ashwathi et al., 2023). These policies comprise a variety of measures, such as incentives, laws, and standards, that are intended to facilitate the incorporation of renewable energy sources and energy-efficient technologies in the design, construction, and operation of buildings (Baronin et al., 2023). Frequently, governmental entities employ tax incentives, grants, and subsidies as a means to promote the widespread adoption of solar panels, wind turbines, energy-efficient appliances, and sustainable building materials.

Furthermore, the implementation of rigorous building codes and certifications plays a crucial role in enforcing the establishment of minimum energy efficiency standards for both new constructions and restorations. These policies involve a range of aspects, including fiscal incentives such as tax credits, rebates, and subsidies, that aim to promote the adoption of renewable energy technology like solar panels, wind turbines, and geothermal systems (Meddah et al., 2023). Moreover, these regulations encompass rigorous building rules and standards that enforce the imposition of minimum energy efficiency prerequisites for both new constructions and repairs. Certain policies also encompass renewable energy mandates, with the objective of augmenting the percentage of energy obtained from environmentally sustainable sources within a specified period (Binega Yemesegen & Memari, 2023). As a result, these regulations have a substantial influence on the design and operational aspects of green buildings.

Infrastructure requirements encompass the tangible components and interconnected systems within structures that are designed to enhance energy efficiency and promote sustainability. The criteria encompass a wide array of elements, which consist of renewable energy systems such as solar panels or geothermal heating, advanced insulation materials aimed at enhancing energy retention, intelligent building automation systems that optimize energy consumption, and water conservation technologies (Assef & Mangold, 2022). Energy-efficient green buildings attempt to decrease dependence on conventional energy sources, mitigate waste generation, and establish a more ecologically sustainable and economically advantageous operational framework by integrating these components into the design and construction of the structure (Bapat et al., 2022). Every constituent aspect of green buildings plays a pivotal role in augmenting their total energy efficiency (presented in Figure 10.2) and environmental footprint, hence adding to their enduring sustainability and ability to withstand environmental adversities.

10.2.1 Impact on green buildings

The policies in question have a far-reaching influence that transcends beyond mere economic advantages, encompassing substantial environmental and societal benefits. In addition to driving innovation within the renewable energy industry, they also contribute to economic growth by facilitating the establishment of employment opportunities within green businesses (Ashwathi et al., 2023). Moreover, through the reduction of greenhouse

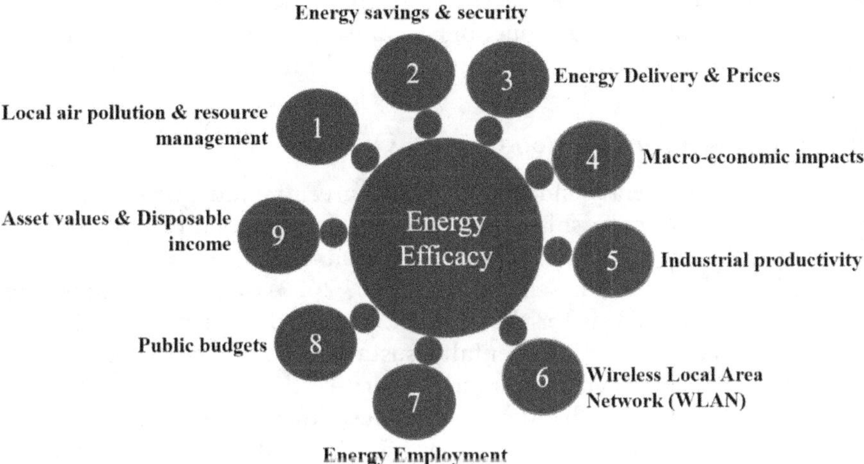

Figure 10.2 Energy efficacy

gas emissions and the decrease in dependence on non-renewable energy sources, these policies make a significant contribution toward the mitigation of climate change, the enhancement of air quality, and the improvement of energy security.

10.2.1.1 Renewable energy incentives

Governments provide financial incentives, such as tax credits and grants, with the aim of promoting the integration of renewable energy sources, such as solar panels and wind turbines, inside green building designs. The provision of these incentives effectively mitigates the initial investment expenses, hence enhancing the accessibility and financial feasibility of sustainable technology for builders and property owners (Baronin et al., 2023). This policy effectively mitigates dependence on traditional energy sources and diminishes carbon emissions in buildings, fostering a more sustainable energy framework through the promotion of renewable energy integration.

10.2.1.2 Energy-efficient building codes and standards

The implementation of rigorous building regulations and energy efficiency standards necessitates the adherence to particular performance criteria for both new constructions and restorations. The laws necessitate the utilization of energy-efficient appliances, lighting fixtures, insulating materials, and heating, ventilation, and air conditioning (HVAC) systems (Meddah et al., 2023). The adherence to compliance measures guarantees that green buildings function at elevated levels of energy efficiency, leading to a drop in energy consumption, a reduction in utility expenses, and mitigation of the environmental consequences throughout the lifespan of the building.

10.2.1.3 Green certification programs

Governments frequently endorse or enforce green building certification initiatives similar to leadership in energy and environmental design (LEED) or building research establishment environmental assessment method (BREEAM) (Binega Yemesegen & Memari, 2023). These programs offer guidelines and standards for the evaluation and acknowledgment of buildings that adhere to environmentally sustainable practices. Through the implementation of incentives or mandatory certification, governing bodies play a crucial role in facilitating the acceptance and implementation of sustainable construction methods (Assef & Mangold, 2022). This approach serves to enhance consciousness regarding environmentally friendly building practices while simultaneously establishing benchmarks and guidelines for the design, construction, and operation of green buildings within the industry.

10.2.1.4 Net zero energy building mandates

Certain governments implement regulations or objectives to promote the construction of net zero energy buildings. These architectural constructions are capable of producing an equivalent amount of energy to the amount they consume, typically achieved by the integration of on-site renewable energy sources in conjunction with extremely efficient building design and technologies (Bapat et al., 2022). The implementation of mandates aimed at achieving net zero energy not only serves as a catalyst for innovation in sustainable construction practices but also yields substantial reductions in environmental impact by promoting energy self-sufficiency (Cooper et al., 2023). Consequently, these mandates play a crucial role in stimulating the development of a more sustainable energy system.

10.2.1.5 Energy performance contracts

Governments have the potential to participate in energy performance contracting, a collaborative effort with private enterprises aimed at retrofitting existing buildings to achieve improved energy efficiency. These contractual agreements serve to guarantee that the costs incurred from implementing enhancements, such as updated lighting or HVAC systems, are gradually repaid by the energy savings generated over a certain period (Kanagaraj et al., 2023). This approach serves the dual purpose of rejuvenating preexisting structures and diminishing energy usage and operational expenses, all the while being in accordance with sustainability objectives.

10.2.1.6 Feed-in tariffs for renewable energy production

Feed-in tariffs (FITs) are a policy mechanism that ensures producers of renewable energy get consistent and extended compensation for the electricity they supply to the grid. The implementation of tariffs catalyzes the advancement of renewable energy infrastructure, thereby motivating property owners to make investments in solar panels or wind turbines for their respective properties (Kayaci & Kanbur, 2023). FITs incentivize the proliferation of decentralized renewable energy generation by providing financial rewards, hence encouraging the incorporation of environmentally friendly technologies into structures and diminishing aggregate carbon emissions.

10.2.1.7 Green building tax credits

Governments offer tax incentives to developers or property owners that undertake the construction or renovation of buildings in accordance with designated green building criteria. Tax incentives provide significant financial advantages for the implementation of sustainable construction practices (Ashwathi et al., 2023). Investments in energy-efficient technology, sustainable materials, and designs are actively promoted, thereby stimulating the expansion of green building efforts and mitigating the environmental consequences associated with construction projects.

10.2.1.8 Public-private partnerships for sustainable infrastructure

The establishment of partnerships between governmental agencies and private organizations facilitates the development of sustainable infrastructure projects. These collaborations utilize resources, knowledge, and financial support from both the public and private sectors to establish extensive programs for environmentally friendly construction (Baronin et al., 2023). Through the collaborative consolidation of resources and specialized knowledge, these alliances facilitate the advancement of novel ideas, expedite the integration of ecologically friendly technology, and establish frameworks for the implementation of expansive, sustainable infrastructure projects that yield positive outcomes for both the local community and the surrounding ecosystem depending on building material presented in Figure 10.3.

10.2.2 Infrastructure requirements for energy-efficient green buildings

The infrastructure prerequisites for energy-efficient green buildings involve a diverse range of components and technologies that jointly contribute to their sustainability and diminished environmental footprint. These

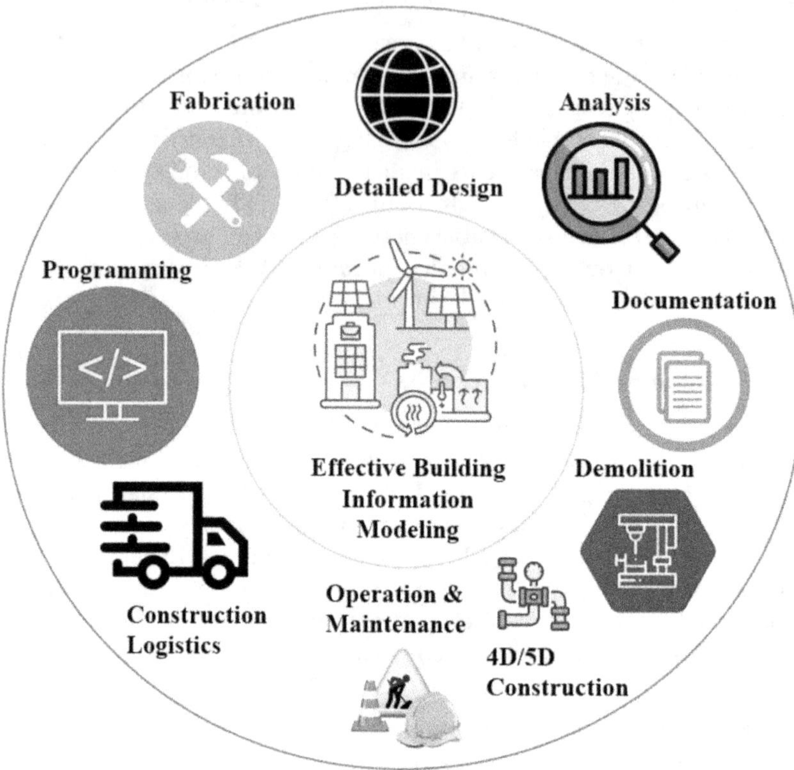

Figure 10.3 Energy efficiency optimization in operating building environment

stipulations are of utmost importance in augmenting energy efficiency, mitigating resource utilization, and fostering a more sustainable constructed milieu.

10.2.2.1 Renewable energy systems

The incorporation of renewable energy systems, including solar panels, wind turbines, and geothermal sets, is a fundamental aspect of energy-efficient green building infrastructure. Solar panels are capable of capturing and utilizing sunlight, wind turbines transform the energy present in the wind, and geothermal systems are designed to extract and utilize the Earth's consistent temperature. These systems have a substantial impact on reducing dependence on conventional power grids and fossil fuels, thereby mitigating carbon emissions and promoting the establishment of a sustainable energy infrastructure (Meddah et al., 2023). Strategically positioned and customized to optimize energy generation in accordance with the specific characteristics of the building's surroundings, these systems offer the potential for substantial cost savings by reducing utility expenditures and potentially enabling the injection of surplus energy into the power grid (Baronin et al., 2023). Consequently, they facilitate the development of a more self-reliant and environmentally sustainable energy framework.

10.2.2.2 Advanced insulation and building materials

Energy-efficient green buildings are established on the foundation of advanced insulation materials, efficient windows, and sustainable building materials. The utilization of high-performance insulation effectively mitigates heat transfer, hence diminishing the necessity for continuous heating or cooling (Meddah et al., 2023). Windows that are equipped with low-emissivity coatings or numerous panes are designed to enhance energy efficiency by effectively retaining heat indoors during the winter season and preventing excessive heat from entering during the summer months. The utilization of sustainable materials not only serves to mitigate conservational impact but also contributes to the augmentation of durability and energy efficiency (Ashwathi et al., 2023). When these materials are selected and utilized with care, they can effectively provide thermal comfort and simultaneously lead to substantial reductions in energy requirements for heating, cooling, and lighting. As a result, this contributes to decreased operational expenses and a reduced environmental impact throughout the lifecycle of the building.

10.2.2.3 Intelligent building automation systems

Intelligent building automation systems, encompassing smart thermostats, sensors, and control systems, have brought about a transformative impact on energy use within buildings. These technologies are responsible for the regulation of heating, cooling, lighting, and other energy-consuming functions, taking into account factors identical to occupancy and ambient conditions (Bapat et al., 2022). By adaptively responding to immediate requirements, they effectively reduce energy inefficiency and enhance utilization, resulting in significant energy conservation without compromising comfort or operational effectiveness. Smart technologies also provide remote monitoring and control, permitting building managers to optimize systems for optimal efficiency, hence leading to decreased operational expenses and less environmental footprint (Cooper et al., 2023).

10.2.2.4 Energy-efficient lighting

Energy-efficient lighting solutions, such as LED bulbs and daylight harvesting systems, play a crucial role in the development of sustainable building infrastructure. LED lighting is known to have a substantial impact on reducing electricity consumption when compared to traditional bulbs

(Kanagaraj et al., 2023). Additionally, LED lights have longer lifespans and provide superior illumination. Sunshine harvesting systems make use of natural light by modulating the intensity of artificial lighting in response to the amount of available sunshine, hence reducing energy consumption to a greater extent (Indunil et al., 2023). The use of these lighting solutions not only results in a reduction in power expenses but also leads to a decrease in maintenance expenditures and environmental harm by limiting energy usage and waste generation. This contributes to the overall enhancement of energy efficiency and sustainability in green buildings.

10.2.2.5 Water conservation technologies

Water conservation technologies, which include water-efficient fixtures, graywater recycling, and rainwater harvesting systems, are integral components of sustainable building infrastructure. Water-efficient fixtures, such as low-flow faucets and toilets, have been shown to substantially decrease water usage while maintaining their intended purpose and operation (Rathnasiri et al., 2021). Graywater recycling systems are designed to gather and process wastewater originating from sinks and showers, with the purpose of reusing it for activities such as irrigation or toilet flushing. This practice contributes to the conservation of freshwater resources (Harmathy, 2021). The practice of rainwater harvesting entails the collection and subsequent storage of rainwater for the use of non-potable applications, hence diminishing dependence on municipal water sources. These technologies reduce water wastage, decrease utility expenses, and alleviate pressure on water resources (Assef & Mangold, 2022). As a result, they play a significant role in promoting sustainable water management and boosting the overall sustainability of green buildings.

10.2.2.6 High-efficiency HVAC systems

Highly efficient heating, ventilation, and air conditioning (HVAC) systems play a crucial role as integral components of energy-efficient, sustainable buildings. These systems utilize sophisticated designs, components for energy recovery, and technologies with variable speeds to maximize the regulation of indoor air quality and temperature while simultaneously decreasing energy usage (Tamur & Erzaij, 2021). High-efficiency HVAC systems substantially decrease operational costs and minimize environmental impact by utilizing less energy for heating or cooling purposes. In addition, the company's utilization of modern filtration systems contributes to the enhancement of indoor air quality, hence promoting increased levels of occupant comfort and well-being (Yigit & Ozorhon, 2018). The implementation of these technologies not only guarantees enhanced energy efficiency but also makes a significant contribution toward creating a building environment that is more sustainable and comfortable.

10.2.2.7 Passive design strategies

Passive design solutions, such as appropriate orientation, utilization of natural ventilation, and implementation of shading systems, constitute essential elements of sustainable building infrastructure. Through the strategic alignment of buildings to optimize solar gain or natural ventilation, these approaches effectively mitigate the reliance on artificial heating or cooling systems (Lambertz et al., 2019). Effectively constructed shading solutions, such as overhangs or louvers, serve to reduce solar heat intake in the summer while permitting sunlight penetration during cooler months, thereby enhancing thermal comfort. Passive methods are employed to optimize temperature and lighting conditions by leveraging natural components (Nandhini et al., 2020). This approach leads to a notable decrease in energy requirements and an improvement in occupant comfort without heavy reliance on mechanical systems. Consequently, passive strategies play a crucial role in promoting energy efficiency and sustainable building design (Tamur & Erzaij, 2021).

10.2.2.8 Monitoring and control systems

These systems provide ongoing monitoring of energy consumption, allowing building managers to spot inefficiencies and make operational adjustments to optimize efficiency. These insights are derived from data analysis and are aimed at understanding energy patterns and trends. Their purpose is to support decision-making processes by providing information that can guide the implementation of energy-saving initiatives (Tamur & Erzaij, 2021). Moreover, the inclusion of remote-control capabilities facilitates the ability to make modifications and precise calibrations to building systems from a centralized position, guaranteeing the attainment of peak performance and enhanced energy efficiency. Monitoring and control systems play a crucial role in minimizing energy waste, decreasing operational expenses, and ensuring the long-term sustainability of building performance (Lambertz et al., 2019). These systems achieve this by furnishing actionable data and facilitating proactive management.

10.3 ENERGY POLICIES FOR GREEN BUILDINGS

Energy policies at both the national and local levels exert significant influence on the development and functioning of green buildings, with the objective of reducing environmental harm and advancing sustainability, as demonstrated within this section.

10.3.1 National energy policies

National governments frequently implement comprehensive energy policies that cover a wide array of laws, incentives, and targets aimed at fostering the use of energy-efficient practices and renewable energy sources in the design and operation of buildings. These policies encompass the implementation of building regulations that require minimum energy performance criteria for both new constructions and restorations, thereby assuring compliance with sustainable building practices (Nandhini et al., 2020). Moreover, governmental entities have the potential to provide economic inducements, such as tax credits, grants, or subsidies, in order to foster the adoption of sustainable energy solutions, including solar panels, wind turbines, and geothermal systems.

10.3.1.1 Energy efficiency standards and labeling programs

The main goal is to develop rigorous energy efficiency standards for appliances, lighting, and building materials, thereby guaranteeing that products adhere to the minimum energy performance criteria. Labeling schemes that are implemented alongside items provide consumers with information regarding the energy efficiency of those products, hence facilitating well-informed purchasing choices (Ashwathi et al., 2023). The ENERGY STAR program in the United States establishes energy efficiency criteria for a range of products, encompassing appliances, electronics, and lighting. The ENERGY STAR label is awarded to products that satisfy the specified standards, indicating their exceptional energy efficiency (Baronin et al., 2023). These requirements provide manufacturers with incentives to make energy-efficient items, hence resulting in a reduction in energy usage within buildings. The implementation of labeling systems provides consumers with the ability to make informed choices on more energy-efficient alternatives, resulting in reduced energy expenses and diminished carbon emissions (Baronin et al., 2023). The extensive implementation of efficient products has a beneficial effect on the total energy efficiency of buildings and appliances throughout the country, thereby making a significant contribution to the development of a more sustainable built environment.

10.3.1.2 Renewable energy incentives and subsidies

In order to promote the widespread utilization of renewable energy sources such as solar, wind, and geothermal systems, it is proposed to provide financial incentives, tax credits, and subsidies to property owners or developers. The Investment Tax Credit (ITC) in the United States offers a federal tax credit with the purpose of encouraging the adoption of solar energy systems in both residential and commercial settings. This credit serves as an incentive for the installation of solar panels (Meddah et al., 2023). The integration of renewable energy technologies into buildings is motivated by these incentives, which serve to decrease reliance on traditional power sources. Through the reduction of installation expenses and the enhancement of payback times, the utilization of sustainable technologies is expedited, resulting in a decrease in greenhouse gas emissions and a shift toward cleaner energy sources within the constructed environment (Ashwathi et al., 2023).

10.3.1.3 Building energy codes and standards

The implementation of mandatory regulations for the minimum energy performance standards for buildings is proposed. These regulations would encompass several aspects like insulation, HVAC systems, and lighting, with the aim of improving the overall energy efficiency of buildings (Baronin et al., 2023). The International Energy Conservation Code (IECC) establishes guidelines for the implementation of energy-efficient construction methods, outlining specific criteria for insulation, windows, lighting, and HVAC systems. Adherence to these norms guarantees that buildings conform to energy efficiency criteria, hence diminishing energy usage and operational expenditures (Meddah et al., 2023). These systems have a role in enhancing indoor comfort, reducing environmental impact, and achieving long-term energy savings, thus fostering the development of a more sustainable and energy-efficient building inventory, as depicted in Figure 10.4.

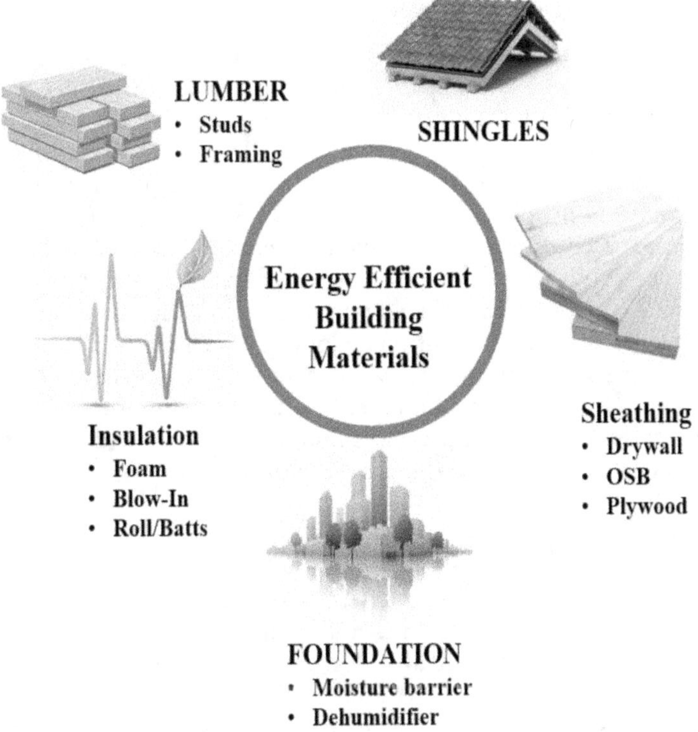

Figure 10.4 Energy-efficient building materials

10.3.1.4 Net zero energy building initiatives

The advancement of net zero energy buildings, which effectively achieve a balance between energy consumption and energy generation by employing a synergistic approach involving renewable energy systems and optimized design strategies, aimed at maximizing efficiency. The Bullitt Center, located in Seattle, Washington, serves as a prime example of a net zero energy structure (Assef & Mangold, 2022). It successfully integrates many sustainable technologies like solar panels, rainwater collecting systems, and energy-efficient design elements to attain complete self-reliance in energy usage. Initiatives promoting the development of net zero buildings facilitate the cultivation of novel ideas and stimulate the acceptance of cutting-edge technologies, thereby establishing standards for the construction of energy-efficient structures (Meddah et al., 2023). These structures exhibit sustainable practices, which lead to a substantial decrease in dependence on external energy sources. Moreover, they serve as a tangible example of the possibility of attaining energy self-sufficiency within the constructed environment.

10.3.1.5 Research and development funding for green technologies

The allocation of resources toward research and development (R&D) endeavors aimed at fostering innovation in the realm of green building technologies, materials, and practices, hence propelling the progress of sustainable solutions (Assef & Mangold, 2022). Government-funded initiatives such as the Advanced Research Projects Agency-Energy (ARPA-E) in the United States provide financial assistance for research and development endeavors focused on the advancement of energy-efficient technology and systems. The provision of financial resources for research and development (R&D) endeavors catalyzes the advancement and implementation of state-of-the-art energy-efficient solutions (Cooper et al., 2023). The investments in question give rise to innovations that result in the development of materials and technologies that are more efficient. These innovations drive progress in sustainable building practices and facilitate the wider acceptance of environmentally friendly alternatives within the construction industry.

10.3.2 Local energy policies

At the sub-national level, local governments and regional authorities frequently augment national policies by implementing localized legislation or programs that are specifically designed to tackle regional difficulties or capitalize on area benefits (Kanagaraj et al., 2023). These initiatives may encompass zoning rules that incentivize sustainable building design, land-use planning strategies that foster the development of energy-efficient communities, or incentive schemes aimed at promoting green building certifications. Local governments may also provide grants or subsidies to incentivize energy-efficient upgrades or facilitate public-private collaborations aimed at the development of sustainable infrastructure (Ashwathi et al., 2023). Through the establishment of alignment with national policies and the customization of strategies to meet specific local demands, these policies effectively cultivate an environment that is conducive to the advancement and functioning of environmentally sustainable structures within communities.

10.3.2.1 Municipal building performance standards

Implementing localized building performance standards that exceed national codes is essential for achieving superior energy efficiency and sustainability in both new constructions and restoration projects. The municipality of Boulder, located in the state of Colorado, has implemented the building efficiency Ordinance, a regulatory measure that mandates commercial and industrial structures to adhere to rigorous energy efficiency standards that surpass those outlined in the state's codes (Tamur & Erzaij, 2021). The aforementioned regional regulations advocate for the use of high-quality, energy-efficient designs and technologies, resulting in a substantial decrease in energy usage within buildings. The promotion of innovative systems and materials is advocated, resulting in cost reductions, less carbon emissions, and enhanced environmental sustainability at the local level (Lambertz et al., 2019).

10.3.2.2 Green building incentive programs

It provides economic incentives, such as financial rewards, tax exemptions, or streamlined approval procedures, to developers or homeowners who integrate environmentally sustainable building methods into their projects (Figure 10.5). The municipality of Austin, Texas, offers financial incentives and streamlined authorization processes for environmentally sustainable construction initiatives that surpass the energy code mandates set forth by the city (Nandhini et al., 2020). Incentive schemes serve as catalysts for promoting the widespread adoption of sustainable building approaches and renewable energy technology. The implementation of these initiatives facilitates the development of more energy-efficient structures, resulting in decreased energy consumption, heightened utilization of renewable energy resources, and a transition toward construction methods that prioritize environmental sustainability (Tamur & Erzaij, 2021).

10.3.2.3 Energy benchmarking and disclosure requirements

The implementation of a regulatory requirement for building owners to monitor and disclose energy consumption serves to foster transparency and

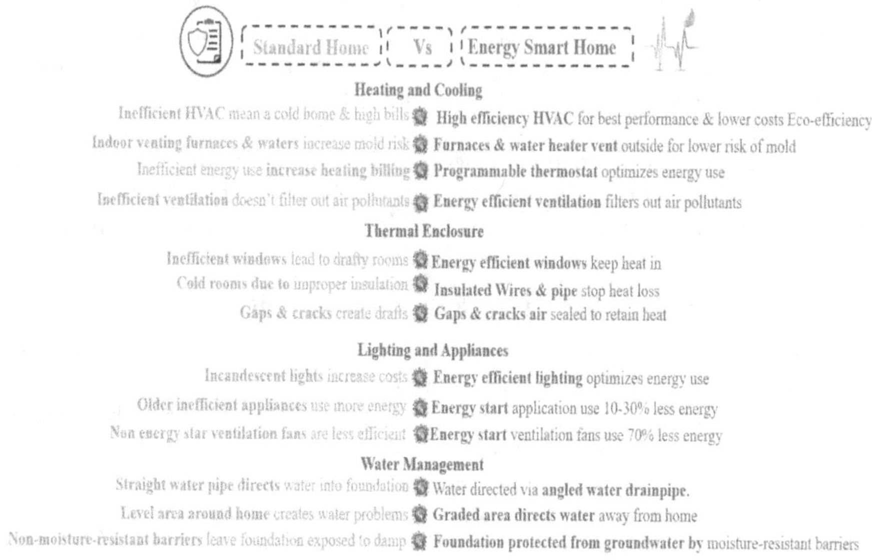

Figure 10.5 Comparison of smart home vs. energy smart home

enhance responsibility in relation to energy performance. Metropolitan areas such as New York and San Francisco have enacted energy benchmarking regulations that mandate building proprietors to submit yearly reports containing information on their energy consumption (Ashwathi et al., 2023). Through the act of divulging energy consumption data, these policies serve to enhance the consciousness of both building proprietors and inhabitants, thereby cultivating a climate conducive to the preservation of energy resources (Baronin et al., 2023). They enable the process of making decisions based on data, promoting the allocation of resources toward renovations that enhance energy efficiency, ultimately leading to a decrease in energy wastage and advancements in building performance.

10.3.2.4 Community renewable energy programs

Promoting the implementation of community-based renewable energy initiatives aimed at facilitating the provision of clean energy from local sources and mitigating reliance on centralized energy systems. Communities, exemplified as Burlington, Vermont, have implemented localized initiatives for renewable energy, wherein they allocate resources toward solar or wind ventures to supply power to public infrastructure and provide environmentally friendly energy alternatives to inhabitants (Meddah et al., 2023). Community-based renewable energy initiatives play a crucial role in enhancing local resilience, mitigating carbon footprints, and fostering active community engagement. Through the allocation of resources toward the development of clean energy generation, these programs actively contribute to the promotion of efficient and sustainable energy practices within the local community, thereby potentially reducing the dependence on fossil fuels (Binega Yemesegen & Memari, 2023).

10.3.2.5 Green infrastructure requirements for development

It is recommended that new buildings or redevelopment projects be mandated to integrate green infrastructure components, such as green roofs, permeable pavement, or rain gardens, in order to manage stormwater and enhance energy efficiency effectively (Assef & Mangold, 2022). Cities such as Portland, Oregon, enforce regulations that require the inclusion of green infrastructure elements in development initiatives, with the aim of promoting sustainability and mitigating adverse environmental effects. The incorporation of green infrastructure elements has been found to promote energy efficiency, mitigate heat island effects, and enhance stormwater management (Bapat et al., 2022). These policies promote the advancement of sustainable urban development, leading to the creation of communities that are more robust to environmental challenges, characterized by decreased energy use and improved preservation of natural resources.

10.4 INFRASTRUCTURE REQUIREMENTS

This section presents a detailed discussion of essential infrastructure elements for energy-efficient green buildings, like renewable energy systems, smart technologies, insulation, and water conservation features, among others.

10.4.1 Monitoring and control systems

Monitoring and control systems play a crucial role in the infrastructure of green buildings, as they provide real-time monitoring and optimization of energy consumption. These systems provide constant monitoring of energy consumption, allowing building managers to spot inefficiencies and make operational adjustments to optimize efficiency (Cooper et al., 2023). The provision of data-driven insights into energy patterns and trends enables the facilitation of informed decision-making for the implementation of energy-saving measures. The incorporation of remote-control capabilities enables the ability to make modifications and precise adjustments to building systems from a centralized position, guaranteeing the attainment of optimal performance and energy efficiency (Kanagaraj et al., 2023). Monitoring and control systems play a substantial role in mitigating energy waste, decreasing operational expenses, and upholding sustainable building performance through the provision of actionable data and facilitation of proactive management.

10.4.2 Passive design strategies

Passive design strategies, such as appropriate orientation, natural ventilation, and shading systems, constitute essential elements of sustainable building infrastructure. By deliberately aligning buildings to optimize solar exposure or facilitate natural airflow, these techniques mitigate the reliance on artificial heating or cooling systems (Kayaci & Kanbur, 2023). Optimal thermal comfort can be achieved by the use of well-designed shading solutions, such as overhangs or louvers, which effectively reduce solar heat gain during the summer season while permitting sunlight penetration during cooler months. Passive techniques are employed to harness natural factors for the purpose of regulating temperature and lighting (Indunil et al., 2023). By doing so, they effectively decrease energy requirements and improve occupant comfort without heavy reliance on mechanical systems. Consequently, these strategies play a crucial role in promoting energy efficiency and sustainability in building design.

10.4.3 High-efficiency HVAC systems

These systems utilize sophisticated designs, components for energy recovery, and technology for variable speed in order to maximize the control of indoor air quality and temperature while simultaneously reducing energy usage. High-efficiency HVAC systems offer a notable reduction in operational expenses and environmental effects by minimizing energy consumption for heating and cooling purposes (Kanagaraj et al., 2023). In addition, the utilization of modern filtration systems by the company contributes to the enhancement of indoor air quality, hence promoting increased levels of occupant comfort and improved health. The implementation of these technologies not only guarantees enhanced energy efficiency but also makes a significant contribution toward creating a building environment that is more sustainable and comfortable (Tamur & Erzaij, 2021).

10.4.4 Water conservation technologies

Water conservation technologies, which include water-efficient fixtures, graywater recycling, and rainwater harvesting systems, are integral components of sustainable building infrastructure. Water-efficient fixtures, such as low-flow faucets and toilets, have been shown to decrease water consumption while maintaining their performance effectively (Yigit & Ozorhon, 2018). Graywater recycling systems are designed to gather and process wastewater originating from sinks and showers, with the purpose of reusing it for activities such as irrigation or toilet flushing. This practice serves to preserve freshwater resources effectively. The practice of rainwater harvesting entails the collection and subsequent storage of rainwater for the use of non-potable applications, hence mitigating dependence on municipal water sources (Nandhini et al., 2020). These technologies reduce water wastage, decrease utility expenses, and alleviate pressure on water resources. As a result, they play a significant role in promoting sustainable water management and boosting the overall sustainability of green buildings.

10.4.5 Energy-well-organized lighting

Energy-efficient lighting technologies, for instance light-emitting diode (LED) bulbs and daylight harvesting systems, play a crucial role in the construction of environmentally sustainable buildings. LED lighting is known to exhibit a substantial decrease in electricity consumption when compared to conventional bulbs, in addition to providing extended lifespans and superior illumination (Ashwathi et al., 2023). Sunshine harvesting systems make use of natural light by modulating the intensity of artificial lighting in response to the amount of available sunshine, hence reducing energy consumption to a greater extent. The use of these lighting solutions not only results in a reduction in power expenses but also leads to decreased maintenance expenditures and a diminished environmental footprint by limiting energy usage and waste generation (Baronin et al., 2023). This contributes to the overall enhancement of energy efficiency and sustainability in green buildings.

10.4.6 Smart technologies and building automation systems

Intelligent building automation systems, encompassing advanced technologies such as smart thermostats, sensors, and control systems, have brought about a transformative impact on energy use within buildings. These technologies are responsible for the regulation of heating, cooling, lighting, and other energy-consuming functions, which are determined by factors identical occupancy and ambient conditions (Meddah et al., 2023). Through the dynamic adjustment to real-time demands, energy waste is minimized, and utilization is optimized, resulting in significant energy savings without compromising comfort or functionality. Smart technologies also provide remote monitoring and control, permitting building managers to optimize systems for optimal efficiency, hence leading to decreased operational expenses and less environmental footprint (Binega Yemesegen & Memari, 2023).

10.4.7 Advanced building materials and insulation

The utilization of advanced insulation materials, efficient windows, and sustainable building materials establishes energy-efficient green buildings.

High-performance insulation effectively mitigates heat transfer, hence decreasing the necessity for continuous heating or cooling (Harmathy, 2021). Windows that are equipped with low-emissivity coatings or numerous panes effectively enhance energy efficiency by retaining heat indoors during the winter season and preventing excessive heat from entering during the summer months. The utilization of sustainable materials such as bamboo or recycled steel not only serves to mitigate environmental consequences but also contributes to the improvement of durability and energy efficiency (Rathnasiri et al., 2021). When these materials are selected and utilized with care, they can effectively provide thermal comfort and lead to substantial reductions in energy consumption for heating, cooling, and lighting. As a result, this contributes to decreased operational expenses and a reduced environmental impact throughout the lifecycle of the building.

10.4.8 Renewable energy systems

The incorporation of on-site renewable energy sources such as solar panels, wind turbines, or geothermal systems plays a crucial role in strengthening the infrastructure of energy-efficient green buildings. Solar panels are capable of capturing and utilizing sunlight, wind turbines convert wind energy into usable forms, and geothermal systems are designed to extract and utilize the Earth's consistent temperature (Indunil et al., 2023). These systems have a substantial impact on reducing dependence on conventional electrical grids and fossil fuels, thereby mitigating carbon emissions and promoting the establishment of a sustainable energy provision (Kanagaraj et al., 2023). Strategically positioned and customized to optimize energy generation in accordance with the specific characteristics of the building's surroundings, these systems offer the potential for substantial cost reductions by significantly reducing utility expenses. Additionally, they potentially contribute surplus energy to the power grid, thereby promoting a more self-reliant and environmentally sustainable energy network (Bapat et al., 2022).

10.5 SELECTED CASE STUDIES

This section presents three in-depth investigations of specific green building projects that showcase the successful integration of energy policies and infrastructure requirements.

10.5.1 One Central Park, Sydney, Australia

One Central Park is an exemplary mixed-use property that effectively demonstrates cutting-edge sustainable design principles and seamless integration of advanced technologies. The building integrates many energy-efficient components, such as photovoltaic panels, high-efficiency lighting, a water recycling system, and abundant vegetation (Ashwathi et al., 2023). The laws implemented by Australia to promote environmentally sustainable building techniques are in accordance with the architectural principles of One Central Park. The project conforms to rigorous building requirements that prioritize energy efficiency and the use of sustainable materials. The system incorporates sustainable energy sources, utilizing photovoltaic panels to produce electricity on the premises (Meddah et al., 2023).

The infrastructure integrates advanced water recycling technologies designed for irrigation purposes, thereby mitigating water use and optimizing the utilization of graywater. In addition, the architectural design of the building has been optimized to enhance the utilization of natural light and ventilation, hence reducing the need for artificial lighting and HVAC systems (Meddah et al., 2023). The project demonstrates the effective harmonization of policy goals with tangible execution. Through the integration of various infrastructural elements, One Central Park was able to attain a 5-star Green Star rating, thereby establishing novel standards for sustainable urban development. The implementation of this measure resulted in a notable decrease in energy consumption, a reduction in carbon emissions, and an improvement in the overall livability and sustainability of the neighboring neighborhood (Bapat et al., 2022).

10.5.2 Bullitt Center, Seattle, USA

The Bullitt Center, located in Seattle, is a six-story commercial office building that is designed with the objective of achieving net zero energy and water consumption. The building incorporates a variety of sustainable elements, such as the utilization of solar panels, rainwater collecting systems, sophisticated insulation techniques, and efficient ventilation systems (Cooper et al., 2023). The project is in accordance with Seattle's lofty sustainability objectives and severe green building regulations. The structure aligns with the principles of the Living Building Challenge, showcasing a dedication to upholding the highest standards of sustainability (Kayaci & Kanbur, 2023).

The system incorporates renewable energy technologies, such as solar panels, to provide on-site electricity alongside rainwater harvesting devices that gather and purify water for on-site utilization. The architectural design integrates natural ventilation and daylighting, thereby diminishing the reliance on artificial lighting and HVAC systems (Indunil et al., 2023). The Bullitt Center has achieved global recognition as a leading exemplar of high-performance green buildings through its adoption of rigorous energy regulations and incorporation of sustainable infrastructure. This study highlights the attainability of net zero energy and water objectives, illustrating that rigorous policies and inventive infrastructure can result in exceptionally sustainable buildings with negligible ecological consequences (Rathnasiri et al., 2021).

10.5.3 The Edge, Amsterdam, Netherlands

The Edge, a prominent office skyscraper located in Amsterdam, serves as a prime example of innovative, sustainable architecture and advanced technological integration. The structure integrates several energy-efficient elements, such as solar panels, intelligent lighting systems, rainwater collection mechanisms, and modern insulating materials (Harmathy, 2021). The energy policies of the Netherlands, which prioritize sustainable construction and renewable energy, are in accordance with the aims of the Edge. The structure incorporates solar panels widely, facilitating the generation of renewable energy to power its many functions. Smart technologies are capable of regulating lighting and heating systems by taking into account factors like occupancy and daylight levels, with the aim of reducing energy inefficiencies (Yigit & Ozorhon, 2018). Rainwater harvesting systems are designed to gather and purify rainwater for non-potable use, hence diminishing dependence on municipal water sources.

The Edge serves as a prime example of effectively incorporating policy-driven sustainable practices into innovative infrastructure. The building has garnered significant recognition for its exceptional sustainability ratings and its status as one of the most energy-efficient structures on a worldwide scale (Yigit & Ozorhon, 2018). The Edge exemplifies the potential of sustainable, high-performing structures in urban environments through the implementation of rigorous energy policies and the integration of sophisticated infrastructure. This serves to highlight the positive impact that efficient building design and technologies can have on the overall sustainability of metropolitan areas (Tamur & Erzaij, 2021). These presented case studies highlight the effective integration of energy policy and innovative infrastructure within green building initiatives. The buildings exemplify the effectiveness of rigorous policies, as well as innovative design and technological integration, in achieving exceptional levels of sustainability and energy efficiency. These buildings serve as notable standards within the industry (Ashwathi et al., 2023).

10.6 ENERGY AND INFRASTRUCTURE POLICIES FUTURE DIRECTIONS

This section details the discussion on the implications of the findings for energy policies and infrastructure development and future trends of this approach to sustainable green building.

10.6.1 Grid modernization and flexibility

Presently, there is a prevailing inclination in energy policy to prioritize the modernization of electrical grids by including intelligent technology that can effectively accept a wide range of renewable energy sources. This entails the improvement of grid flexibility through the implementation of storage systems and demand-side management techniques (Baronin et al., 2023). It is anticipated that future developments will involve a persistent emphasis on enhancing grid flexibility in order to handle the intermittent nature of renewable energy sources effectively. Anticipated developments encompass the integration of AI in grid management, the development of novel storage systems, and the implementation of demand response mechanisms (Baronin et al., 2023). Policies are designed with the objective of establishing flexible grids that can effectively accommodate higher levels of renewable energy integration while also ensuring the reliability of the system.

10.6.2 Decentralized energy systems

Currently, energy regulations endorse the implementation of decentralized systems, which aim to support the production of energy at a local level and facilitate the development of microgrids. According to projected future developments, there will be a sustained emphasis on cultivating community-oriented renewable initiatives and prosumer frameworks (Meddah et al., 2023). Anticipated developments encompass the development of streamlined regulatory frameworks that facilitate the integration of dispersed energy resources, as well as the implementation of novel finance methods that promote local energy autonomy (Binega Yemesegen & Memari, 2023). Policies are expected to create incentives for the adoption of peer-to-peer energy trading and grid interconnectivity, hence facilitating the development of a decentralized and resilient energy environment.

10.6.3 Electrification and clean transportation

The present policies are vital in propelling the electrification endeavors in the transportation sector, as they provide incentives for the adoption of electric vehicles (EVs) and facilitate the expansion of charging infrastructure. It is anticipated that there will be a growing inclination toward the adoption of electric vehicles (EVs) in the future (Assef & Mangold, 2022). The development of improved infrastructure and a wider range of available EV models will facilitate this movement. Expected developments in this field encompass the development of standardized charging methods, the creation of batteries with increased energy density, and the introduction of unique charging systems (Bapat et al., 2022). It is anticipated that policies will be implemented to enhance further initiatives aimed at gradually eliminating internal combustion engines and establishing extensive charging infrastructure, thereby promoting the broad adoption of sustainable electric mobility.

10.6.4 Energy efficiency and building standards

Currently, there is a prioritization of policies that emphasize the implementation of rigorous energy efficiency standards for both buildings and appliances. Contemporary trends place significant emphasis on the promotion of retrofitting programs and the provision of incentives for the adoption of sustainable designs (Cooper et al., 2023). Future directions encompass the pursuit of even greater efficiency goals, which will be facilitated by technical breakthroughs such as smart materials and enhanced energy modeling tools. Anticipated advancements encompass the implementation of zero-energy building standards and the enhanced integration of IoT devices to optimize energy consumption (Kayaci & Kanbur, 2023). It is anticipated that policies will be implemented to impose stricter rules, with the aim of promoting zero-carbon construction and the adoption of circular economy concepts. These measures are intended to mitigate environmental harm and improve the overall sustainability of buildings (Indunil et al., 2023).

10.6.5 Resilient and adaptive infrastructure

The existing policies prioritize the enhancement of infrastructure resilience in the face of climate hazards, advocating for the implementation of adaptive measures and the development of robust methods for disaster response. It is projected that the next trends will involve a heightened emphasis on the development of adaptive infrastructure design, which will incorporate nature-based solutions and utilize AI-driven risk assessments (Rathnasiri et al., 2021). Anticipated developments encompass the implementation of resilient urban planning strategies, the development of flood-resistant infrastructure, and the establishment of early warning systems. Policies are expected to emphasize prioritizing investments in resilient infrastructure and incorporating creative designs that include the impacts of changing climate patterns and environmental uncertainties (Harmathy, 2021). These measures aim to ensure the long-term durability and functionality of infrastructure systems.

10.6.6 Smart and connected infrastructure

Current trends indicate the implementation of policies that support the integration of smart technology for the purpose of effectively managing infrastructure. The present endeavors primarily focus on the integration of IoT and the utilization of data analytics to achieve enhanced resource allocation (Ashwathi et al., 2023). The future trajectory involves the integration of AI-driven systems and blockchain technology on a broader scale, with the aim of improving security measures and facilitating interoperability. The expected progressions encompass predictive maintenance, autonomous systems, and city-wide connection (Baronin et al., 2023). It is anticipated that policies will create incentives for fostering collaborations between the public and private sectors in order to facilitate the widespread implementation of sophisticated technologies. The objective of such collaborations would be to achieve seamless integration and optimization across various infrastructure sectors (Meddah et al., 2023).

10.6.7 Green and sustainable infrastructure

Presently, prevailing policies endorse the adoption of sustainable infrastructure practices, which entail the promotion of green design aspects and the utilization of eco-friendly construction methods. Current developments in urban planning and development are centered around the promotion and prioritization of renewable materials, the creation of green spaces, and the implementation of low-impact development strategies (Assef & Mangold, 2022). It is anticipated that future endeavors will further enhance these initiatives, placing greater emphasis on the integration of circular economy principles and the application of biomimicry in the design of infrastructure. Anticipated developments encompass a broader utilization of recycled materials, the implementation of nature-based solutions to enhance climate resilience, and the integration of urban greenery (Bapat et al., 2022). Policies are anticipated to require rigorous sustainability certifications and provide incentives for the use of novel green technology. These policies aim to encourage the development of infrastructure that not only reduces environmental impact but also enhances ecosystem services and improves human well-being (Cooper et al., 2023).

10.6.8 Investment in digital infrastructure

The existing policies place a high priority on the expansion of digital infrastructure, with the goal of achieving extensive connectivity and enhanced accessibility to digital services. Current trends are focused on improving the availability of broadband connectivity and strengthening steps to ensure cybersecurity (Kayaci & Kanbur, 2023). The forthcoming endeavors are expected to enhance these initiatives further, with a particular emphasis on the growth of 5G technology, promoting digital inclusivity, and implementing rules to safeguard data privacy. Anticipated advancements encompass the integration of augmented reality, the development of sophisticated cybersecurity frameworks, and the promotion of equal access to digital resources (Meddah et al., 2023). Policies are anticipated to prioritize the implementation of digital literacy initiatives and the allocation of resources toward developing technologies, with the aim of cultivating an inclusive and safe digital environment that facilitates societal and economic advancement (Baronin et al., 2023).

10.7 LESSONS LEARNED FROM THE CHAPTER AND CONCLUSION

As we draw to the conclusion of the study, present the insights from the chapter based on the literature we will review and finally conclude the chapter with key takeaways and their importance.

10.7.1 Lessons learned

These lessons combined highlight the complex interplay between energy policies and infrastructure needs, underlining the necessity of a unified approach to promote effective and biologically friendly growth in the constructed surroundings.

- The need for integration to achieve success is paramount. The chapter emphasizes the significance of integrating energy policies with infrastructure needs in order to achieve optimal sustainability outcomes. Policies that are in line with the development of infrastructure contribute to the creation of a favorable environment for the implementation of efficient green buildings.
- The significance of innovation in infrastructure components such as renewable energy systems, smart technologies, and sustainable materials underscores the pivotal role that innovation plays in promoting energy-efficient construction practices and driving progress.
- The significance of regulatory frameworks: The adoption of energy-efficient technologies is heavily influenced by the presence of strong energy regulations and construction standards. Robust regulatory frameworks play a pivotal role in fostering compliance with sustainability requirements.
- The importance of collaboration cannot be overstated. Collaboration among politicians, industry stakeholders, and communities is frequently observed in the achievement of successful green-building initiatives. Collaborative endeavors propel the adoption of sustainable infrastructure and regulations.
- The adoption of a long-term vision leads to positive outcomes. The findings demonstrate that emphasizing future trends underscores the necessity of adopting a forward-thinking approach in the development of policies and the planning of infrastructure. The anticipation of future requirements and improvements is of utmost importance in the context of sustainable development.
- The chapter underscores the significance of resilience in infrastructure, highlighting its non-negotiable nature. The establishment of resilient and adaptive infrastructure is of utmost importance in order to effectively withstand the potential impacts of climate threats and guarantee long-lasting durability.

- The integration of smart technology and digital infrastructure plays a crucial role in shaping the future. The utilization of technology breakthroughs plays a pivotal role in enhancing the efficiency and connectedness of infrastructure.
- The concept of sustainability is multifaceted, encompassing multiple elements within green and sustainable infrastructure. These elements include but are not limited to energy efficiency, decentralization, electrification, and resilient designs. Comprehensive sustainability necessitates the adoption of a holistic strategy.

10.7.2 The conclusion

The investigation into energy regulations and the necessary infrastructure for promoting the efficiency of ecologically friendly buildings highlights a significant convergence of policy development, technological advancements, and the establishment of sustainable infrastructure. The success of green construction efforts is contingent upon the harmonization of progressive regulations and inventive infrastructure solutions. The lessons gleaned underscore the importance of including integration, innovation, and collaboration as key drivers of sustainable growth. To effectively navigate the dynamic energy landscape, it is imperative to establish robust regulatory frameworks accompanied by long-term visions and resilient infrastructure designs. The prioritization of technology as a facilitator of advancement and the complex characteristics of sustainability indicate the necessity for a comprehensive methodology. In conclusion, the chapter elucidates a trajectory for progress that entails the implementation of integrated approaches, collaborative partnerships, and a steadfast dedication to sustainability in order to establish energy-efficient, resilient, and environmentally conscious constructed environments for subsequent cohorts.

REFERENCES

Ashwathi, R., Yuvaraj, S., & Jeyanth, B. (2023). Enhancing strength properties by incorporating fly ash for sustainable environment. *Materials Today: Proceedings*. https://doi.org/10.1016/j.matpr.2023.03.560

Assef, D., & Mangold, G. (2022). Building on change, ESG considerations for the construction industry. *Civil Engineering*, *30*(6).

Bapat, H., Sarkar, D., & Gujar, R. (2022). A sustainable approach to reduce embodied and operational cooling energy for an elevated metro rail station of Ahmedabad, India, using Building Information Modelling (BIM) and Factor Comparison Method. *Journal of The Institution of Engineers (India): Series A*, *103*(1). https://doi.org/10.1007/s40030-021-00594-1

Baronin, S. A., Guschina, E. S., & Romanova, A. I. (2023). Integrated green construction as a prerequisite for sustainable urban development. *E3S Web of Conferences*, *403*. https://doi.org/10.1051/e3sconf/202340302013

Binega Yemesegen, E., & Memari, A. M. (2023). A review of experimental studies on Cob, Hempcrete, and bamboo components and the call for transition towards sustainable home building with 3D printing. In *Construction and building materials* (Vol. 399). https://doi.org/10.1016/j.conbuildmat.2023.132603

Cooper, N., Galasiu, A., & Bahiraei, F. (2023). Evaluating sustainable and green building designs using human factor approaches. *Sustainable construction in the era of the fourth industrial revolution* (p. 107). https://doi.org/10.54941/ahfe1003087

Harmathy, N. (2021). Investigation of decarbonization potential in green building design to accelerate the utilization of renewable energy sources. *Thermal Science*, *26*(6 Part A). https://doi.org/10.2298/TSCI200324195H

Indunil, G. M., Kumari, T. M. C. P., & Subasinghe, C. W. (2023). Infrastructure, curriculum delivery and service provision improvements made by the University of Kelaniya, Sri Lanka in transforming education towards sustainability in the midst of economic crisis, in post-covid 19 pandemic period. *IOP Conference Series: Earth and Environmental Science*, *1194*(1). https://doi.org/10.1088/1755-1315/1194/1/012028

Kanagaraj, B., Kiran, T., Anand, A., Al Jabri, K., & S., J. (2023). Development and strength assessment of eco-friendly geopolymer concrete made with natural and recycled aggregates. *Construction Innovation*, *23*(3). https://doi.org/10.1108/CI-08-2021-0157

Kayaci, N., & Kanbur, B. B. (2023). Numerical and economic analysis of hydronic-heated anti-icing solutions on underground park driveways. *Sustainability*, *15*(3). https://doi.org/10.3390/su15032564

Kovalchuk, J. A., & Stepnov, I. M. (2023). Achieving energy efficiency goals in the public procurement. In *Environmental footprints and eco-design of products and processes*. https://doi.org/10.1007/978-3-031-28457-1_31

Lambertz, M., Theißen, S., Höper, J., & Wimmer, R. (2019). Importance of building services in ecological building assessments. *E3S Web of Conferences*, *111*. https://doi.org/10.1051/e3sconf/201911103061

Meddah, M. S., Al Owaisi, M., Abedi, M., & Hago, A. W. (2023). Mortar and concrete with lime-rich calcined clay pozzolana: A sustainable approach to enhancing performances and reducing carbon footprint. *Construction and Building Materials*, *393*. https://doi.org/10.1016/j.conbuildmat.2023.132098

Nandhini, V., Ambika, D., Kumar, V. S., Priya, S. D., Poovizhi, G., & Rubini, V. S. (2020). An analysis on low cost and energy efficient materials for sustainable housing. *International Journal of Engineering Trends and Technology*, *68*(2). https://doi.org/10.14445/22315381/IJETT-V68I2P215S

Rathnasiri, P., Jayasena, S., & Siriwardena, M. (2021). Assessing the applicability of green building information modelling for existing green buildings. *International Journal of Design and Nature and Ecodynamics*, *15*(6). https://doi.org/10.18280/ijdne.150601

Tamur, S. T., & Erzaij, K. R. (2021). Selecting appropriate delivery system for implementation of green buildings. *E3S Web of Conferences*, *318*. https://doi.org/10.1051/e3sconf/202131802002

Yigit, S., & Ozorhon, B. (2018). A simulation-based optimization method for designing energy efficient buildings. *Energy and Buildings*, *178*. https://doi.org/10.1016/j.enbuild.2018.08.045

Chapter 11

Building performance prediction and optimization

Mira Chitt, Sivasakthivel Thangavel, Ashwani Kumar, Varun Pratap Singh, and Chandan Swaroop Meena

11.1 INTRODUCTION

In the ever-evolving landscape of urban development, the quest for sustainable and efficient building performance stands as a pivotal challenge. The intricate interplay of architectural design, technological advancements, and environmental considerations has given rise to a fascinating field known as building performance prediction and optimization (BPPO). This domain seeks not only to understand the intricacies of how buildings function but also to harness that knowledge for creating spaces that are energy-efficient, comfortable for occupants, and environmentally conscious (Wang & Liu, 2019).

11.2 THE COMPLEX WEB OF BUILDING PERFORMANCE

At the heart of BPPO lies a recognition of the complex web of factors influencing building performance. These factors extend beyond the traditional architectural considerations of form and function, venturing into the realms of energy consumption, thermal dynamics, occupant behavior, and environmental impact. A holistic understanding of these elements is imperative for constructing buildings that go beyond mere structures to become dynamic, responsive entities within their urban ecosystems. The realm of building performance is a complex tapestry woven from a myriad of interconnected factors, each influencing the other in a delicate dance that shapes the efficiency, sustainability, and functionality of built environments. From the macro-scale considerations of architectural design to the micro-scale intricacies of energy consumption and occupant behavior, the complex web of building performance defies simple categorization. At its core, building performance encompasses a multifaceted understanding of how structures interact with their surroundings and inhabitants. One of the primary pillars within this intricate web is energy efficiency—a linchpin

DOI: 10.1201/9781003496656-11

that not only dictates the economic viability of buildings but also holds profound implications for environmental sustainability. Energy efficiency in buildings involves a nuanced interplay of design choices, technological integrations, and operational strategies. Architects and engineers strive to create structures that maximize natural lighting, optimize thermal insulation, and incorporate renewable energy sources (Verma, 2023). The goal is not merely to construct buildings that exist in isolation but to seamlessly integrate them into the broader urban ecosystem, minimizing their carbon footprint and energy demand. Dynamic simulation tools emerge as indispensable allies in deciphering the complexities of building performance. These tools allow for the virtual modeling of structures, enabling architects and engineers to simulate various scenarios and assess the impact of design decisions on energy consumption, thermal dynamics, and overall functionality. From assessing the effectiveness of passive design strategies to fine-tuning HVAC systems, these simulations provide invaluable insights that guide the optimization of building performance (Turner et al., 2014). Yet, within this tapestry of technology and design, the human factor emerges as a pivotal thread. Occupant comfort, influenced by factors such as temperature, air quality, and lighting, transcends the realm of mere luxury to become an integral component of building performance. Designing spaces that cater to the diverse needs and preferences of occupants involves a delicate balancing act—one that seeks to optimize physical comfort while minimizing energy consumption. Human-centric design principles underscore the importance of tailoring-built environments to the occupants' well-being. The integration of smart sensors, adaptive lighting systems, and personalized climate control not only enhances comfort but also contributes to energy efficiency by aligning building operations with real-time occupancy patterns. Occupant behavior, a variable as unpredictable as it is influential, adds another layer of complexity to the web of building performance. Understanding how individuals interact with and utilize spaces is essential for optimizing building operations. Real-time data collection through sensors and Internet of Things (IoT) devices provides a window into occupant behavior, enabling building managers to adapt and optimize systems based on actual usage patterns (Williams et al., 2011). Technological frontiers further complicate this intricate tapestry. Artificial intelligence (AI), with its capacity to process vast datasets and identify patterns, transforms building management systems into intelligent entities capable of learning and adapting. Advanced materials, designed with enhanced thermal properties and sustainability in mind, reshape the physical fabric of buildings, further influencing their performance. However, navigating the complex web of building performance is not without its challenges. Interdisciplinary collaboration becomes imperative as architects, engineers, data scientists, and behavioral psychologists converge to address the multifaceted nature of building performance. Bridging the communication gaps between these diverse disciplines is essential for devising holistic solutions that consider both the technical and human dimensions of building design and operation (Yang & Wu, 2019).

11.3 ENERGY EFFICIENCY AS THE CORNERSTONE

Energy efficiency emerges as a cornerstone in the pursuit of optimal building performance. Buildings are significant contributors to global energy consumption, and their efficiency directly impacts both economic and environmental sustainability. Predicting and optimizing energy usage involves a sophisticated interplay of design principles, technological integration, and behavioral insights. In the intricate landscape of building performance, energy efficiency stands tall as the cornerstone, shaping the very foundation of sustainable and resilient structures. Beyond being a buzzword, energy efficiency is a dynamic principle that goes beyond mere conservation—it is a holistic approach that influences design, construction, and operation, promising a future where buildings seamlessly integrate with the environment while minimizing their impact on valuable resources.

At its essence, energy efficiency in buildings involves a strategic orchestration of architectural elements, technological advancements, and operational practices to optimize energy usage. This optimization is not just a matter of financial prudence, it is an ethical imperative and a commitment to environmental stewardship. Buildings, responsible for a significant share of global energy consumption, play a pivotal role in the quest for a more sustainable future. Architects and engineers embark on a journey to design structures that are not only aesthetically pleasing but also inherently efficient in their use of energy. Passive design strategies become key players in this endeavor, utilizing natural elements to enhance comfort and reduce the reliance on mechanical systems. From orientation that maximizes natural light to innovative shading devices that mitigate heat gain, these strategies form the first layer of the energy efficiency framework. Dynamic simulation tools, sophisticated and powerful, come into play to test and refine these design choices. Architects leverage these tools to create virtual prototypes, simulating various scenarios to understand how different elements impact the building's energy performance. The goal is not only to meet regulatory standards but to surpass them, creating structures that operate at the intersection of sustainability and innovation. The integration of renewable energy sources emerges as a pivotal element in the pursuit of energy efficiency. Solar panels, wind turbines, and other renewable technologies transform buildings from passive consumers to active contributors to the energy grid. The symbiotic relationship between the built environment and renewable energy sources not only reduces reliance on conventional power but also positions buildings as potential sources of clean energy (Ahmadizadeh, 2024). In the operational phase, building management systems equipped with advanced sensors and automation technologies play a crucial role in maintaining and enhancing energy efficiency. These systems go beyond traditional HVAC controls, adapting to real-time conditions and occupant behaviors. Smart lighting that adjusts based on natural light availability, intelligent climate control that learns from usage patterns, and energy management systems that optimize consumption become integral components of a building's energy efficiency arsenal. The economic benefits of energy efficiency are undeniable. Reduced energy consumption translates into lower utility bills, offering tangible savings for building owners and occupants alike. However, the impact extends beyond immediate financial gains. Energy-efficient buildings also enjoy increased market value, reflecting the growing awareness and demand for sustainable living and working spaces.

Environmental sustainability is perhaps the most compelling argument for prioritizing energy efficiency. The reduction in greenhouse gas emissions, decreased reliance on finite resources, and a smaller ecological footprint position energy-efficient buildings as catalysts for positive change. As the global community grapples with the challenges of climate change, the role of energy-efficient structures becomes increasingly critical in mitigating environmental impact. Challenges, however, accompany the pursuit of energy efficiency. The initial costs of implementing advanced technologies and sustainable practices may pose a barrier for some. Convincing stakeholders of the long-term benefits and return on investment requires a nuanced understanding of the economic implications of energy efficiency. Yet, as technologies mature and become more accessible, the economic barriers are gradually diminishing, paving the way for widespread adoption.

11.4 ROLE OF PREDICTIVE ANALYTICS

Predictive analytics, fueled by advancements in data science, plays a pivotal role in anticipating energy needs and consumption patterns. By analyzing historical data, weather patterns, and occupant behavior, predictive models can forecast future energy demands with remarkable accuracy. This foresight empowers building managers and designers to implement preemptive measures, such as load shedding or the integration of renewable energy sources, to mitigate energy wastage and enhance overall efficiency. The integration of predictive analytics in the realm of building performance prediction and optimization (BPPO) heralds a new era in the way we conceive, design, and manage built environments. Predictive analytics plays a pivotal role in forecasting, adapting, and enhancing various facets of building performance, contributing to energy efficiency, occupant comfort, and overall sustainability (Chitt, 2024 and Khashehchi, 2024).

At the core of its contribution to BPPO is the ability to foresee and model future trends based on historical data and relevant variables. Energy consumption, a cornerstone of building performance, benefits significantly from the predictive prowess of analytics. By analyzing past energy usage patterns, weather conditions, and occupant behavior, predictive models can accurately anticipate future energy demands. This forward-thinking approach empowers building managers to implement proactive measures, such as load optimization, demand response strategies, and the integration of renewable energy sources, to enhance overall energy efficiency.

The role of predictive analytics extends beyond energy considerations to predictive maintenance—a critical component of building optimization. By leveraging sensor data and historical performance records, predictive maintenance models can forecast when equipment is likely to fail. This enables building managers to schedule maintenance activities efficiently, reducing downtime, minimizing disruptions, and extending the lifespan of crucial building systems. In the context of BPPO, this predictive approach to maintenance not only enhances reliability but also contributes to cost savings and operational efficiency (Zhang & Wang, 2016). Occupant comfort, a dynamic and subjective aspect of building performance, is another arena where predictive analytics proves invaluable. By analyzing data on occupant behavior, preferences, and responses to environmental conditions, predictive models can optimize building parameters in real-time. From personalized climate control to adaptive lighting, the application of predictive analytics ensures that building environments are responsive to the unique needs of occupants, thereby enhancing comfort and satisfaction. This personalized approach not only contributes to the well-being of occupants but also aligns with the broader goals of sustainable and human-centric design in BPPO. The real-time adaptability provided by predictive analytics transforms decision-making processes in building management. Instead of reacting to issues as they arise, building managers gain the ability to make informed decisions based on anticipated trends and patterns. This dynamic responsiveness ensures that building performance remains optimized in the face of evolving variables, contributing to a more resilient and adaptive built environment. While the benefits of predictive analytics in BPPO are substantial, challenges exist, particularly in the realms of data quality, privacy, and ethical considerations. The accuracy of predictions hinges on the availability and relevance of high-quality data. Additionally, the responsible and transparent handling of sensitive occupant data is paramount to address privacy concerns and foster trust in the use of predictive analytics.

11.5 DYNAMIC SIMULATION TOOLS

In the realm of building performance, dynamic simulation tools stand out as indispensable assets. These tools allow architects and engineers to virtually model and simulate various scenarios, assessing how different design elements and technologies impact the building's performance. From daylighting simulations that optimize natural light usage to thermal simulations that evaluate insulation effectiveness (Chang & Wang, 2012), these tools provide invaluable insights into the potential performance of a building before it even breaks ground.

In the quest for sustainable and efficient built environments, dynamic simulation tools (DSTs) stand as indispensable assets, reshaping the landscape of building performance prediction and optimization (BPPO). These sophisticated tools offer a virtual playground for architects, engineers, and designers, enabling them to model and simulate diverse scenarios to understand, predict, and optimize the performance of buildings. From assessing energy consumption to fine-tuning thermal comfort, DSTs play a pivotal role in enhancing the overall sustainability and functionality of built environments.

11.5.1 Power of virtual prototyping

At the heart of DSTs lies the ability to create virtual prototypes of buildings, mimicking their physical characteristics and behavior. This virtual prototyping allows for the dynamic simulation of various scenarios, offering insights into how different design choices and operational strategies impact building performance. Architects and engineers can experiment with different configurations, materials, and systems, observing the potential outcomes before a single brick is laid.

11.5.2 Energy efficiency optimization

One of the primary applications of DSTs in BPPO is the optimization of energy efficiency. These tools simulate the energy performance of buildings under different conditions, considering factors such as orientation, insulation, and the integration of renewable energy sources. By providing a visual representation of energy flows and consumption patterns, DSTs empower designers to make informed decisions that enhance energy efficiency and minimize environmental impact.

11.5.3 Thermal comfort analysis

Thermal comfort is a critical aspect of building performance, influencing occupant well-being and productivity. DSTs facilitate detailed thermal comfort analysis by simulating the thermal behavior of spaces under various conditions. This includes evaluating the impact of natural ventilation, shading devices, and HVAC systems on indoor temperatures (Chen & Li, 2014). The insights gained from these simulations inform design choices that not only optimize comfort but also contribute to energy-efficient climate control strategies (Garcia et al., (2015)).

11.5.4 Daylighting simulations

Natural light is a valuable resource in building design, impacting both energy usage and occupant satisfaction. DSTs excel in daylighting simulations, allowing designers to assess how natural light penetrates and interacts with interior spaces throughout the day and across seasons. This information guides the placement of windows, the design of shading devices, and the selection of materials to maximize daylight while minimizing glare and heat gain.

11.5.5 Indoor air quality modeling

Beyond energy and thermal considerations, DSTs also contribute to the optimization of indoor air quality (IAQ). By simulating airflow patterns, pollutant dispersion, and ventilation strategies, these tools help designers create environments that promote health and well-being (Johnson & Williams, 2010). This is particularly crucial in the context of BPPO, where holistic optimization involves not only energy efficiency but also the creation of spaces that foster occupant health.

11.5.6 Adaptive building design

DSTs enable the exploration of adaptive building design, where structures respond dynamically to changing conditions. This includes the integration of smart building technologies that adjust lighting, HVAC systems, and other parameters based on real-time data and occupant behavior (Kim & Lee, 2013). The dynamic simulation of these adaptive features ensures that buildings remain responsive to evolving environmental and user-driven factors.

11.5.7 Challenges and advancements

While DSTs offer immense potential in BPPO, challenges exist, including the need for accurate input data, computational complexity, and the requirement for specialized expertise. Advancements in technology, however, are addressing these challenges. Improved algorithms, cloud computing, and user-friendly interfaces are making DSTs more accessible and efficient, broadening their applicability across the architectural and engineering spectrum.

Thus, dynamic simulation tools are transformative assets in the realm of building performance prediction and optimization. Their ability to create virtual prototypes, simulate diverse scenarios, and provide actionable insights positions them as essential contributors to the creation of sustainable, efficient, and occupant-centric built environments (Kim & Park, 2015). As technology continues to advance, the seamless integration of DSTs will play a pivotal role in shaping the future of architecture and engineering, ensuring that buildings not only meet but exceed performance expectations in an ever-evolving urban landscape.

11.5.8 Human factor: Occupant comfort and behavior

Beyond the technical aspects, BPPO recognizes the human factor as a critical dimension of building performance. Occupant comfort, influenced by factors such as temperature, air quality, and lighting, directly impacts well-being and productivity. Understanding occupant behavior is equally crucial, as it shapes how spaces are utilized and how systems are operated.

11.5.9 Occupant comfort: A multifaceted dimension

Occupant comfort transcends the conventional boundaries of temperature control and lighting; it encompasses a nuanced understanding of the psychological, physiological, and perceptual aspects that contribute to a positive and productive environment. Architects and designers strive to create spaces that go beyond mere functionality, aiming for a harmonious balance between aesthetics and occupant well-being.

11.5.10 Thermal comfort

Thermal comfort, a cornerstone of occupant satisfaction, involves creating indoor environments where individuals feel neither too hot nor too cold (Lee & Chen, 2016). Dynamic simulation tools (DSTs) are often employed to simulate and optimize thermal conditions, allowing designers to fine-tune HVAC systems and insulation strategies. This attention to thermal comfort not only enhances the well-being of occupants but also contributes to energy efficiency by aligning climate control with actual user needs (Liu & Li, 2011).

11.5.11 Visual comfort

Beyond temperature, visual comfort plays a crucial role in occupant satisfaction. Daylighting simulations, facilitated by DSTs, help designers understand how natural light interacts with interior spaces. Strategic placement of windows, intelligent shading solutions, and the use of reflective materials contribute to visual comfort, reducing reliance on artificial lighting and fostering a connection with the outdoor environment (Smith et al., 2018).

11.5.12 Acoustic comfort

The auditory environment significantly influences occupant comfort and productivity. The design and layout of spaces, as well as the selection of materials, impact sound transmission. Acoustic simulations within DSTs assist in optimizing room configurations and materials to create spaces that mitigate noise and enhance acoustic comfort.

11.5.13 Occupant behavior: A variable of complexity

Understanding occupant behavior adds another layer of complexity to the optimization of building performance. Occupants interact with spaces in diverse ways, influenced by personal preferences, cultural factors, and even the time of day. Predictive analytics, drawing insights from historical data and real-time monitoring, contribute to the modeling of occupant behavior. This data-driven approach allows for the creation of responsive environments that adapt to the dynamic needs of occupants.

11.5.14 Adaptive building systems

Smart building technologies leverage occupant behavior insights to create adaptive systems. These systems adjust lighting, temperature, and other parameters based on real-time data, creating environments that align with user preferences. Adaptive buildings not only enhance occupant comfort but also contribute to energy efficiency by avoiding unnecessary resource consumption when spaces are unoccupied.

11.5.15 User-centric design

Human-centric design principles underscore the importance of tailoring-built environments to the needs and preferences of occupants. This involves not only providing physical comfort but also creating spaces that cater to the diverse activities and interactions that occur within them. Flexible layouts, multipurpose spaces, and amenities that enhance user experience contribute to a holistic approach to occupant-centric design.

11.5.16 Challenges and considerations

However, the human factor in BPPO is not without its challenges. Occupant behavior is inherently unpredictable, influenced by individual choices, cultural backgrounds, and evolving societal norms. Striking a balance between personalized comfort and overall energy efficiency requires a nuanced approach, and the responsible handling of occupant data raises privacy considerations that must be addressed ethically and transparently.

In the evolving landscape of BPPO, occupant comfort and behavior emerge as critical dimensions that demand attention and understanding. As technology advances, the fusion of predictive analytics, adaptive building systems, and user-centric design principles holds the key to creating environments that not only meet performance standards but also elevate the human experience within built spaces. The future of building performance lies in the delicate equilibrium between technological innovation and the intricate nuances of human well-being, ensuring that buildings become not just structures but nurturing and responsive habitats for the diverse individuals who inhabit them.

11.6 HUMAN-CENTRIC DESIGN

Human-centric design principles underscore the importance of tailoring building environments to the needs and preferences of their occupants. This involves not only optimizing physical comfort but also leveraging technologies like smart sensors and adaptive HVAC systems to create personalized and responsive spaces. By considering the diverse needs and behaviors of occupants, BPPO aims to strike a delicate balance between energy efficiency and user satisfaction.

Human-centric design, a philosophy that places the needs, preferences, and experiences of individuals at the forefront of the creative process, is a transformative approach that transcends traditional design paradigms. In the realm of architecture and product development, human-centric design marks a departure from a focus solely on functionality or aesthetics, embracing a holistic understanding of users and their interactions with the designed environment or product.

11.6.1 Empathy as the foundation

At the core of human-centric design lies empathy—an intimate understanding of the users' perspectives, motivations, and challenges. Designers immerse themselves in the lived experiences of the individuals for whom they are creating, seeking to uncover unmet needs and latent desires. By empathizing with users, designers can develop solutions that resonate on a deeper level, addressing not just functional requirements but also the emotional and psychological aspects of the human experience.

11.6.2 User-centered design principles

Human-centric design adheres to user-centered design principles, emphasizing collaboration, iteration, and feedback throughout the design process. Users are not passive recipients of a final product; they are active participants in the creation journey. Continuous feedback loops allow designers to refine and improve their creations based on real-world usage and evolving user needs.

11.6.3 Personalization and diversity

One of the hallmarks of human-centric design is its recognition of the diversity among users. People have unique preferences, abilities, and cultural backgrounds. A human-centric approach acknowledges this diversity and seeks to create designs that are inclusive and adaptable. Personalization features, such as customizable interfaces or adjustable ergonomic elements, ensure that the designed solutions cater to a broad spectrum of users.

11.6.4 Enhancing user experience in built environments

In the context of architecture, human-centric design transforms space into more than just functional structures. It involves a meticulous consideration of how individuals navigate and experience a space. Lighting, acoustics, and spatial layouts are tailored to enhance well-being and comfort. The result is not only aesthetically pleasing environments but also spaces that support the diverse activities and interactions of their occupants.

11.6.5 Balancing technology with human needs

As technology becomes increasingly integrated into our daily lives, human-centric design becomes a crucial compass for navigating this intersection. Technological innovations are harnessed not merely for their own sake but to enhance and augment the human experience. Whether it's the design of user-friendly interfaces, intuitive controls, or seamless integrations, human-centric design ensures that technology serves as a facilitator rather than an obstacle.

11.6.6 Design thinking in problem-solving

Human-centric design is closely aligned with the principles of design thinking—a problem-solving approach that prioritizes understanding users, defining challenges, ideating solutions, prototyping, and testing. This iterative process fosters innovation by encouraging designers to continually refine their ideas based on user feedback and real-world observations.

11.6.7 Challenges and ethical considerations

While human-centric design brings immense benefits, it also poses challenges and ethical considerations. Striking the right balance between customization and standardization, addressing biases in design, and safeguarding user privacy are critical considerations. Responsible designers recognize the potential impact of their creations on individuals and society at large, working to minimize unintended consequences.

11.6.8 The future landscape of design

As we look toward the future, human-centric design is poised to play a pivotal role in shaping the landscape of design across various disciplines. From sustainable architecture that prioritizes occupant well-being to user-friendly technologies that empower individuals, the principles of human-centric design offer a blueprint for creating solutions that resonate with the diverse and evolving needs of humanity.

11.7 BEHAVIORAL INSIGHTS AND FEEDBACK LOOPS

Incorporating behavioral insights into building design and management involves establishing feedback loops that continuously learn from occupant interactions. Smart building systems, equipped with sensors and IoT devices, can collect real-time data on occupant behavior. Analyzing this data provides a deeper understanding of how spaces are used, enabling adjustments in real-time to optimize both energy efficiency and occupant comfort.

11.7.1 Technological frontiers: Innovations in building performance

The technological landscape is evolving rapidly, offering innovative solutions that redefine the possibilities within BPPO. From the integration of artificial intelligence (AI) in building management systems to the advent of advanced materials with enhanced thermal properties, these technologies are reshaping the way we approach building performance.

11.7.2 AI and building management systems

AI algorithms, with their ability to process vast amounts of data and learn from patterns, have found a natural home in building management systems. These systems can predict maintenance needs, optimize energy consumption, and even adapt to changing environmental conditions in real-time. The result is buildings that operate with a level of intelligence and efficiency previously thought unattainable.

11.7.3 Advanced materials and sustainable construction

The materials used in construction play a pivotal role in building performance. Innovations in materials science are yielding sustainable alternatives that not only reduce the environmental impact of construction but also enhance the thermal and acoustic properties of buildings. From smart glass that adjusts its tint based on external conditions to bio-based insulation materials, these advancements contribute to the overall efficiency and sustainability of built environments.

11.7.4 Challenges on the path to optimization

While the promises of BPPO are tantalizing, the journey is not without its challenges. The complexity of integrating diverse technologies, the need for interdisciplinary collaboration, and the inherent unpredictability of human behavior pose hurdles that demand innovative solutions.

11.7.5 Interdisciplinary collaboration

The optimization of building performance requires a convergence of expertise from diverse fields—architecture, engineering, data science, and behavioral psychology, to name a few. Bridging the communication gaps between these disciplines and fostering a collaborative mindset is crucial for successfully navigating the intricate landscape of BPPO (Brown & Wilson, 2017).

11.7.6 DATA PRIVACY AND SECURITY

The reliance on data-driven technologies raises concerns about data privacy and security. As buildings become increasingly connected and sensor-laden, the need to protect sensitive occupant data becomes paramount. Striking a balance between harnessing the benefits of data-driven optimization and safeguarding privacy is a delicate yet essential aspect of BPPO.

11.7.7 Cost implications

Implementing cutting-edge technologies and sustainable design practices often comes with an initial cost. Convincing stakeholders of the long-term benefits and return on investment requires a comprehensive understanding of the economic implications of BPPO. However, as the technologies mature and become more widespread, the initial costs are likely to decrease, making sustainable practices more accessible.

11.8 TOWARD A SUSTAINABLE FUTURE

Building performance prediction and optimization is not merely a theoretical concept; it is a roadmap to a sustainable future. As urbanization continues to accelerate, the imperative to construct buildings that contribute positively to the environment and the well-being of their occupants becomes ever more urgent.

In the rapidly evolving landscape of urbanization and construction, the imperative to create sustainable and resilient built environments has never been more pressing. Building performance prediction and optimization (BPPO) emerges as a crucial paradigm that not only envisions but actively paves the way toward a future where structures harmonize with the environment, enhance occupant well-being, and contribute positively to the global quest for sustainability.

11.9 THE ESSENCE OF SUSTAINABILITY IN BPPO

Sustainability, in the context of BPPO, extends beyond the mere reduction of environmental impact. It embodies a holistic approach that seeks to balance ecological considerations with economic viability and social well-being. At its core, sustainability in building performance involves creating structures that not only meet the immediate needs of their occupants but also contribute positively to the health of the planet.

11.9.1 Energy efficiency as a pillar

Energy efficiency stands as a foundational pillar in the pursuit of sustainability within BPPO. Buildings, responsible for a significant share of global energy consumption, become focal points for optimization. Predictive analytics, fueled by historical data and advanced modeling techniques, enable the anticipation and optimization of energy consumption. From forecasting demand to dynamically adjusting operational parameters, the integration of energy-efficient practices ensures that buildings operate with minimal environmental impact.

Dynamic simulation tools (DSTs) play a pivotal role in this quest for energy efficiency. These tools allow architects and engineers to virtually model and simulate various scenarios, assessing the impact of design choices on energy consumption. Whether it's optimizing passive design strategies, fine-tuning HVAC systems, or integrating renewable energy sources, DSTs provide a dynamic platform for architects to visualize and enhance the energy performance of their designs.

11.9.2 Occupant well-being and human-centric design

Sustainability in BPPO extends beyond energy metrics to encompass the well-being of occupants. Human-centric design principles become integral to creating spaces that prioritize comfort, health, and productivity. Thermal comfort simulations, acoustic analyses, and lighting optimization within DSTs contribute to the creation of environments that not only meet environmental standards but also enhance the quality of life for those who inhabit them.

Moreover, occupant behavior becomes a key variable in the sustainability equation. Predictive analytics not only anticipates energy needs but also models how occupants interact with and utilize spaces. This data-driven approach allows for the implementation of adaptive building systems that respond to real-time conditions, ensuring a dynamic equilibrium between sustainability and occupant comfort.

11.9.3 Materials innovation and sustainable construction

The materials used in construction represent another frontier in the pursuit of sustainability. Innovations in materials science are yielding alternatives that reduce environmental impact without compromising structural integrity. From recycled and low-impact materials to bio-based alternatives, the choices available to designers are expanding, allowing for the creation of structures that are not only efficient but also environmentally responsible. Sustainable construction practices, guided by the principles of BPPO, go beyond the use of eco-friendly materials. DSTs enable designers to assess the life cycle impacts of construction materials and methodologies. This comprehensive understanding allows for informed decisions that minimize waste, reduce carbon footprints, and contribute to the longevity and sustainability of built structures.

11.9.4 Policy, regulation, and community engagement

The journey toward a sustainable future in BPPO is not solely a technological endeavor. Policy and regulatory frameworks play a pivotal role in shaping the trajectory of sustainable building practices. Governments and municipalities can incentivize green building initiatives, establish energy efficiency standards, and create a conducive environment for the adoption of innovative technologies. Community engagement is equally vital. Educating stakeholders—from architects and builders to residents and businesses—fosters a collective understanding of the importance of sustainability. Awareness campaigns, incentives for sustainable practices, and collaborative initiatives can contribute to a cultural shift toward more conscious and responsible building and living practices.

11.9.5 Challenges and future innovations

Despite the strides made in the realm of BPPO, challenges persist. The initial costs associated with adopting sustainable technologies and materials can be a deterrent for some. Overcoming these challenges requires a holistic perspective that considers not only short-term costs but also long-term benefits, both economic and environmental.

11.10 CONCLUSIONS AND FUTURE SCOPE

In the intricate dance between form and function, aesthetics and efficiency, buildings play a central role in shaping the urban landscape. Building performance prediction and optimization represents a paradigm shift in how we conceive, design, and manage built environments. By embracing advanced technologies, interdisciplinary collaboration, and a human-centric approach, BPPO charts a course towards sustainable urban development—one where buildings are not just static structures but dynamic, responsive entities that contribute positively to the well-being of occupants and the health of the planet. As we navigate this path, the challenges may be formidable, but the rewards—a sustainable and resilient built environment—are worth the journey.

REFERENCES

Ahmadizadeh, M., Heidari, M., Thangavel, S., Naamani, E. A., Khashehchi, M., Verma, V., & Kumar, A. (2024) 'Technological Advancements in Sustainable and Renewable Solar Energy Systems', in *Highly Efficient Thermal Renewable Energy Systems: Design, Design, Optimization and Applications*; V. Verma, S. Thangavel, N. Dutt, A. Kumar, R. Weerasinghe Editors; CRC Press: Boca Raton, FL; Chapter 02; ISBN 9781032595641.

Brown, P. J., & Wilson, D. S. (2017) 'Optimization of Building Envelope Materials for Enhanced Thermal Performance', *Sustainable Architecture and Design*, 30(1), 245–260.

Chang, L., & Wang, Y. (2012) 'A Comparative Study of Genetic Algorithms for Building Energy Performance Optimization', *Journal of Sustainable Building Design*, 15(4), 120–135.

Chen, Q., & Li, X. (2014) 'Simulation-Based Optimization for HVAC System Design in High-Performance Buildings', *Energy Optimization Quarterly*, 18(3), 210–225.

Chitt, M., Thangavel, S., Verma, V., & Kumar, A. (2024) 'Green Hydrogen Productions: Methods, Designs and Smart Applications', in *Highly Efficient Thermal Renewable Energy Systems: Design, Design, Optimization and Applications*; V. Verma, S. Thangavel, N. Dutt, A. Kumar, R. Weerasinghe Editors; CRC Press: Boca Raton, FL; Chapter 16; ISBN 9781032595641.

Garcia, M. A., et al. (2015) 'An Integrated Framework for Building Performance Prediction and Control', *Journal of Building Performance Analysis*, 24(2), 112–128.

Garcia, R. B., & Patel, S. K. (2018) 'A Comparative Study of Optimization Algorithms for Building Performance Enhancement', *Journal of Sustainable Engineering*, 32(1), 45–58.

Johnson, M. P., & Williams, R. S. (2010) 'Predictive Modeling for Energy Efficiency in Buildings', *Energy Optimization Quarterly*, 15(4), 120–135.

Khashehchi, M., Thangavel, S., Rahmanivahid, P., Heidari, M., Moazzeni, T., Verma, V., & Kumar, A. (2024) 'Solar Desalination Techniques: Challenges and Opportunities', in *Highly Efficient Thermal Renewable Energy Systems: Design, Design, Optimization and Applications*; V. Verma, S. Thangavel, N. Dutt, A. Kumar, R. Weerasinghe Editors; CRC Press: Boca Raton, FL; Chapter 19; ISBN 9781032595641.

Kim, H., & Lee, J. (2013) 'Advanced Control Strategies for Energy-Efficient HVAC Systems in Office Buildings', *Automation in Construction*, 25(3), 55–68.

Kim, S., & Park, J. (2015) 'Machine Learning-Based Predictive Control for Adaptive Lighting in Commercial Buildings', *Automation in Construction*, 22(3), 112–125.

Lee, K. W., & Chen, H. M. (2016) 'Predictive Modeling of Thermal Comfort in Office Buildings: A Case Study in Urban Environments', *Building Simulation*, 20(4), 365–378.

Liu, X., & Li, Z. (2011) 'Optimizing Thermal Mass in Passive Solar Design for Residential Buildings', *Energy and Buildings*, 14(2), 78–93.

Patel, N. R., & Gupta, A. K. (2016) 'Machine Learning Approaches for Predictive Maintenance in Smart Buildings', *International Journal of Sustainable Engineering*, 28(2), 78–93.

Smith, J. A. (2005) 'Optimizing Building Performance: A Comprehensive Approach', *Building Science Journal*, 10(2), 45–60.

Smith, T. L., et al. (2018) 'Optimizing Daylighting Strategies for Energy-Efficient Building Design: A Case Study in Commercial Buildings', *Renewable Energy and Sustainable Development*, 21(1), 45–60.

Thompson, H. C., & Brown, L. E. (2013) 'Integration of Machine Learning Techniques for Building Energy Performance Prediction', *Sustainable Architecture and Design*, 25(3), 78–92.

Turner, M. J., et al. (2014) 'A Framework for Multi-Objective Optimization of Building Energy Performance', *Energy and Sustainable Development*, 18(1), 55–68.

Verma, V., Meena, C. S., Thangavel, S., Kumar, A., Choudhary, T., & Dwivedi, G. (2023) 'Ground and solar assisted heat pump systems for space heating and cooling applications in the northern region of India – A study on energy and CO2 saving potential', *Sustainable Energy Technologies and Assessments*, 59, 103405, ISSN 2213-1388, https://doi.org/10.1016/j.seta.2023.103405.

Wang, Z., & Liu, H. (2019) 'Predictive Modeling of Indoor Air Quality for Occupant Comfor in Green Buildings', *Building Simulation*, 32(4), 365–378.

Williams, A. P., et al. (2011) 'Data-Driven Approaches to Predictive Maintenance in HVAC Systems', *International Journal of Building Performance*, 14(2), 210–225.

Yang, L., & Wu, F. (2019) 'Optimizing Building Envelope Design for Energy Efficiency: A Case Study of Residential Buildings', *Energy and Buildings*, 40(5), 112–128.

Zhang, Q., & Li, Y. (2017) 'Advanced Simulation Techniques for Building Performance Prediction: A Review', *Journal of Building Performance Simulation*, 28(3), 245–260.

Zhang, Y., & Wang, L. (2016) 'A Data-Driven Approach to Predictive Maintenance in Building Systems: A Case Study in Educational Facilities', *Journal of Building Performance*, 19(4), 112–125.

Chapter 12

Energy planning and thermal comfort in buildings

Morteza Khashehchi, Sivasakthivel Thangavel, Pooyan Rahmanivahid, and Milad Heidari

12.1 INTRODUCTION

In the contemporary era of escalating environmental concerns and an ever-growing awareness of the need for sustainable practices, the nexus between energy planning and thermal comfort in buildings has become a focal point in architectural and engineering discourse. The global imperative to mitigate climate change has spurred a reevaluation of traditional approaches to building design, construction, and operation (Ifiok Enobong Mfon & Luna Bassey, 2023). Energy planning, a multifaceted discipline encompassing efficient architectural design, advanced heating, ventilation, and air conditioning (HVAC) systems, and the integration of renewable energy sources, has emerged as a linchpin in the quest for reduced energy consumption and diminished environmental impact (B. Becerik-Gerber et al., 2022). Concurrently, the significance of thermal comfort, once relegated to a mere aspect of climate control, has evolved into a pivotal consideration in building design and operation. Acknowledging the profound influence of thermal conditions on occupant well-being, productivity, and satisfaction, architects and engineers are now compelled to navigate the intricate balance between energy efficiency and human-centric design (Annamaria Buonomano, 2024).

This evolving landscape underscores the imperative for a holistic approach that transcends the conventional boundaries between energy planning and thermal comfort considerations. The challenge lies in developing synergistic strategies that not only minimize energy consumption and carbon footprints but also enhance the livability and satisfaction of building occupants. The interplay between these two domains demands a nuanced understanding of building physics, occupant behaviors, and the dynamic interactions within the built environment (Robert Lowe, 2018). Successful navigation of this intersection involves harnessing cutting-edge technologies, such as smart building systems and data analytics, to optimize energy use while ensuring thermal conditions that promote occupant well-being.

 DOI: 10.1201/9781003496656-12

Furthermore, the exploration of energy planning and thermal comfort in buildings is enriched by a wealth of interdisciplinary insights, drawing from fields as diverse as architecture, engineering, environmental science, and human psychology (C.S. Meena, 2022). As we stand at the intersection of ecological imperatives and the human experience within the built environment, the pursuit of sustainable, energy-efficient, and occupant-centric building solutions becomes not only a technical challenge but a holistic endeavor that integrates scientific rigor with a profound understanding of human needs and aspirations (Mansor & Sheau-Ting, 2021). This comprehensive background establishes the context for a deeper exploration of the symbiotic relationship between energy planning and thermal comfort in buildings, offering a pathway to innovative and transformative practices that align with the imperatives of a sustainable and human-centered future (B. Becerik-Gerber et al., 2022).

12.2 IMPORTANCE OF ENERGY EFFICIENCY IN BUILDINGS

The importance of energy efficiency in buildings cannot be overstated in the contemporary global context marked by environmental concerns, resource scarcity, and the imperative to mitigate climate change. Buildings are substantial consumers of energy, contributing significantly to carbon emissions and resource depletion. Energy-efficient practices in building design, construction, and operation play a pivotal role in reducing the environmental impact of the built environment. Beyond environmental considerations, enhancing energy efficiency is economically prudent, as it reduces operational costs and fosters long-term sustainability (C.S. Meena, 2022). Energy-efficient buildings not only consume less energy but also contribute to the overall resilience of energy infrastructure by alleviating stress on the power grid. Moreover, energy efficiency is inseparable from the pursuit of occupant comfort and well-being, ensuring that buildings are not only environmentally responsible but also conducive to a high quality of life. This approach aligns with the broader goal of creating sustainable, smart cities that prioritize resource conservation and provide healthier, more comfortable living spaces (Ruben Sánchez-Corcuera, 2019). Therefore, the importance of energy efficiency in buildings extends beyond immediate environmental and economic benefits, serving as a cornerstone for the creation of resilient, sustainable, and livable urban environments.

12.3 SIGNIFICANCE OF THERMAL COMFORT FOR OCCUPANTS

The significance of thermal comfort for occupants within the built environment is a crucial dimension that transcends the boundaries of mere climate control, deeply influencing the overall quality of life, productivity, and well-being of individuals. At its core, thermal comfort refers to the state where occupants experience a sense of satisfaction with the thermal conditions in their surroundings, neither feeling too hot nor too cold (S.K. Sansaniwal et al., 2019). This concept is grounded in a complex interplay of physiological, psychological, and environmental factors, making it a cornerstone in contemporary building design, construction, and operational considerations (A. Kumar, 2023).

Physiologically, the impact of thermal comfort on occupants is profound. Human performance and cognitive functions are intricately tied to the ambient temperature. Studies have consistently shown that individuals working in environments where thermal conditions are optimal exhibit enhanced concentration, improved problem-solving abilities, and increased overall productivity. Conversely, discomfort arising from extreme temperatures can lead to distractions, irritability, and a decrease in work efficiency. In educational settings, the thermal environment significantly influences students' ability to focus and engage in learning activities. A well-maintained thermal comfort level is, therefore, not just a matter of occupant preference but a crucial factor in creating environments that support peak cognitive performance and educational outcomes.

Beyond the physiological realm, the psychological dimensions of thermal comfort are equally significant. Individuals have distinct preferences for temperature conditions, and these preferences are influenced by factors such as personal experiences, cultural background, and individual comfort thresholds. Spaces that accommodate and cater to these preferences contribute to a positive psychological experience for occupants. Moreover, the psychological impact extends to mood and overall satisfaction, with people generally reporting higher levels of contentment in spaces where they can maintain a comfortable body temperature. This connection between thermal comfort and psychological well-being highlights the importance of considering occupants' subjective experiences and preferences in the design and operation of buildings.

In the context of workplace environments, where a significant portion of an individual's daily life is spent, the importance of thermal comfort becomes even more pronounced (Zhe Wang et al., 2018). Employers increasingly recognize the impact of the built environment on employee satisfaction, engagement, and, consequently, retention. A workplace that prioritizes thermal comfort fosters a positive organizational culture, contributes to employee well-being, and can even influence recruitment efforts. In the era of remote work, as more individuals design home offices, the significance of thermal comfort extends into residential spaces, impacting the work-life balance and overall satisfaction of individuals.

As the world grapples with the challenges of climate change and the imperative to create sustainable, resilient cities, the significance of thermal comfort gains additional prominence. Striking a balance between energy efficiency and occupant well-being becomes pivotal. Human-centric approaches, such as adaptive comfort models, which consider occupants' ability to adapt to varying thermal conditions, and personalized thermal comfort solutions are becoming essential considerations in modern architectural and engineering practices.

In fact, the significance of thermal comfort for occupants within the built environment is multifaceted and goes beyond the immediate physical experience of temperature. It is intricately linked to physiological health, cognitive performance, psychological well-being, and overall satisfaction. Recognizing and prioritizing thermal comfort in building design and operation is not only an investment in the immediate comfort of occupants but also a strategic move toward creating environments that foster productivity, enhance well-being, and contribute to the overall quality of life.

12.4 RELATIONSHIP BETWEEN ENERGY PLANNING AND THERMAL COMFORT

The primary purpose of this chapter is to systematically address the intricate interplay between energy planning and thermal comfort within the built environment. Recognizing the symbiotic relationship between these two critical aspects of building design and operation, the chapter aims to provide a comprehensive understanding of how energy planning decisions impact the thermal comfort experienced by building occupants. By delving into the complex dynamics of this interconnection, the chapter seeks to elucidate the challenges and opportunities inherent in optimizing both energy efficiency and occupant well-being. Through an exploration of design principles, technological interventions, and case studies, the chapter aims to offer insights and practical strategies for achieving a harmonious balance between energy planning objectives and the provision of optimal thermal conditions. Ultimately, the purpose is to guide architects, engineers, researchers, and policymakers in making informed decisions that not only enhance the sustainability and energy efficiency of buildings but also prioritize the comfort and satisfaction of those who inhabit these spaces.

12.4.1 Identifying key strategies for optimizing both aspects of building design and operation

The central purpose of this chapter is to identify and elucidate key strategies for optimizing both energy planning and thermal comfort in building design and operation. Recognizing the essential interconnectedness of these elements, the chapter seeks to offer practical insights and actionable approaches that architects, engineers, and stakeholders can employ to achieve a synergistic balance. Through a thorough examination of best practices, innovative technologies, and successful case studies, the chapter aims to distill a set of tangible strategies that promote energy efficiency while simultaneously enhancing thermal comfort for building occupants. By pinpointing these key strategies, the chapter aims to provide a valuable resource for professionals and decision-makers in the field, offering guidance on how to navigate the complex landscape of energy planning and thermal comfort considerations. Ultimately, the overarching purpose is to contribute to the development of sustainable, occupant-centric built environments by outlining effective pathways and methodologies for the optimization of both energy performance and thermal conditions in buildings.

12.5 ENERGY PLANNING IN BUILDINGS

Energy planning serves as the foundational pillar in the pursuit of sustainable and efficient building design and operation (Figure 12.1). This section provides a comprehensive overview of energy planning, delineating its

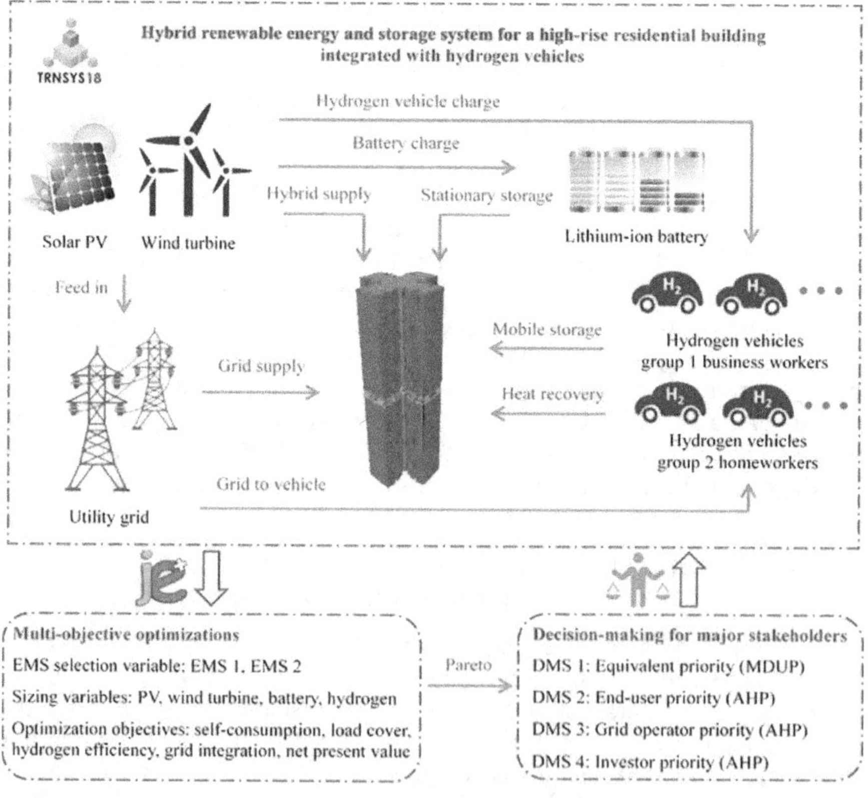

Figure 12.1 Multi-objective optimizations in buildings (Jia Liu et al., 2021)

definition, scope, objectives, and the multifaceted factors that influence its formulation and implementation.

12.5.1 Definition and scope of energy planning

Energy planning, in the context of the built environment, refers to a strategic and systematic approach to managing energy resources throughout the lifecycle of a building. It encompasses a broad spectrum of considerations, from the initial design and construction phases to the operational and maintenance stages. The overarching goal is to optimize energy use, minimize waste, and reduce the environmental impact of a building's energy consumption. Energy planning is not merely a technical endeavor but a holistic process that integrates environmental, economic, and social considerations.

The scope of energy planning extends across various dimensions, including architectural design, HVAC systems, lighting, and the integration of renewable energy sources. It involves a careful examination of energy needs and patterns specific to the building type, location, and purpose. Moreover, energy planning must adapt to the dynamic nature of energy technologies and evolving environmental standards, making flexibility a key aspect of its scope (M. Ahmadizadeh, 2024).

12.5.2 Objectives and goals of energy planning

The objectives of energy planning are multifaceted and align with broader sustainability goals. Key objectives include the reduction of energy consumption, the optimization of energy efficiency, and the integration of renewable energy sources. Energy planning aims to mitigate the environmental impact of buildings by decreasing carbon emissions, lowering reliance on non-renewable resources, and fostering resilience in the face of fluctuating energy markets (Z. Razaviyn, 2024).

Setting specific goals within the energy planning framework is essential for measuring success and progress. These goals may include achieving a certain level of energy efficiency, implementing renewable energy systems to meet a percentage of the building's energy needs, or aligning with recognized sustainability certifications and standards.

12.5.3 Factors influencing energy planning

Several interrelated factors influence the energy planning process, necessitating a holistic and interdisciplinary approach. Building design and construction materials play a pivotal role, influencing the thermal characteristics and overall energy performance of a structure. HVAC (heating, ventilation, and air conditioning) systems are critical components, impacting both energy consumption and occupant comfort. Additionally, the geographical location of a building, local climate conditions, and regional energy availability influence energy planning decisions. Energy planning also intersects with emerging technologies and innovations, such as smart building systems and the Internet of Things (IoT). These technologies offer opportunities to optimize energy use through real-time data analysis, predictive modeling, and adaptive control systems.

12.5.4 Energy modeling and simulation

The importance of accurate energy modeling and simulation cannot be overstated in the energy planning process. These tools enable designers and engineers to assess the potential impact of various strategies before implementation. Energy modeling allows for the evaluation of different design scenarios, HVAC system configurations, and renewable energy integrations, providing valuable insights into the expected energy performance of a building.

Simulation tools, coupled with advanced computational methods, enable dynamic analyses that consider variations in occupancy, climate conditions, and operational parameters. This predictive capability empowers decision-makers to make informed choices early in the design phase, minimizing the need for costly retrofits and optimizing energy efficiency from the outset.

This overview of energy planning sets the stage for a deeper exploration of its intricate components and their role in achieving the dual objectives of energy efficiency and occupant comfort. The subsequent sections will delve into specific factors, strategies, and case studies that exemplify the integration of energy planning into building design and operation. Through this comprehensive exploration, the chapter aims to provide a robust foundation for professionals and stakeholders seeking to navigate the complex landscape of energy planning in the pursuit of sustainable and comfortable built environments.

12.5.5 Factors influencing energy planning

The successful execution of energy planning within the built environment hinges on a nuanced understanding of the multifaceted factors that exert influence on this intricate process. This section delves into the key determinants that shape energy planning decisions, emphasizing the interconnected nature of building design, construction, and operation.

12.5.5.1 Building design and construction materials

The design and construction of a building lay the foundation for its energy performance. The choice of materials, architectural design, and overall building form significantly impact the structure's thermal characteristics, insulation properties, and response to external environmental conditions. Energy planning, therefore, necessitates a meticulous consideration of these elements to optimize both energy efficiency and occupant comfort.

Innovations in sustainable building materials, such as high-performance insulation and low-emissivity coatings, offer avenues for improving thermal resistance and reducing heat transfer. Passive design strategies, such as proper orientation, shading devices, and the incorporation of natural ventilation, further enhance a building's energy performance by minimizing the reliance on mechanical systems. Integrating these considerations into the early stages of the design process is critical for aligning architectural intent with energy efficiency goals.

12.5.5.2 HVAC systems

Heating, ventilation, and air conditioning (HVAC) systems constitute a pivotal component in the energy planning paradigm, influencing both energy consumption and occupant comfort. The selection of HVAC systems, their efficiency ratings, and the integration of advanced control mechanisms play a central role in achieving optimal thermal conditions while minimizing energy use. Energy-efficient HVAC systems, including variable refrigerant flow (VRF) systems, geothermal heat pumps, and demand-controlled ventilation, offer solutions for reducing energy consumption and operational costs. Additionally, the implementation of smart controls, occupancy sensors, and predictive analytics enhances the adaptive capacity of HVAC systems, aligning energy use with actual building requirements. Balancing the trade-off between energy efficiency and the need for a comfortable indoor environment is a key challenge in the realm of HVAC systems, and energy planning strategies must carefully navigate this equilibrium.

12.5.5.3 Geographical location and climate conditions

The geographical location of a building introduces a set of contextual factors that profoundly influence energy planning decisions. Local climate conditions, including temperature ranges, humidity levels, and prevailing wind patterns, play a crucial role in determining the energy needs for heating, cooling, and ventilation. Energy planning must adapt to the unique challenges posed by different climates, emphasizing region-specific strategies to maximize efficiency. In colder climates, for instance, considerations may revolve around effective insulation, high-efficiency heating systems, and strategies to capture and utilize solar heat. Conversely, in warmer climates, a focus on shading, natural ventilation, and cooling strategies becomes paramount. Climate-responsive energy planning ensures that buildings are not only energy-efficient but also resilient to the specific environmental challenges of their location.

12.5.5.4 Regional energy availability and policies

The availability and cost of energy resources within a region influence the overall energy planning strategy. Regions with abundant renewable energy resources may prioritize the integration of solar, wind, or geothermal systems. In contrast, areas with limited access to renewable sources may lean toward optimizing energy efficiency through advanced building technologies and energy conservation measures (M. Chitt, 2024 and M. Khashehchi, 2024).

Moreover, regional energy policies and regulations shape the framework within which energy planning operates. Incentives for renewable energy adoption, energy efficiency standards, and carbon reduction initiatives impact the decision-making process. Energy planners must navigate this regulatory landscape, ensuring compliance with local standards while seeking opportunities to exceed them.

12.5.5.5 Emerging technologies and innovations

The rapid evolution of technology introduces dynamic opportunities and challenges for energy planning. Smart building technologies, Internet of Things (IoT) devices, and advanced analytics provide avenues for real-time monitoring, adaptive control, and data-driven decision-making. Energy planning that incorporates these innovations can optimize energy use, enhance system reliability, and respond dynamically to changing environmental conditions and occupant behaviors (N. Dutt, 2024).

In summary, understanding and navigating the intricate interplay of these factors are essential for effective energy planning in the built environment. The subsequent sections will delve into specific strategies and case studies that showcase the successful integration of these considerations into comprehensive energy planning initiatives. By addressing these factors collectively, energy planners can devise strategies that not only optimize energy efficiency but also create environments that prioritize occupant comfort and well-being.

12.6 ENERGY MODELING AND SIMULATION

Energy modeling and simulation stand as indispensable tools within the realm of energy planning, offering a proactive approach to understanding and optimizing the complex dynamics of a building's energy performance (Dhruvi Brijeshbhai Shah et al., 2022). This section explores the significance of these tools, their methodologies, and the transformative impact they wield on the decision-making process in building design and operation.

12.6.1 Importance of energy modeling

Energy modeling serves as a predictive instrument, allowing architects, engineers, and planners to assess and optimize a building's energy performance before construction begins. This anticipatory capability is crucial in achieving energy efficiency goals and ensuring that the designed systems align with sustainability objectives. By providing insights into the potential energy consumption and thermal behavior of a building under various scenarios, energy modeling aids in making informed decisions early in the design phase, reducing the need for costly retrofits and adjustments.

The importance of energy modeling extends beyond the realm of efficiency—it is a foundational element in achieving occupant comfort. Modeling tools consider variables such as thermal comfort indices, indoor air quality, and daylighting, providing a holistic perspective on building performance. This ensures that energy planning not only addresses the reduction of environmental impact but also prioritizes the well-being of those inhabiting the space.

12.6.2 Tools and methodologies

A myriad of tools and methodologies exist to facilitate energy modeling and simulation, ranging from simplified tools for early-stage conceptualization to sophisticated software for detailed analysis. Each tool caters to specific aspects of energy planning, offering a spectrum of capabilities to meet the diverse needs of professionals in the field.

Building Information Modeling (BIM) software integrates architectural, structural, and MEP (mechanical, electrical, plumbing) systems into a cohesive digital model. BIM enables the visualization of design concepts and facilitates the identification of energy performance opportunities from the early stages of design. Simulation software, such as EnergyPlus, OpenStudio, and DesignBuilder, goes further by providing detailed energy analyses based on the building's geometry, materials, occupancy patterns, and system specifications. These tools utilize algorithms to simulate the dynamic interactions between various building components, climate conditions, and occupant behaviors, offering a granular understanding of energy consumption and thermal comfort.

Advanced computational methods, including parametric modeling and machine learning algorithms, enhance the predictive capabilities of energy models. Parametric modeling enables the exploration of multiple design iterations by systematically varying input parameters, while machine learning algorithms leverage historical data to make predictions and optimize energy performance based on real-time information.

12.6.3 Integration into design and decision-making

The integration of energy modeling into the design and decision-making process is critical for realizing its full potential. Early-stage conceptual modeling helps architects explore design alternatives that balance aesthetic, functional, and energy performance considerations. Iterative simulations guide decision-making by quantifying the energy impact of design choices, allowing for the identification of optimal solutions that achieve both efficiency and comfort.

During the design development phase, detailed energy simulations inform the selection of building systems, materials, and operational strategies. By evaluating different HVAC configurations, insulation levels, and renewable energy integrations, energy modeling enables a comprehensive understanding of trade-offs and synergies. This iterative process aligns energy planning with the broader goals of sustainability and occupant well-being, fostering a balanced approach that considers both technical and human-centric aspects (V. Verma, 2024).

12.6.4 Benefits and challenges

The benefits of incorporating energy modeling into the energy planning process are manifold. By predicting energy consumption and thermal performance, modeling tools empower stakeholders to make informed decisions that align with sustainability goals, regulatory requirements, and occupant comfort standards. The iterative nature of modeling allows for continuous refinement and optimization, contributing to the development of high-performance, environmentally responsible buildings.

However, challenges exist in the successful implementation of energy modeling. Accurate modeling requires precise input data, including building geometry, material properties, and occupancy patterns. Obtaining this data can be challenging, and inaccuracies in the input can lead to discrepancies between predicted and actual performance. Moreover, the complexity of simulation tools demands a level of expertise that may not be readily available to all practitioners. Addressing these challenges requires a commitment to data accuracy, ongoing training, and collaboration among architects, engineers, and energy modelers.

In conclusion, energy modeling and simulation play a pivotal role in the energy planning process, offering a forward-looking approach that integrates technical precision with the broader goals of sustainability and occupant well-being. The subsequent sections will delve into specific case studies that highlight the successful application of energy modeling, showcasing its transformative impact on building design and operation. Through a nuanced exploration of these tools, the chapter aims to provide a comprehensive understanding of their role in optimizing both energy efficiency and thermal comfort within the built environment.

12.7 THERMAL COMFORT: FUNDAMENTALS AND CONSIDERATIONS

Thermal comfort is a cornerstone of human-centered design within the built environment, influencing occupant satisfaction, productivity, and overall well-being. This section delves into the fundamentals of thermal comfort, exploring its definition, metrics, and the myriad factors that contribute to creating environments where occupants feel optimally at ease.

12.7.1 Definition and metrics

At its core, thermal comfort refers to the state in which individuals perceive the thermal environment of a space as acceptable, neither feeling too hot nor too cold. Achieving thermal comfort involves striking a delicate balance between various environmental factors, personal variables, and the dynamic nature of human physiology. The American Society of Heating, Refrigerating, and Air Conditioning Engineers (ASHRAE) defines thermal comfort as "that condition of mind which expresses satisfaction with the thermal environment."

Metrics such as the predicted mean vote (PMV) and predicted percentage dissatisfied (PPD) are commonly employed to quantify thermal comfort. PMV quantifies the mean thermal sensation of a group of occupants, while PPD represents the percentage of individuals within that group who may find the thermal environment dissatisfactory. These metrics provide a standardized framework for assessing and designing thermal conditions that align with the expectations and preferences of building occupants.

12.7.2 Factors influencing thermal comfort

Achieving thermal comfort is a multifaceted challenge influenced by an array of factors that extend beyond air temperature alone. The combination of air temperature, humidity, air velocity, and personal factors such as clothing and activity level collectively shapes the perceived thermal environment (Ghogare Abhijeet Ganesh et al., 2021).

12.7.2.1 Air temperature

Perhaps the most apparent factor, air temperature sets the foundation for thermal comfort considerations. However, it is crucial to note that individual preferences and perceptions of comfort can vary widely, and a fixed temperature may not universally satisfy all occupants.

12.7.2.2 Humidity

The moisture content in the air significantly influences thermal comfort. High humidity levels can impede the body's ability to cool itself through evaporation, leading to discomfort even at moderate temperatures.

Conversely, excessively low humidity can result in dry skin and respiratory discomfort.

12.7.2.3 Air velocity

The movement of air, whether through natural ventilation or mechanical means, impacts thermal comfort. A gentle breeze can enhance perceived comfort by promoting evaporative cooling, while stagnant air may lead to a perception of warmth.

12.7.2.4 Clothing and activity level

Personal factors, such as the type and amount of clothing worn and the level of physical activity, are integral to thermal comfort. Individuals engaged in strenuous activities may prefer cooler conditions, while those clad in heavy clothing may seek a warmer environment.

12.7.3 Human-centric design principles

As the understanding of thermal comfort has evolved, so too has the approach to design within the built environment. Human-centric design principles acknowledge the individuality of comfort preferences and the dynamic nature of thermal sensation. Adaptive comfort models, such as those developed by Fanger and de Dear, recognize that occupants can acclimate to a range of thermal conditions over time, challenging the notion of a universal comfort standard.

Personalized thermal comfort solutions, often facilitated by smart building technologies, allow occupants to exert a degree of control over their immediate environment. From adjustable HVAC settings to personalized climate zones, these technologies enable a more nuanced and responsive approach to meeting diverse comfort requirements.

12.7.4 Challenges in achieving thermal comfort

While the science of thermal comfort has advanced significantly, challenges persist in consistently achieving optimal conditions for all occupants. Variability in individual preferences, transient factors like metabolic rate, and the subjective nature of comfort pose challenges for designers and engineers seeking to create universally satisfying environments.

Balancing thermal comfort with energy efficiency adds another layer of complexity. Designing for comfort often involves increased energy consumption, particularly in climates that demand extensive heating or cooling. Striking an equilibrium between energy efficiency goals and occupant satisfaction remains an ongoing challenge, necessitating innovative solutions and a holistic approach to building design and operation.

12.7.5 Emerging trends and technologies

In the pursuit of enhanced thermal comfort, emerging trends and technologies are reshaping the landscape of building design. Advanced materials with improved insulation properties and phase-change materials that can store and release energy based on environmental conditions offer novel approaches to temperature regulation. Additionally, wearable technologies that monitor individual physiological responses and preferences contribute to the personalization of thermal comfort solutions.

The integration of artificial intelligence (AI) and machine learning into building management systems allows for real-time adjustments based on occupant behavior and preferences. These technologies, coupled with IoT devices and data analytics, enable a more responsive and adaptive approach to thermal comfort management, aligning building conditions with the ever-evolving needs and expectations of occupants.

In conclusion, the exploration of thermal comfort fundamentals and considerations provides a foundational understanding of the complexities inherent in creating environments that prioritize occupant well-being. From the definition and metrics to the multifactorial influences and challenges, the chapter underscores the dynamic nature of thermal comfort within the context of human-centric design. As the field continues to evolve, with advancements in both theory and technology, the quest for achieving thermal comfort remains central to the overarching goal of creating built environments that are not only sustainable and energy-efficient but also inherently responsive to the diverse and evolving needs of their occupants.

12.8 INTEGRATION OF ENERGY PLANNING AND THERMAL COMFORT

The seamless integration of energy planning and thermal comfort represents a pivotal paradigm shift in the field of building design and operation. This section explores the synergies, challenges, and innovative strategies that underpin the convergence of these two critical aspects, emphasizing the importance of creating environments that are not only energy-efficient but also conducive to optimal occupant well-being (R. Mansor & L. Sheau-Ting, 2020).

12.8.1 Symbiotic relationship: Energy planning and thermal comfort

The traditional dichotomy between energy efficiency and occupant comfort is giving way to a more holistic approach that recognizes the symbiotic relationship between these two imperatives. Energy planning, which historically focused primarily on reducing consumption and environmental impact, is increasingly embracing a human-centric perspective. Simultaneously, considerations of thermal comfort are expanding beyond individual comfort preferences to encompass broader sustainability goals.

In this integrated paradigm, the design and operation of buildings are approached with a balanced consideration of both technical and human factors. Architectural decisions, HVAC system configurations, and renewable energy integrations are evaluated not only for their energy performance implications but also for their impact on indoor environmental quality and occupant experience. The goal is to achieve a harmonious equilibrium where energy efficiency complements rather than compromises thermal comfort, and vice versa (P. K. Kushwaha, 2023).

12.8.2 Design principles for integration

The successful integration of energy planning and thermal comfort begins at the design phase. Human-centric design principles prioritize the creation of spaces that promote occupant well-being while optimizing energy use. Passive design strategies, such as natural ventilation, daylighting, and the strategic placement of thermal mass, contribute to both energy efficiency and thermal comfort.

Considerations of building orientation become paramount, leveraging natural elements to harness or deflect solar heat based on seasonal variations. High-performance building envelopes, with enhanced insulation and glazing technologies, not only minimize heat transfer but also contribute to a more stable and comfortable indoor environment. The integration of these principles into the early stages of design lays the groundwork for a built environment that seamlessly marries energy efficiency with thermal comfort.

12.8.3 Adaptive systems and smart technologies

The advent of adaptive building systems and smart technologies has transformed the landscape of energy planning and thermal comfort integration. HVAC systems equipped with advanced sensors and controls can dynamically respond to changing occupancy patterns, weather conditions, and individual comfort preferences. Smart thermostats, often coupled with machine learning algorithms, learn from occupant behaviors to optimize temperature settings, achieving a personalized and energy-efficient balance.

Building management systems, powered by data analytics and the Internet of Things (IoT), enable real-time monitoring and adjustment of various parameters affecting both energy use and thermal comfort. From lighting systems that adjust based on daylight availability to automated shading devices responding to solar angles, these technologies contribute to a more responsive and adaptable built environment.

12.8.4 Performance metrics and feedback loops

The integration of energy planning and thermal comfort is reinforced by robust performance metrics and feedback loops. Continuous monitoring of energy consumption, indoor air quality, and occupant feedback creates a data-driven framework for refining building performance. Post-occupancy evaluations, often utilizing occupant surveys and comfort assessments, provide valuable insights into the effectiveness of design and operational strategies.

Performance metrics such as the total equivalent temperature difference (TETD) and the building performance index (BPI) offer quantifiable measures of both energy efficiency and thermal comfort. These metrics serve as benchmarks for assessing the success of integration strategies and guiding improvements over time. The establishment of feedback loops ensures that the built environment remains responsive to changing conditions, evolving occupant preferences, and advancements in technology.

12.8.5 Challenges and future directions

Despite the promising strides in integrating energy planning and thermal comfort, challenges persist in achieving a seamless and universally applicable approach. Balancing competing priorities, such as the need for energy conservation versus the desire for personalized thermal conditions, requires careful consideration. Designers and planners must navigate the trade-offs between energy efficiency and thermal comfort without compromising either aspect.

The future of integration lies in further advancements in technology, increased interdisciplinary collaboration, and a deeper understanding of occupant behavior and preferences. Emerging technologies, such as occupant-centric environmental controls and wearable devices that interface with building systems, hold the potential to create more adaptive and responsive environments. Interdisciplinary collaborations that bring together experts in architecture, engineering, psychology, and data science can contribute to a more nuanced and comprehensive approach to integration.

The integration of energy planning and thermal comfort represents a transformative approach to building design and operation, transcending the conventional boundaries between energy efficiency and occupant well-being. By adopting human-centric design principles, leveraging adaptive systems and smart technologies, and establishing robust performance metrics, the built environment can evolve into a harmonious space that optimally balances energy use with occupant comfort. As challenges are addressed and innovative solutions continue to emerge, the integration of energy planning and thermal comfort stands as a cornerstone in the pursuit of sustainable, resilient, and occupant-centric built environments.

12.9 CONCLUSION

In the exploration of the integration of energy planning and thermal comfort, it becomes evident that this synergy represents a fundamental shift in the philosophy of building design and operation. The conventional dichotomy between energy efficiency and occupant comfort is evolving into a holistic approach that recognizes the interconnectedness of these two crucial aspects. This concluding section encapsulates the key insights, implications, and transformative potential of seamlessly integrating energy planning and thermal comfort within the built environment.

12.9.1 Human-centric design

Human-centric design principles are at the forefront of this integration, emphasizing the importance of understanding and prioritizing occupant needs. Passive design strategies, adaptive systems, and smart technologies all converge to create spaces that align with both technical energy efficiency goals and the subjective experiences of occupants.

12.9.2 Adaptive technologies

The emergence of adaptive building systems and smart technologies has ushered in a new era of responsiveness in building design. HVAC systems that adjust based on real-time occupancy, smart thermostats that learn from user preferences, and data-driven building management systems exemplify the adaptability and dynamism that define this integration.

12.9.3 Data-driven decision-making

Robust performance metrics and feedback loops play a pivotal role in refining building performance over time. The continuous monitoring of energy consumption, indoor environmental quality, and occupant feedback establishes a data-driven framework for making informed decisions and improvements.

12.9.4 Continuous adaptation

The built environment is not static, and successful integration is contingent on continuous adaptation. Emerging technologies, advancements in materials, and evolving occupant expectations all underscore the need for an agile and responsive approach to building design and operation.

As the field continues to evolve, the integration of energy planning and thermal comfort stands as a beacon for future-oriented building design. The journey toward harmonizing energy efficiency and occupant comfort is dynamic, multifaceted, and holds the promise of creating spaces that are not only environmentally responsible but also enriching, comfortable, and conducive to the flourishing of those who inhabit them. The chapter's exploration serves as an invitation to embrace this transformative potential, recognizing that the pursuit of sustainable and human-centric built environments is an ever-evolving and collaborative endeavor.

REFERENCES

A. Buonomano, C. Forzano, V. M. Gnecco, I. Pigliautile, A. L. Pisello, G. Russo, 2024, "Enhancing energy efficiency and comfort with a multi-domain approach: Development of a novel human thermoregulatory model for occupant-centric control," *Energy and Buildings*, Volume 303, 15 January 2024, 113771.

A. Kumar, V. P. Singh, C. S. Meena, N. Dutt, 2023, *Thermal Energy Systems: Design, Computational Techniques and Applications*, Taylor & Francis (CRC Press), https://doi.org/10.1201/9781003395768.

B. Becerik-Gerber, G. Lucas, A. Aryal, M. Awada, M. Bergés, S. Billington, O. Boric-Lubecke, A. Ghahramani, A. Heydarian, C. Höelscher, F. Jazizadeh, A. Khan, J. Langevin, R. Liu, F. Marks, M. L. Mauriello, E. Murnane, H. Noh, M. Pritoni, S. Roll, D. Schaumann, M. Seyedrezaei, J. E. Taylor, J. Zhao, R. Zhu, 2022, "The field of human building interaction for convergent research and innovation for intelligent built environments," *Scientific Reports*, Volume 12, Article number: 22092.

C. S. Meena, A. Kumar, S. Jain, A. U. Rehman, S. Mishra, 2022, "Innovation in green building sector for sustainable future," *Energies*, Volume 15, 6631, https://doi.org/10.3390/en15186631.

C. S. Meena, A. N. Prajapati, A. Kumar, M. Kumar, 2022, "Utilization of solar energy for water heating application to improve building energy efficiency: An experimental study," *Buildings*, Volume 12, 2166, https://doi.org/10.3390/buildings12122166.

D. Shah, H. Kathiriya, H. Suthar, P. Pandya, J. Soni, 2022, "Enhancing the building's energy performance through building information modelling—A review," in *Recent Trends in Construction Technology and Management: Select Proceedings of ACTM*; Ranadive, M.S., Das, B.B., Mehta, Y.A., Gupta, R. Editors; Lecture Notes in Civil Engineering, Volume 260. Springer, Singapore, 247–253.

G. A. Ganesh, S. L. Sinha, T. N. Verma, S. K. Dewangan, 2021, "Investigation of indoor environment quality and factors affecting human comfort: A critical review," *Building and Environment*, Volume 204, 15 October 2021, 108146.

I. E. Mfon, L. Bassey, 2023, "Sustainable materials and construction practices in industrial buildings," *International Journal of Developmental Studies and Environmental Monitoring*, Volume 2, Issue 1, https://doi.org/10.6084/m9.figshare.24514870.

J. Liu, S. Cao, X. Chen, H. Yang, J. Peng, 2021, "Energy planning of renewable applications in high-rise residential buildings integrating battery and hydrogen vehicle storage," *Applied Energy*, Volume 281, 116038.

M. Ahmadizadeh, M. Heidari, S. Thangavel, E. A. Naamani, M. Khashehchi, V. Verma, A. Kumar, 2024, "Technological advancements in sustainable and renewable solar energy systems," in *Highly Efficient Thermal Renewable*

Energy Systems: Design, Design, Optimization and Applications; V. Verma, S. Thangavel, N. Dutt, A. Kumar, R. Weerasinghe Editors; CRC Press: Boca Raton, FL, Chapter 02; ISBN 9781032595641.

M. Chitt, S. Thangavel, V. Verma, A. Kumar, 2024, "Green hydrogen productions: Methods, designs and smart applications," in *Highly Efficient Thermal Renewable Energy Systems: Design, Design, Optimization and Applications*; V. Verma, S. Thangavel, N. Dutt, A. Kumar, R. Weerasinghe Editors; CRC Press: Boca Raton, FL; Chapter 16; ISBN 9781032595641.

M. Khashehchi, S. Thangavel, P. Rahmanivahid, M. Heidari, T. Moazzeni, V. Verma, A. Kumar, 2024, "Solar desalination techniques: Challenges and opportunities," in *Highly Efficient Thermal Renewable Energy Systems: Design, Design, Optimization and Applications*; V. Verma, S. Thangavel, N. Dutt, A. Kumar, R. Weerasinghe Editors; CRC Press: Boca Raton, FL; Chapter 19; ISBN 9781032595641.

N. Dutt, A. Hedau, A. Kumar, M. K. Awasthi, S. Hedau, C. S. Meena, 2024, "Thermo-hydraulic performance investigation of solar air heater duct having staggered D-shaped ribs: Numerical approach," *Heat Transfer*, 1–31, https://doi.org/10.1002/htj.22998.

P. K. Kushwaha, N. K. Sharma, A. Kumar, C. S. Meena, 2023, "Recent advancements in augmentation of solar water heaters using nanocomposites with PCM: Past, present, and future," *Buildings*, 13, 79, https://doi.org/10.3390/buildings13010079.

R. Lowe, L. F. Chiu, T. Oreszczyn, 2018, "Socio-technical case study method in building performance evaluation," *Building Research & Information*, Volume 46, Issue 5, 469–484.

R. Mansor, L. Sheau-Ting, 2021, "Criteria for occupant well-being: A qualitative study of Malaysian office buildings," *Building and Environment*, Volume 186, 107364.

R. Sánchez-Corcuera, A. Nuñez-Marcos, J. Sesma-Solance, J. Bilbao-Jayo, R. Mulero, U. Zulaika, A. Almeida, 2019, "Smart cities survey: Technologies, application domains and challenges for the cities of the future," *International Journal of Distributed Sensor Networks*, Volume 15, Issue 6, 1550147719853984.

S. K. Sansaniwal, J. Mathur, S. Mathur, 2019, "Review of practices for human thermal comfort in buildings: Present and future perspectives," *International Journal of Ambient Energy*, Volume 43, Issue 1, 2097–2123. https://doi.org/10.1080/01430750.2020.1725629.

V. Verma, S. Thangavel, N. Dutt, A. Kumar, R. Weerasinghe, 2024, "Recent development of thermal energy storage: Solar, geothermal and hydrogen energy," in *Highly Efficient Thermal Renewable Energy Systems: Design, Design, Optimization and Applications*; V. Verma, S. Thangavel, N. Dutt, A. Kumar, R. Weerasinghe Editors; CRC Press: Boca Raton, FL; Chapter 01; ISBN 9781032595641.

Z. Razaviyn, M. Heidari, V. Verma, S. Thangavel, A. Kumar, K. K. Saxena, G. Dwivedi, 2024, "Numerical simulation of a marine energy convertor based on vortex induced vibrations," *Proceedings of the Institution of Mechanical Engineers, Part E: Journal of Process Mechanical Engineering* (In Press).

Z. Wang, R. de Dear, M. Luo, B. Lin, Y. He, A. Ghahramani, Y. Zhu, 2018, "Individual difference in thermal comfort: A literature review," *Building and Environment*, Volume 138, 15 June 2018, 181–193.

Chapter 13

Renewable energy technologies application in sustainable buildings

Jitendra Yadav, Varun Pratap Singh, and Ashwani Kumar

13.1 INTRODUCTION TO SUSTAINABLE BUILDINGS

The opening of this chapter provides a thorough exploration of the evolution of renewable energy technologies, connecting their past uses to their current importance in sustainable construction methods. Examining the historical context, the chapter demonstrates the long-standing connection between renewable energy sources such as windmills, water wheels, and biomass, and the advancement of human civilization. Shifting to the present, it offers a comprehensive examination of the worldwide market situation, providing valuable observations on the current status of renewable energy technology and predicting future developments. The review encompasses the use of various technologies in sustainable structures, exploring the incorporation of solar, wind, geothermal, and biomass systems. The debate highlights the factors that are driving the adoption of technology in response to the changing landscape. These factors include regulatory backing, economic incentives, and the need to overcome technical obstacles. In addition, the chapter examines predictions made by end-users, differentiating between the viewpoints of industrial and residential sectors, highlighting the historical development, and outlining the significant technical possibilities.

The text explores the theoretical foundations of renewable energy technologies, including the principles of photovoltaics, the dynamics of wind turbines, the mechanics of geothermal heat transmission, and the processes of biomass energy conversion. The discussion encompasses approaches to increase performance and characteristics, spanning from increasing efficiency in solar panels to introducing novel design concepts for wind turbines, hence emphasizing key topics being examined. The significance of technology is emphasized through talks on the integration of architecture, selection of building materials, and implementation of smart technologies. Additionally, an analysis of intellectual property rights (IPRs) on a

DOI: 10.1201/9781003496656-13

worldwide scale offers valuable insights into the inventive environment. The chapter explores many developing study areas globally, including interdisciplinary efforts and advancements where architecture intersects with renewable technology [1].

The narrative predicts future advancements and highlights areas for more study and difficulties, including undiscovered opportunities and technology deficiencies in storage and grid integration. A critical summary provides a concise overview of the main discoveries and patterns while identifying areas of study that have not been explored and problems that remain unsolved in the current body of literature. The chapter concludes by concisely summarizing its goals—offering a comprehensive examination of historical origins, analyzing the worldwide market, investigating applications, drivers, and challenges, exploring perspectives from end-users and industrial history, evaluating technological fundamentals and significance, understanding intellectual property rights and emerging research areas, and outlining future research directions [2].

13.1.1 Overview of sustainable design

Sustainable design is a significant change in the architectural and construction business. It involves incorporating renewable energy technology to create environmentally friendly and energy-efficient structures. Sustainable design aims to minimize the environmental effect of structures during their entire lifespan by prioritizing resource efficiency, carbon footprint reduction, and the utilization of eco-friendly materials. Within the realm of renewable energy technology, the primary objective is to utilize nature's plentiful resources to fulfill energy demands while reducing dependence on non-renewable sources. Solar energy, which is highly prevalent, encompasses the integration of photovoltaic systems, solar thermal technologies, and passive solar design concepts. Wind power is harnessed by using wind turbines, whereas geothermal systems utilize the Earth's heat stores for heating and cooling. Biomass technologies, which make use of organic resources, contribute to the expansion of the renewable energy portfolio. The review examines the symbiotic relationship between sustainable design concepts and renewable energy applications, which not only tackles environmental issues but also brings economic advantages by lowering operational expenses and enhancing energy self-sufficiency. Sustainable design utilizes creative architectural integration and a comprehensive approach to building planning to convert structures into dynamic entities that actively contribute to a more sustainable future. This chapter provides a comprehensive exploration of the practical uses, difficulties, and potential future developments of incorporating renewable energy technology into sustainable building design. It offers useful knowledge for professionals, policymakers, and scholars alike [3].

13.1.2 Importance of renewable energy integration

The growing acknowledgment of the integration of renewable energy as a crucial component in worldwide efforts to address environmental issues, strengthen energy stability, and promote sustainable development is emphasized by many significant elements. One of the most important benefits of renewable energy is its significant beneficial effect on the environment. Renewable sources produce very little greenhouse gas emissions, which helps address climate change and decreases air pollution. In order to reduce carbon emissions and limit global temperature increase, the implementation of renewable energy technology is crucial in mitigating climate change. Moreover, the plentiful and extensive dispersion of renewable sources reduces reliance on limited and politically sensitive fossil fuel supplies, therefore improving resource security. Furthermore, the renewable energy industry acts as a powerful driver for economic expansion, employment generation, and technical advancement [4]. Due to technology breakthroughs and economies of scale, the costs of renewable sources are decreasing, making them more competitive and appealing. Diversifying energy sources promotes both energy independence and the development of robust and self-reliant energy infrastructures. Moreover, renewable projects that prioritize the community have immediate social advantages, enabling local inhabitants to take control and promote sustainable habits. The integration of renewable energy goes beyond environmental considerations and includes economic, social, and geopolitical aspects. It is a crucial and revolutionary approach as the globe moves toward a sustainable energy future [5].

13.2 RENEWABLE ENERGY TECHNOLOGIES SYSTEMS APPLICATION IN SUSTAINABLE BUILDINGS

Within the context of sustainable buildings, the comprehensive examination of renewable energy technologies comprises a wide array of inventive solutions designed to diminish environmental harm and improve energy efficiency. These technologies play a crucial role in converting traditional building methods into sustainable and environmentally friendly structures. The main elements consist of solar energy systems, which incorporate photovoltaic (PV) and solar thermal systems. These systems directly convert sunshine into electricity and absorb solar heat to fulfill various building requirements. Wind power technologies, which utilize wind turbines to generate energy, and geothermal systems, which employ heat pumps to access the Earth's constant temperature, provide sustainable options for power and HVAC needs [6]. Biomass cogeneration is the process of converting biological waste into electricity. Architectural design integration involves including solar components and wind-responsive features, which combines renewable energy generation with aesthetic concerns. Energy storage systems, such as battery storage, improve dependability by storing surplus energy, while smart building technologies, such as energy management systems and IoT integration, optimize energy use [7]. The continuous development of sustainable construction solutions is demonstrated by emerging technologies such as transparent solar cells and piezoelectric flooring [8]. As the demand for environmentally conscious practices increases, it is crucial to include these technologies in building design and operations to achieve energy efficiency, decrease carbon emissions, and create durable and sustainable buildings. Figure 13.1 and Table 13.1 discuss the various aspects

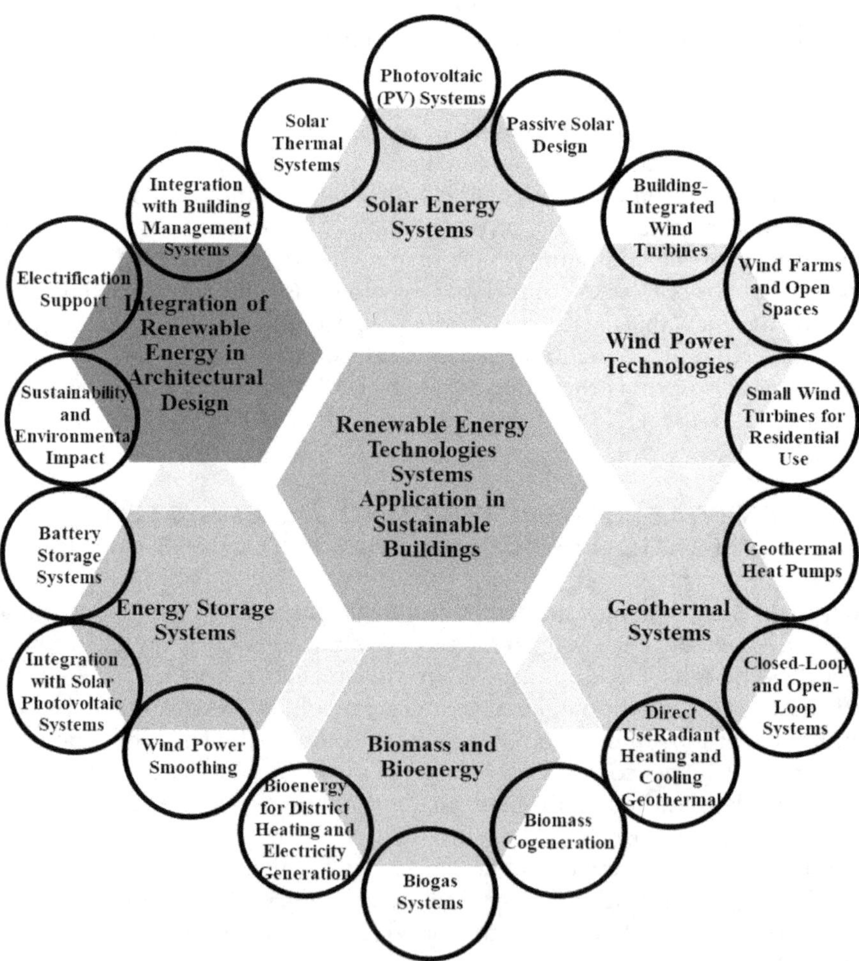

Figure 13.1 Application of different renewable energy technologies systems in sustainable buildings

Table 13.1 Renewable energy technologies systems application in sustainable buildings

Technology	*Parameters*	*Solar Thermal*	*Solar PV*	*Wind Power*		*Geothermal*	*Biomass*	*Energy Storage*
Solar Thermal Collectors	Efficiency (%)	30–70%	15–25%					
	Temperature Range (°C)	60–250				90–200		
	Cost per kWh	$0.05–$0.20	$0.03–$0.15					
Solar PV Systems	Efficiency (%)	15–22%	18–25%					
	Capacity Factor (%)	15–25%	20–30%					
	Cost per kWh	$0.03–$0.15	$0.02–$0.10					
Wind Power Systems	Capacity Factor (%)			25–40%	30–45%			
	Turbine Type			Horizontal	Vertical			
	Cost per kWh			$0.02–$0.08	$0.03–$0.12			
Geothermal Systems	Efficiency (%)					10–25%		
	Temperature Range (°C)					90–200		
	Cost per kWh					$0.04–$0.12		
Biomass Systems	Efficiency (%)						25–45%	
	Feedstock Type						Wood, Crop Res	
	Cost per kWh						$0.05–$0.15	
Energy Storage Systems	Efficiency (%)							70–90%
	Cycle Life							5,000–15,000
	Cost per kWh							$0.05–$0.20

of the application of different renewable energy technologies systems in sustainable buildings [9].

13.2.1 Solar energy systems

The integration of solar energy systems in sustainable buildings signifies a fundamental change in how energy is obtained, bringing a revolutionary and ecologically aware method of powering structures. Solar energy, derived from the sun's plentiful and renewable beams, presents itself as a pristine and enduring substitute for traditional power sources [10]. The implementation of solar energy systems in sustainable buildings involves a diverse range of technologies and design concepts to maximize energy output and usage [11].

Photovoltaic (PV) systems are crucial for generating renewable power on-site. Solar panels, which consist of photovoltaic cells, transform sunlight into electricity, providing a decentralized and eco-friendly energy source. Roof-mounted solar arrays, strategically positioned on the rooftops of buildings, optimize sunshine exposure without requiring further land, demonstrating a highly effective utilization of the existing area. Building-integrated photovoltaics (BIPV) involves the integration of solar components with architectural design, allowing for the seamless incorporation of solar windows, glass facades, or solar roof tiles into the building's structure [12].

Solar thermal systems enhance the range of solar technologies, therefore contributing to sustainable solutions for heating and cooling. Solar water heating, accomplished by utilizing solar collectors, diminishes reliance on conventional water heating techniques. In addition, solar air conditioning, which utilizes solar heat as its power source, offers a sustainable alternative to traditional air conditioning systems. Passive solar design measures, such as optimizing the direction of the structure, using shading components, and including daylighting features like skylights, improve energy efficiency by reducing the need for artificial lighting and heating [13].

Solar energy is used not only for generating electricity but also for powering electric cars, among other inventive alternatives. Solar carports, equipped with solar panels on the roof, serve the dual purpose of producing power and offering shaded parking spots for electric vehicles, so encouraging eco-friendly mobility. Off-grid and hybrid solar systems promote energy autonomy by including energy storage capabilities and integrating solar energy with other renewable sources or conventional grid electricity. These systems are particularly beneficial in distant or off-grid regions [14].

Energy storage options, such as battery storage, mitigate the problem of intermittent energy supply by storing surplus solar energy for use during periods of reduced sunshine or increased demand. Intelligent technologies, such as solar tracking systems and energy management systems, enhance the efficiency of solar energy utilization by altering the alignment of solar panels and monitoring consumption patterns, respectively. The use of solar energy systems is economically practical and contributes to the larger aims of sustainability and environmental stewardship due to government subsidies, tax credits, and green building certifications that incentivize its use [15].

The use of solar energy systems in sustainable buildings embodies a varied and comprehensive strategy for sourcing and using energy. In addition to its positive impact on the environment, this integration improves energy self-sufficiency, adaptability, and economic effectiveness, establishing solar energy systems as a crucial element in creating a sustainable and ecologically aware constructed environment. Given the progress of technology and the decreasing costs, it is crucial to widely implement solar energy systems in order to achieve a more sustainable future.

13.2.2 Wind power technologies

The integration of wind power technologies into sustainable buildings signifies a dynamic and forward-thinking approach to harnessing renewable energy for on-site power generation. Wind power, stemming from the kinetic energy of the wind, presents a clean and efficient alternative to conventional energy sources. Building-integrated wind turbines exemplify innovation in sustainable energy solutions. These compact turbines, strategically mounted on building rooftops, and vertical axis wind turbines (VAWT) designed for urban environments open avenues for structures to capture wind energy and convert it into electricity. Roof-mounted turbines, efficient and space-conscious, capture wind energy at elevated levels, while vertical axis wind turbines adapt to variable wind directions in urban settings.

Beyond individual buildings, the utilization of wind farms and open spaces emerges as an opportunity for larger structures or campuses to implement on-site wind farms. The strategic placement of multiple turbines in wind farms efficiently captures wind energy, contributing significantly to the overall energy needs of the building. Moreover, landscaping around buildings can be purposefully designed to channel and accelerate wind flow, optimizing the efficiency of wind turbines. A decentralized approach is also seen in small wind turbines for residential use, where scaled-down turbines integrated into homes or small buildings provide a supplementary source of electricity, contributing to a more sustainable energy mix [16].

Microgrid and hybrid systems further exemplify the integration of wind power into broader energy solutions. Wind-solar hybrid systems combine the benefits of wind and solar energy, ensuring a consistent and reliable renewable energy supply. Microgrids, incorporating wind power, enhance energy resilience and provide off-grid capabilities, particularly in remote areas. Offshore wind power introduces the concept of offshore wind farms for buildings near coastlines, capitalizing on the strong and consistent winds over bodies of water. Smart wind power systems leverage advanced technologies, including wind forecasting and predictive technologies, to analyze weather patterns for optimized turbine operation. Smart grid integration ensures the seamless incorporation of wind power into the broader energy infrastructure, maintaining stability and reliability [17].

The application of wind power technologies not only reduces dependence on traditional energy sources but also contributes to a more resilient and environmentally friendly energy infrastructure. As technology advances and our understanding of wind dynamics improves, the integration of wind power into building design continues to evolve, playing a pivotal role in the ongoing transition to a more sustainable built environment [18].

13.2.3 Geothermal systems

Geothermal systems emerge as a sustainable and efficient solution to address the heating, cooling, and hot water requirements of buildings by harnessing the Earth's internal heat. At the forefront of these systems are geothermal heat pumps, specifically ground source heat pumps (GSHPs), which utilize the constant temperature of the ground to transfer heat into or out of a building. During winter, GSHPs extract heat from the ground for heating, while in summer, excess heat is transferred back into the ground for cooling. The distinction between closed-loop and open-loop systems lies in the manner of heat exchange. Closed-loop systems involve a network of pipes with a heat-transfer fluid buried in the ground, while open-loop systems extract groundwater for heat exchange and then discharge it back to the ground or surface water body [19].

Direct use geothermal systems, such as district heating systems, provide an innovative approach where hot water from underground reservoirs is distributed to nearby buildings for space heating and domestic hot water. Additionally, geothermal finds application in radiant floor heating, circulating warm water through pipes in the floor for efficient and comfortable heating, and radiant cooling, adapting geothermal systems to circulate cooled water through the floor or ceiling in warmer climates. Hybrid systems, combining geothermal with solar energy, present a strategic approach to optimize energy efficiency, with geothermal providing a constant baseline and solar energy contributing during peak demand. Geothermal heat recovery mechanisms, specifically waste heat recovery, further enhance overall energy efficiency by capturing waste heat from industrial processes or power generation [20].

Geothermal retrofitting demonstrates the adaptability of these systems to existing buildings, seamlessly integrating with traditional heating, ventilation, and air conditioning (HVAC) systems. Environmental and economic benefits are paramount, with geothermal systems minimizing carbon emissions compared to conventional systems, contributing significantly to climate change mitigation [21]. Despite potentially higher upfront installation costs, the long-term cost savings associated with lower energy consumption and maintenance requirements make geothermal systems financially attractive. Regulatory support and incentives, such as government incentives, tax credits, and grants, further bolster the adoption of geothermal systems, positioning them as a pivotal element in the transition toward sustainable building practices. As technology advances and awareness of their benefits grows, geothermal systems continue to play a crucial role in shaping the future of sustainable building practices [22].

13.2.4 Biomass and bioenergy

Biomass and bioenergy systems represent a sustainable and renewable avenue for meeting the energy needs of buildings, contributing to an environmentally friendly and circular approach. These systems leverage organic materials to generate various forms of energy, offering a versatile solution for a range of applications. Biomass cogeneration, notably combined heat and power (CHP), exemplifies a key aspect where organic materials like wood chips or agricultural residues are burned to produce steam, driving a turbine for electricity generation while utilizing residual heat for space heating or industrial processes. Biogas systems, achieved through anaerobic digestion of organic waste, yield methane-rich gas suitable for electricity generation, heating, or as a vehicle fuel [23].

Pellet stoves and biomass boilers emerge as practical solutions for residential heating, utilizing biomass pellets made from compressed sawdust or agricultural residues. These systems provide a renewable alternative to traditional fossil fuel-based heating, contributing to decentralized and sustainable energy sources. Biomass for district heating involves centralized biomass-fired systems that generate hot water or steam distributed through a network of pipes to heat multiple buildings. Biomass also plays a vital role in electricity generation through dedicated biomass power plants, offering a renewable and carbon-neutral energy source to the grid [24].

Combined biomass and solar systems showcase hybrid approaches, enhancing overall energy reliability and efficiency. While biomass provides a consistent energy source, solar technologies contribute during periods of sunlight, offering a complementary and sustainable energy mix. Bioenergy derived from agricultural residues, including crop stalks and husks, reduces the environmental impact of agricultural waste, showcasing the versatility of biomass applications [25]. Additionally, bioenergy in waste-to-energy systems, specifically from municipal solid waste (MSW), presents a sustainable solution for waste management through incineration or anaerobic digestion [26].

13.2.5 Energy storage systems

Energy storage systems play a pivotal role in enhancing the reliability, efficiency, and sustainability of buildings, mitigating the intermittency and variability inherent in renewable energy sources. Among the key aspects of their application, battery storage systems stand out, with lithium-ion batteries being widely utilized across residential, commercial, and industrial settings due to their efficient storage and release of electrical energy. Emerging technologies such as flow batteries and solid-state batteries offer additional scalability, flexibility, and improved safety, presenting promising alternatives to traditional lithium-ion batteries [27].

The integration of energy storage systems with solar photovoltaic systems is a critical strategy for sustainable buildings. These systems facilitate peak shaving, storing excess solar energy during sunny periods to reduce dependency on grid electricity during peak demand [28]. Additionally, time shifting ensures a continuous power supply by utilizing stored solar energy during non-sunlight hours, optimizing self-consumption. In the context of wind power, energy storage systems play a key role in stabilizing wind energy output by mitigating variability. They store excess energy during periods of high wind and release it when wind speeds are low, providing a stable and reliable energy supply. Furthermore, energy storage systems contribute to grid stabilization and ancillary services, offering rapid responses to fluctuations in grid frequency and aiding peak load management, thereby enhancing overall grid stability [29].

Microgrid resilience is another critical application, where energy storage enables buildings to operate independently as a microgrid during grid outages, ensuring the continuity of critical functions and services. Demand response and load-shifting strategies are facilitated by energy storage, allowing for optimized energy use during periods of high or low electricity prices and shifting electricity consumption to off-peak hours [30]. Integration with building management systems enables smart building integration, optimizing energy use, reducing peak demand charges, and enhancing overall energy efficiency [31]. Sustainability and environmental impact considerations highlight the role of energy storage systems in reducing the carbon footprint of buildings, optimizing renewable energy use, and extending the lifespan of renewable assets, contributing significantly to the broader transition toward a more sustainable and reliable energy infrastructure [32].

13.2.6 Integration of renewable energy in architectural design

The use of renewable energy in architectural design exemplifies a forward-thinking and comprehensive strategy for producing environmentally friendly and energy-efficient structures. Architects are progressively integrating cutting-edge technology and architectural features to use the potential of renewable sources, while concurrently improving the visual and practical aspects of buildings. Building-integrated photovoltaics (BIPV) exemplifies the seamless integration of solar panels into facades and rooftops, effectively converting surfaces into both energy-generating sources and visually attractive architectural elements. Wind-responsive designs, such as compact and visually appealing wind turbines and optimized building structures, harness wind energy while maintaining the overall aesthetic attractiveness [33]. Daylighting methods involve the strategic placement of windows and the use of light-diverting devices to optimize the amount of natural light, hence minimizing the reliance on artificial lighting. Green roofs and walls enhance insulation and energy efficiency while simultaneously improving the overall environmental quality. Passive solar design maximizes building orientation and incorporates high thermal mass materials to efficiently regulate temperature. Rainwater harvesting technologies and sustainable water features integrate harmoniously into building designs, fostering water conservation. Architects are designing structures that minimize environmental effects and actively involve inhabitants in energy conservation efforts by focusing on sustainable materials, energy-efficient building envelopes, and incorporating smart technology such as IoT integration and energy management systems. The combination of a progressive mindset, educational elements, and intuitive controls promotes a seamless integration of practicality, visual appeal, and ecological consciousness in the constructed surroundings [34].

13.3 BENEFITS OF RENEWABLE ENERGY INTEGRATION IN SUSTAINABLE BUILDINGS

The integration of renewable energy into sustainable buildings offers a myriad of benefits that extend across environmental, economic, and resilience dimensions. A primary advantage is the substantial reduction in the carbon footprint of buildings. Renewable energy sources such as solar, wind, and biomass produce electricity without emitting greenhouse gases, aligning with global efforts to mitigate climate change and transition to a low-carbon economy. This reduction in carbon emissions contributes significantly to environmental conservation and sustainability. Additionally, sustainable buildings that harness renewable energy enjoy energy cost savings by generating their electricity, reducing dependence on conventional energy sources, and mitigating the impact of fluctuating energy prices. Over time, this leads to substantial cost savings for building owners and occupants, fostering economic sustainability [35].

The integration of renewable energy systems enhances energy independence for sustainable buildings. By relying less on traditional grid power, these buildings become more resilient during power outages or disruptions, ensuring a continuous energy supply. Lower operating costs are another key advantage, as renewable energy systems often entail lower maintenance costs compared to conventional sources. Once installed, technologies like solar panels and wind turbines provide long-term, reliable energy with minimal ongoing expenses. Moreover, the incorporation of renewable energy solutions enhances the overall property value and marketability of sustainable buildings, making them increasingly attractive to environmentally conscious investors and tenants. This aligns with the growing demand for green technologies and sustainable practices in the real estate market. Additionally, the integration of renewable energy supports job creation and economic stimulus, contributing to the growth of local industries and communities within the renewable energy sector. Overall, the integration of renewable energy in sustainable buildings represents a holistic approach that addresses environmental concerns while offering economic, social, and resilience benefits, aligning with the principles of sustainable development [36].

The adoption of renewable energy in sustainable buildings is further incentivized by various factors such as government incentives, rebates, and compliance with environmental regulations and certifications. Many governments globally offer financial incentives to encourage the adoption of renewable energy technologies, making their integration into sustainable buildings economically viable. Achieving certifications such as LEED (Leadership in Energy and Environmental Design) or BREEAM (Building Research Establishment Environmental Assessment Methodology) not only showcases a commitment to sustainable practices but also aligns with increasingly stringent environmental regulations. This integration also contributes to community engagement and education by serving as an educational tool and fostering discussions about clean energy within communities [37]. Moreover, the technological advancements and innovation driven by the integration of renewable energy in buildings contribute to the evolution of more efficient and cost-effective technologies in the renewable energy sector. Sustainable building practices, including the use of green roofs and reflective surfaces, mitigate the urban heat island effect, reducing overall energy demand for cooling in urban environments. By diversifying energy sources, sustainable buildings become less vulnerable to price fluctuations and supply chain disruptions associated with conventional energy resources. The positive impact on public image and the contribution to global environmental conservation further emphasize the far-reaching benefits of integrating renewable energy into sustainable buildings [38].

13.4 CHALLENGES AND CONSIDERATIONS

The integration of renewable energy into sustainable buildings, while offering substantial benefits, is accompanied by various challenges and considerations that necessitate careful attention to ensure successful implementation. One primary challenge is the intermittency and variability inherent in renewable energy sources, particularly solar and wind. Factors such as cloud cover, nightfall, and fluctuations in wind speed can impact energy production, requiring effective strategies for energy storage or backup systems to maintain a consistent power supply. The associated costs of high-capacity energy storage systems represent another significant hurdle, necessitating advancements in storage technologies and cost reductions to facilitate widespread adoption [39].

Grid integration poses additional challenges, including stability concerns, voltage fluctuations, and the management of bi-directional power flow. As renewable energy systems become more decentralized, smart grid technologies and regulatory frameworks must evolve to accommodate these changes effectively. Land use and aesthetic concerns arise with the large-scale deployment of renewable energy systems, such as solar farms or wind turbines, prompting the need to strike a delicate balance between energy production and environmental preservation. The upfront capital costs associated with installing renewable energy systems, such as solar panels or wind turbines, can be prohibitive, highlighting the importance of access to financing options and financial incentives to enhance accessibility [40]. The rapid pace of technological advancements in the renewable energy sector may lead to concerns about system obsolescence, emphasizing the need for regular updates and considerations for future upgrades [41].

Inconsistent or outdated regulatory frameworks and policies can create uncertainties for investors and hinder the adoption of renewable energy in buildings. Clear and supportive policies are necessary to incentivize sustainable practices effectively. The success of renewable energy integration also depends on the cooperation and engagement of building occupants [42]. Education and awareness programs are needed to encourage energy-efficient behaviors and ensure the optimal use of renewable energy systems. Supply chain concerns, resource availability, local climate conditions, building design, maintenance, social acceptance, cultural considerations, and the need for a skilled workforce further underscore the multifaceted nature of challenges associated with the integration of renewable energy in sustainable buildings. A holistic and collaborative approach, involving policymakers, architects, engineers, building owners, and the broader community, is crucial to proactively address these issues and facilitate the efficient and successful integration of renewable energy technologies in sustainable building projects [1].

13.5 OPTIMIZING RENEWABLE ENERGY TECHNOLOGIES

Optimizing the integration of renewable energy technologies in sustainable buildings necessitates a comprehensive and multifaceted approach encompassing various key strategies. A foundational step involves conducting site analyses and energy audits to comprehend local conditions and assess the current energy usage of buildings. Smart building design, emphasizing optimal orientation, architectural features for passive heating, and energy-efficient building envelopes, plays a critical role in enhancing overall energy efficiency. Additionally, the deployment of advanced solar technologies, such as bifacial solar panels and solar tracking systems, along with wind turbine optimization and the integration of energy storage solutions, ensures a balanced and reliable energy supply [43].

The incorporation of hybrid renewable energy systems that combine sources like solar and wind, along with intelligent energy management systems and demand response strategies, contributes to overall efficiency. The flexibility of design for future upgrades, involving modular and scalable systems, is pivotal in adapting to evolving energy needs. Collaboration and stakeholder engagement, including architects, engineers, energy experts, and building occupants, facilitate a holistic approach to renewable energy integration. Life cycle assessments guide decision-making from manufacturing to disposal, considering environmental impacts [44]. Continuous monitoring and maintenance, along with exploring incentives, tax credits, and financing options, contribute to the economic viability of renewable energy technologies. Lastly, community education and outreach foster a culture of energy conservation and sustainability, aligning with the principles of sustainable development and creating resilient, energy-efficient structures [36].

13.6 FUTURE TRENDS AND EMERGING TECHNOLOGIES

As the demand for sustainable and energy-efficient buildings continues to rise, several key trends and emerging technologies are shaping the landscape of sustainable construction. Advanced solar technologies, such as building-integrated photovoltaics (BIPV), are evolving to seamlessly integrate solar cells into various building materials, including windows, facades, and roofing [45]. This not only enhances functionality but also contributes to the aesthetics of the building. The development of innovative wind turbines, particularly vertical axis wind turbines (VAWT), is gaining attention for their potential to capture wind energy more efficiently, especially in urban environments where wind patterns can be unpredictable [46].

Next-generation energy storage technologies, including breakthroughs in battery technologies like solid-state batteries, are crucial for enhancing energy storage efficiency, safety, and longevity. These advancements play a pivotal role in balancing the supply and demand for renewable energy. The integration of artificial intelligence (AI) into smart building systems for predictive energy management is becoming more prevalent. AI algorithms analyze data from various sources, such as weather forecasts and occupancy patterns, to optimize the usage of renewable energy in buildings [47].

Renewable hydrogen production is advancing with the use of electrolysis technologies powered by renewable energy, leading to the development of green hydrogen. This serves as a clean energy carrier with applications across various sectors [48]. Additionally, blockchain technology is being explored for transparent and decentralized transactions in renewable energy markets, enabling peer-to-peer energy trading and fostering a more distributed and resilient energy ecosystem. Urban energy harvesting, through innovations like piezoelectric tiles on sidewalks and roads, captures energy from human movement and transportation, contributing to sustainable power generation in densely populated areas [49]. Circular economy principles are gaining prominence in construction practices, emphasizing recycling and repurposing materials to minimize waste. Biogenic photovoltaics, inspired by natural processes like photosynthesis, hold potential for sustainable energy generation by mimicking nature in innovative and environmentally friendly designs. The rise of renewable energy microgrids, socially responsible design emphasizing equitable access to renewable energy benefits, and other cutting-edge trends collectively pave the way for a more resilient and environmentally conscious future in sustainable building practices [50].

13.7 CONCLUSION

In conclusion, the exploration of renewable energy technologies in sustainable buildings reveals a dynamic landscape marked by innovation and transformative possibilities. The integration of these technologies not only addresses the imperative for environmentally conscious construction but also unlocks economic, social, and resilience benefits. Here are key points that encapsulate the findings and provide a pathway forward:

1. The evolution of building-integrated photovoltaics (BIPV) is redefining the aesthetics of sustainable buildings by seamlessly incorporating solar cells into architectural elements.
2. Advanced wind turbine designs, notably vertical axis wind turbines (VAWT), are emerging as efficient solutions for capturing wind energy, especially in urban environments.
3. Breakthroughs in energy storage, including next-generation battery technologies, are pivotal in addressing the intermittent nature of renewable sources, ensuring a continuous and reliable energy supply.
4. Artificial intelligence (AI) is playing a central role in predictive energy management, optimizing renewable energy usage through real-time data analysis and smart building technologies.
5. The production of green hydrogen through electrolysis powered by renewable energy signifies a promising avenue for clean energy carriers and contributes to a more sustainable energy mix.
6. Flexible and organic photovoltaic technologies are expanding design possibilities, allowing solar panels to adapt to various surfaces and architectural forms.
7. The exploration of blockchain technology in energy transactions promotes transparent and decentralized markets, fostering peer-to-peer energy trading and increased energy autonomy.

8. Urban energy harvesting, particularly through piezoelectric innovations, is transforming urban infrastructure into sources of sustainable power, capturing energy from human movement and transportation.

13.7.1 Path forward: A call to action

To propel the sustainable building revolution forward, a multifaceted approach is essential. Firstly, embracing collaborative research is paramount; architects, engineers, researchers, and policymakers must engage in interdisciplinary collaboration to drive innovation in renewable energy technologies. Secondly, prioritizing investment in renewable energy infrastructure is critical. Governments and private sectors should channel resources toward incentivizing the adoption of advanced technologies and sustainable building practices. Thirdly, education becomes a key catalyst for change. Implementing educational programs to raise awareness among building owners, occupants, and communities about the benefits of renewable energy integration is vital for fostering a culture of sustainability. Supporting policy frameworks is equally essential; advocating for and developing clear, supportive policies encourages the widespread adoption of renewable energy technologies in construction. Ensuring equitable access to these technologies is a moral imperative, considering the social and economic diversity of communities and fostering inclusive sustainability. Furthermore, promoting green building certification, such as LEED and BREEAM, serves as a tangible acknowledgment and reward for sustainable building practices and renewable energy integration. Lastly, continuous research and development must be prioritized, with ongoing investment aimed at exploring emerging technologies, addressing challenges, and enhancing the efficiency and affordability of renewable energy solutions. This comprehensive strategy lays the foundation for a resilient, sustainable, and regenerative future in the realm of sustainable building practices.

As we navigate the future of sustainable construction, the convergence of renewable energy technologies with innovative building practices presents an exciting frontier. By implementing these key findings and heeding the call to action, we can collectively shape a built environment that not only meets the demands of the present but also ensures a resilient, sustainable, and regenerative future.

REFERENCES

1. A. GhaffarianHoseini, N. D. Dahlan, U. Berardi, A. GhaffarianHoseini, N. Makaremi, and M. GhaffarianHoseini, "Sustainable energy performances of green buildings: A review of current theories, implementations and challenges," *Renew. Sustain. Energy Rev.*, vol. 25, pp. 1–17, 2013, doi: 10.1016/j.rser.2013.01.010.

2. D. Chwieduk, "Towards sustainable-energy buildings," *Appl. Energy*, vol. 76, no. 1, pp. 211–217, 2003, doi: 10.1016/S0306-2619(03)00059-X.
3. A. Chel and G. Kaushik, "Renewable energy technologies for sustainable development of energy efficient building," *Alexandria Eng. J.*, vol. 57, no. 2, pp. 655–669, 2018, doi: 10.1016/j.aej.2017.02.027.
4. A. Kundu, A. Kumar, V. P. Singh, C. S. Meena, and N. Dutt, "Chapter 1: Introduction to thermal energy resources and their smart applications," in *Thermal Energy Systems: Design, Computational Techniques, and Applications*, 2023, 1st ed., A. Kumar, V. P. Singh, C. S. Meena, and N. Dutt, Eds., Boca Raton, CRC Press, pp. 1–15.
5. P. A. Stergaard, N. Duic, Y. Noorollahi, H. Mikulcic, and S. Kalogirou, "Sustainable development using renewable energy technology," *Renew. Energy*, vol. 146, pp. 2430–2437, 2020, doi: 10.1016/j.renene.2019.08.094.
6. V. P. Singh, A. Karn, G. Dwivedi, T. Alam, and A. Kumar, "Experimental assessment of variation in open area ratio on thermohydraulic performance of parallel flow solar air heater," *Arab. J. Sci. Eng.*, vol. 41, no. 12, pp. 1–17, 2022, doi: 10.1007/s13369-022-07525-7.
7. N. Dutt, A. Jageshwar, H. Ashwani, K. Mukesh, K. Awasthi, and V. Pratap, "Thermo Hydraulic performance of solar air heater having discrete D Shaped ribs as artificial roughness," *Environ. Sci. Pollut. Res.*, 2023, doi: 10.1007/s11356-023-28247-9.
8. G. S. Rathore, S. Rathor, A. Nazeer, A. Karn, and V. P. Singh, "Recent progress in solar dryers using phase changing material: A review," *Int. J. Energy Resour. Appl.*, vol. 2, no. 1, pp. 57–77, 2023, doi: 10.56896/IJERA.2023.2.1.005.
9. B. Rezaie, E. Esmailzadeh, and I. Dincer, "Renewable energy options for buildings: Case studies," *Energy Build.*, vol. 43, no. 1, pp. 56–65, 2011, doi: 10.1016/j.enbuild.2010.08.013.
10. V. P. Singh *et al.*, "Heat transfer and friction factor correlations development for double pass solar air heater artificially roughened with perforated multi-V ribs," *Case Stud. Therm. Eng.*, vol. 39, no. September, p. 102461, 2022, doi: 10.1016/j.csite.2022.102461.
11. Y. Lu, Z. A. Khan, M. S. Alvarez-Alvarado, Y. Zhang, Z. Huang, and M. Imran, "A critical review of sustainable energy policies for the promotion of renewable energy sources," *Sustainability*, vol. 12, no. 12, 2020, doi: 10.3390/su12125078.
12. A. G. Hestnes, "Building integration of solar energy systems," *Sol. Energy*, vol. 67, no. 4, pp. 181–187, 1999, doi: 10.1016/S0038-092X(00)00065-7.
13. V. P. Singh, S. Jain, and A. Kumar, "Establishment of correlations for the Thermo-Hydraulic parameters due to perforation in a multi-V rib roughened single pass solar air heater," *Exp. Heat Transf.*, vol. 35, no. 5, pp. 1–20, 2022, doi: 10.1080/08916152.2022.2064940.
14. N. Dutt, A. Binjola, A. J. Hedau, A. Kumar, V. P. Singh, and C. S. Meena, "Comparison of CFD results of smooth air duct with experimental and available equations in literature," *Int. J. Energy Resour. Appl.*, vol. 1, no. 1, pp. 40–47, 2022, doi: 10.56896/IJERA.2022.1.1.006.
15. M. Temiz and I. Dincer, "Design and assessment of a solar energy based integrated system with hydrogen production and storage for sustainable buildings," *Int. J. Hydrog. Energy*, vol. 48, no. 42, pp. 15817–15830, 2023.

16. A. S. Darwish, "Green buildings to approach sustainable buildings by integrating wind and solar systems with smart technologies," in *Green Buildings and Renewable Energy. Innovative Renewable Energy*, A. Sayigh, Ed. Cham: Springer International Publishing, 2020, pp. 433–446.
17. K. Kaygusuz, "Wind power for a clean and sustainable energy future," *Energy Sources B: Econ. Plan. Policy*, vol. 4, no. 1, pp. 122–133, Jan. 2009, doi: 10.1080/15567240701620390.
18. P. Škvorc and H. Kozmar, "Wind energy harnessing on tall buildings in urban environments," *Renew. Sustain. Energy Rev.*, vol. 152, p. 111662, 2021, doi: 10.1016/j.rser.2021.111662.
19. T. Sivasakthivel, V. Verma, R. Tarodiya, C. S. Meena, V. P. Singh, and R. Kumar, "Chapter 11: Analysis of optimum operating parameters for ground source heat pump system for different cases of building heating and cooling mode operations," in *Thermal Energy Systems: Design, Computational Techniques, and Applications*, 1st ed., A. Kumar, N. Dutt, V. Singh, and C. Meena, Eds. CRC Press, 2023, pp. 183–207.
20. F. Trieb *et al.*, "A cost-benefit analysis of power generation from commercial reinforced concrete solar chimney power plant," *Renew. Sustain. Energy Rev.*, vol. 15, no. 1, pp. 104–113, 2013, doi: 10.1016/j.enconman.2013.11.046.
21. A. Saxena, A. Prajapati, G. Pant, C. Meena, A. Kumar, and V. Singh, "Water consumption optimization of hybrid heat pump water heating system," in *Lecture Notes in Mechanical Engineering- Recent Advances in Mechanical Engineering Select Proceedings of FLAME 2022*, A. Shukla, B. Sharma, A. Arabkoohsar, and P. Kumar, Eds. Springer Nature Singapore, 2023, pp. 721–732.
22. T. A. H. Ratlamwala, M. A. Gadalla, and I. Dincer, "Thermodynamic analyses of an integrated PEMFC–TEARS-geothermal system for sustainable buildings," *Energy Build.*, vol. 44, pp. 73–80, 2012, doi: 10.1016/j.enbuild.2011.10.017.
23. A. R. Singh, S. K. Singh, and V. P. Singh, "Process parameters optimization of carbon nano tube based catalytic transesterification of algal oil," *Mater. Today Proc.*, no. xxxx, 2023, doi: 10.1016/j.matpr.2023.01.418.
24. A. Datta, A. Kumar, A. Kumar, A. Kumar, and V. P. Singh, "Advanced materials in biological implants and surgical tools," in *Advanced Materials for Biomedical Applications*, 1st ed., N. D. Ashwani Kumar, Y. Gori, A. Kumar, and C. S. Meena, Eds., Boca Raton, CRC Press, 2022, pp. 283–298.
25. S. Jain, N. Kumar, V. P. Singh, S. Mishra, and N. K. Sharma, "Transesterification of algae oil and little amount of waste cooking oil blend at low temperature in the presence of NaOH," *Energies*, vol. 16, no. 1, pp. 1–12, 2023, doi: 10.3390/ en16031293.
26. M. D. Brenes, *Biomass and bioenergy: New research*. Nova Publishers, 2006.
27. G. Pant, C. S. Meena, A. Saxena, A. Kumar, V. P. Singh, and N. Dutt, "Study the temperature variation in alternate coils of insulated condenser cum storage tank: Experimental study," in *Lecture Notes in Mechanical Engineering*. Springer, 2023, pp. 627–638, doi: 10.1007/978-981-99-2382-3_52.
28. R. K. Shubham Srivastava, D. Verma, S. Thusoo, A. Kumar, and V. P. Singh, "Nanomanufacturing for energy conversion and storage devices," in *Nanomanufacturing and Nanomaterials Design: Principles and Applications*,

1st ed., edited by S. Singh, S. K. Behura, A. Kumar, K. Verma, Boca Raton, CRC Press, 2022, pp. 165–174, doi: 10.1201/9781003220602.
29. R. Parameshwaran, S. Kalaiselvam, S. Harikrishnan, and A. Elayaperumal, "Sustainable thermal energy storage technologies for buildings: A review," *Renew. Sustain. Energy Rev.*, vol. 16, no. 5, pp. 2394–2433, 2012, doi: 10.1016/j.rser.2012.01.058.
30. S. Bawankar *et al.*, "Environmental impact assessment of lithium ion battery employing cradle to grave," *Sustain. Energy Technol. Assessments*, vol. 60, no. May, 2023, doi: 10.1016/j.seta.2023.103530.
31. A. Kumar, V. P. Singh, C. S. Meena, and N. Dutt, "Preface of thermal energy systems," in *Thermal Energy Systems: Design, Computational Techniques, and Applications*, Technical Education Department, Kanpur: CRC Press, 2023, pp. x–xiii.
32. Y. Elaouzy and A. El Fadar, "Potential of thermal, electricity and hydrogen storage systems for achieving sustainable buildings," *Energy Convers. Manag.*, vol. 294, p. 117601, 2023, doi: 10.1016/j.enconman.2023.117601.
33. A. Kundu, A. Kumar, N. Dutt, V. P. Singh, and C. S. Meena, "Chapter 7: Modelling and simulation of thermal energy system for design optimization," in *Thermal Energy Systems: Design, Computational Techniques, and Applications*, A. Kumar, N. Dutt, V. Singh, and C. Meena, Eds. CRC Press, 2023, pp. 103–137.
34. V. P. Singh, S. Jain, A. Karn, A. Kumar, and G. Dwivedi, "Mathematical modeling of efficiency evaluation of double pass parallel flow solar air heater," *Sustainability*, vol. 14, no. 17, pp. 1–22, 2022, doi: 10.3390/su141710535.
35. V. P. Singh, S. Jain, and J. M. L. Gupta, "Analysis of the effect of perforation in multi-v rib artificial roughened single pass solar air heater: - Part A," *Exp. Heat Transf.*, pp. 1–20, Oct. 2021, doi: 10.1080/08916152.2021.1988761.
36. A. A. Hassan and K. El-Rayes, "Optimizing the integration of renewable energy in existing buildings," *Energy Build.*, vol. 238, p. 110851, 2021, doi: 10.1016/j.enbuild.2021.110851.
37. V. P. Singh and G. Dwivedi, "Technical analysis of a large-scale solar updraft tower power plant," *Energies*, vol. 16, no. 1, p. 103325, 2023, doi: 10.3390/en16010494.
38. R. Kumar *et al.*, "Experimental and RSM-based process-parameters optimisation for turning operation of EN36B steel," *Materials*, vol. 16, no. 1, pp. 1–18, 2023, doi: 10.3390/ma16010339.
39. G. Mancò, E. Guelpa, A. Colangelo, A. Virtuani, T. Morbiato, and V. Verda, "Innovative renewable technology integration for nearly zero-energy buildings within the Re-COGNITION project," *Sustainability*, vol. 13, no. 4, 2021, doi: 10.3390/su13041938.
40. Y. G. Ashwani Kumar, A. K. Singh Gangwar, A. Kumar, C. Swaroop Meena, V. P. Singh, N. Dutt, A. Prasad, "Biomedical study of femur bone fracture and healing," in *Advanced Materials for Biomedical Applications*, 2022, 1st ed., A. Kumar, Y. Gori, A. Kumar, C. S. Meena, N. Dutt, Eds., Boca Raton, CRC Press, pp. 212–235.
41. V. P. Singh *et al.*, "Nanomanufacturing and design of high-performance piezoelectric nanogenerator for energy harvesting," in *Nanomanufacturing and Nanomaterials Design: Principles and Applications*, 1st ed., S. Singh,

S. K. Behura, A Kumar, K. Verma, Eds., Boca Raton, CRC Press, 2022, pp. 241–272.

42. M. Nayal, A. K. Sharma, S. Jain, and V. P. Singh, "Chapter 06 design and modelling of solar, geothermal and hybrid energy systems.pdf," in *Highly Efficient Thermal Renewable Energy Systems Design, Optimization and Applications*, 1st ed., R. W. V. Verma, S. Thangavel, N. Dutt, and A. Kumar, Eds. CRC Press, 2024, pp. 1–15.
43. S. A. Kalogirou, "Building integration of solar renewable energy systems towards zero or nearly zero energy buildings," *Int. J. Low-Carbon Technol.*, vol. 10, no. 4, pp. 379–385, Dec. 2015, doi: 10.1093/ijlct/ctt071.
44. F. Ascione, N. Bianco, R. F. De Masi, C. De Stasio, G. M. Mauro, and G. P. Vanoli, "Multi-objective optimization of the renewable energy mix for a building," *Appl. Therm. Eng.*, vol. 101, pp. 612–621, 2016, doi: 10.1016/j.applthermaleng.2015.12.073.
45. V. P. Singh and S. Jain, "Economic analysis of a large scale solar updraft tower power plant," *Sustain. Energy Technol. Assessments*, vol. 58, p. 103325, 2023, doi: 10.1016/j.seta.2023.103325.
46. H. Zhao, *The economics and politics of china's energy security transition*, Academic Press. 2018, doi: 10.1016/C2017-0-01392-7.
47. A. Hussain, S. M. Arif, and M. Aslam, "Emerging renewable and sustainable energy technologies: State of the art," *Renew. Sustain. Energy Rev.*, vol. 71, pp. 12–28, 2017, doi: 10.1016/j.rser.2016.12.033.
48. M. Nayal, A. K. Sharma, S. Jain, and V. P. Singh, "Chapter 11 green hydrogen production methods, designs and applications," in *Highly Efficient Thermal Renewable Energy Systems Design, Optimization and Applications*, 1st ed., R. W. V. Verma, S. Thangavel, N. Dutt, and A. Kumar, Eds. CRC Press, 2024, pp. 178–193.
49. A. K. Sharma, M. Nayal, S. Jain, and V. P. Singh, "Chapter 07 optimization techniques of solar thermal and hybrid energy systems," in *Highly Efficient Thermal Renewable Energy Systems Design, Optimization and Applications*, 1st ed., R. W. V. Verma, S. Thangavel, N. Dutt, and A. Kumar, Eds. CRC Press, 2024, pp. 1–14.
50. N. Wang, P. E. Phelan, C. Harris, J. Langevin, B. Nelson, and K. Sawyer, "Past visions, current trends, and future context: A review of building energy, carbon, and sustainability," *Renew. Sustain. Energy Rev.*, vol. 82, pp. 976–993, 2018, doi: 10.1016/j.rser.2017.04.114.

Chapter 14

Social, economic, and environmental issues associated with sustainable buildings

Pooyan Rahmanivahid, Morteza Khashehchi, Sivasakthivel Thangavel, and Milad Heidari

14.1 INTRODUCTION

In the face of escalating global environmental challenges, the imperative to redefine our approach to building design and operation has never been more pressing. The built environment, a nexus of human activity, energy consumption, and resource utilization, plays a pivotal role in the broader context of sustainability. This chapter embarks on a journey to explore the significance of sustainable buildings, emphasizing the necessity for an integrated approach that spans social, economic, and environmental dimensions.

14.1.1 Significance of sustainable buildings

The environmental challenges confronting the world today—climate change, resource depletion, and ecosystem degradation—underscore the need for transformative actions in every sector, with the built environment standing as a linchpin for change. Buildings, responsible for a substantial share of global energy consumption and greenhouse gas emissions, wield immense influence on the trajectory of environmental sustainability.

Sustainable buildings emerge as a beacon of hope in this landscape of challenges (Figure 14.1). They represent a change in thinking in design and operation, aiming to minimize environmental impact, optimize resource use, and create spaces that nurture the well-being of occupants. Beyond the reduction of carbon footprints, sustainable buildings contribute to resource conservation, biodiversity preservation, and the creation of resilient and adaptable urban ecosystems.

14.1.2 Integrated approach: Social, economic, and environmental dimensions

A pivotal realization driving the discourse on sustainability is the interconnectedness of social, economic, and environmental dimensions. Traditional approaches often focused on isolated solutions, neglecting the intricate web

DOI: 10.1201/9781003496656-14

Figure 14.1 Dimensions of sustainability of green buildings

of relationships that define the built environment. An integrated approach, one that harmonizes the three pillars of sustainability, emerges as the linchpin for transformative change (M. Ahmadizadeh, 2024).

- Social dimension: Sustainable buildings are not merely structures; they are vibrant, inclusive spaces that prioritize the well-being and comfort of their occupants (Flavia Grey, 2019). Integrating social considerations entails creating environments that foster health, equity, and community engagement. The chapter delves into how sustainable design can enhance indoor air quality, natural lighting, and thermal comfort, contributing to the physical and mental health of building occupants.
- Economic dimension: The economic viability of sustainable buildings is integral to their widespread adoption. Beyond the initial costs, the chapter explores how sustainable practices lead to long-term economic benefits, including reduced operational costs, increased property

value, and enhanced occupant productivity. The economic perspective is not relegated to financial considerations alone but extends to broader concepts of resilience and adaptability in the face of economic uncertainties.

- Environmental dimension: Central to the discussion is the environmental impact of buildings. Sustainable practices aim to mitigate this impact by incorporating energy-efficient technologies, renewable energy sources, and eco-friendly materials (Z. Razaviyn, 2024). The chapter investigates how sustainable buildings contribute to a circular economy, minimize waste generation, and play a crucial role in achieving global climate and sustainability targets.

14.1.3 The urgency of integration

As the urgency to address global challenges intensifies, an integrated approach to sustainable buildings becomes not just a choice but a necessity. The compartmentalization of social, economic, and environmental considerations proves inadequate in the face of complex, interdependent challenges. The chapter asserts that to create a lasting impact, an integrated approach must become the cornerstone of our endeavors to build a sustainable future (V. Verma, 2023).

In the chapters that follow, we will delve into specific strategies, case studies, and emerging trends that exemplify the integrated approach to sustainable buildings. From the design phase to operational practices, from individual structures to urban ecosystems, the exploration will illuminate the transformative potential of adopting a holistic perspective. As we navigate the intricate web of sustainability, this chapter serves as a compass, guiding us toward an understanding of the integrated strategies needed to shape a sustainable built environment for generations to come.

14.2 SOCIAL IMPACTS OF SUSTAINABLE BUILDINGS

Sustainable buildings, beyond their tangible environmental benefits, wield a profound influence on the social fabric of communities. This section delves into the myriad ways in which sustainable building practices shape and enhance the well-being, health, and connectivity of occupants. From indoor air quality to community engagement, the social impacts of sustainable buildings resonate far beyond the physical structures, contributing to a more inclusive and resilient built environment (Julia & Lars, 2023).

14.2.1 Indoor environmental quality and health

One of the primary social impacts of sustainable buildings lies in the prioritization of indoor environmental quality. Sustainable design practices focus on enhancing air quality, maximizing natural light, and optimizing thermal comfort. Adequate ventilation, low-emission materials, and efficient HVAC systems contribute to healthier indoor environments, reducing the prevalence of respiratory issues, allergies, and other health concerns. This emphasis on occupant well-being is not only a testament to ethical design but also translates into increased productivity, reduced absenteeism, and improved overall satisfaction among building occupants (V. Verma, 2024).

- Biophilic design and connection to nature: Biophilic design principles, integral to sustainable building practices, recognize the innate human connection to nature. Incorporating elements such as green roofs, living walls, and ample daylight, sustainable buildings create spaces that mimic natural environments. This intentional integration of nature into the built environment has been shown to reduce stress, enhance cognitive function, and foster a sense of well-being among occupants. Beyond individual spaces, the promotion of biodiversity in urban settings through sustainable landscaping contributes to the creation of vibrant, ecologically conscious communities.
- Occupant productivity and comfort: Sustainable buildings go beyond the mere provision of shelter; they are designed to support and enhance the daily activities and productivity of their occupants. From ergonomic workspaces to comfortable shared areas, sustainable design principles prioritize the creation of environments that facilitate occupant comfort and productivity. Research indicates that well-designed, sustainable workplaces lead to increased job satisfaction, creativity, and overall job performance. This section explores the principles of biophilic design, flexible layouts, and ergonomic considerations that underpin the creation of workplaces that are not only energy-efficient but also conducive to occupant well-being.
- Community engagement and social cohesion: Sustainable buildings extend their impact beyond their immediate occupants to foster community engagement and social cohesion. Incorporating communal spaces, shared resources, and participatory design processes, sustainable buildings become focal points for community activities. The chapter explores how sustainable design principles can contribute to the creation of inclusive public spaces, promoting social interaction, and strengthening community bonds. Additionally, sustainable buildings often serve as educational tools, raising awareness about environmental issues and inspiring collective action within communities.
- Equitable access and inclusion: An often-overlooked aspect of social impacts is equitable access to sustainable buildings. Sustainable design strives to create environments that are accessible to all, irrespective of age, ability, or socio-economic status. Universal design principles,

such as barrier-free access and sensory considerations, are integrated to ensure that sustainable buildings are inclusive and welcoming to diverse user groups. This section explores how sustainable buildings can address issues of social equity, providing equal opportunities for all individuals to benefit from the positive social impacts associated with sustainable design.

- Resilience and social equity: The social impacts of sustainable buildings extend to considerations of resilience, especially in the face of environmental and socio-economic uncertainties. Sustainable buildings, designed with adaptability and resilience in mind, contribute to the overall resilience of communities. The section delves into how sustainable practices, such as energy-efficient technologies and resource conservation, play a role in enhancing the capacity of communities to withstand and recover from challenges, thereby promoting social equity in the face of adversity (Vanessa, 2022). The social impacts of sustainable buildings underscore their role as catalysts for positive change within communities. From fostering health and well-being to strengthening social bonds and promoting equity, sustainable buildings exemplify a comprehensive approach that extends far beyond the physical structures. As we navigate the complex terrain of sustainability, understanding and leveraging the social impacts of sustainable buildings becomes integral to the overarching goal of creating resilient, inclusive, and thriving built environments. The subsequent sections will delve into case studies, strategies, and emerging trends that further illuminate the transformative potential of integrating social considerations into sustainable building practices.

14.3 ECONOMIC PERSPECTIVES ON SUSTAINABLE CONSTRUCTION

Sustainable construction practices, often lauded for their environmental benefits, also yield significant economic advantages. This section delves into the economic dimensions of sustainable construction, exploring how such practices contribute to long-term financial viability, operational efficiency, and broader economic resilience. From initial costs to life-cycle analysis, the economic perspectives on sustainable construction shed light on the multifaceted advantages that extend beyond the traditional considerations of building design and operation.

14.3.1 Initial costs and return on investment

A common misconception surrounding sustainable construction is the perception of higher initial costs. While it is true that some sustainable materials and technologies may have a higher upfront price, the focus on long-term value and efficiency often translates into a favorable return on investment (ROI). This section explores how sustainable construction, with careful planning and strategic decision-making, can result in cost savings over the life cycle of a building. Energy-efficient systems, durable materials, and reduced operational costs contribute to a positive economic impact that extends far beyond the construction phase.

14.3.2 Operational efficiency and reduced costs

The economic benefits of sustainable construction manifest prominently during the operational phase of a building's life cycle. Sustainable practices, such as energy-efficient lighting, heating, ventilation, and air conditioning (HVAC) systems, contribute to reduced utility bills and operational costs. The integration of smart building technologies further enhances operational efficiency, allowing for real-time monitoring, adaptive controls, and predictive maintenance. This section delves into how sustainable construction fosters long-term economic gains through streamlined operations, reduced resource consumption, and lower maintenance costs.

14.3.3 Increased property value and market demand

Sustainable construction practices can significantly enhance the market value of a property. The growing awareness of environmental issues and the emphasis on sustainable living have increased the demand for green buildings. This section explores how certifications like LEED (Leadership in Energy and Environmental Design) and BREEAM (Building Research Establishment Environmental Assessment Method) serve not only as indicators of environmental responsibility but also as markers of increased property value. Investors and tenants alike are increasingly prioritizing sustainable features, contributing to a robust market demand for green buildings.

14.3.4 Government incentives and regulatory compliance

Governments worldwide are recognizing the importance of sustainable construction and have introduced a plethora of incentives to encourage its adoption. From tax credits and grants to expedited permitting processes, these incentives significantly offset initial costs and create a more favorable economic environment for sustainable construction. Additionally, compliance with evolving environmental regulations and building codes becomes smoother when incorporating sustainable practices. This section explores how navigating regulatory landscapes and leveraging government incentives can further enhance the economic viability of sustainable construction projects.

14.3.5 Resilience to market volatility and climate risks

The economic perspectives on sustainable construction extend to considerations of resilience in the face of market volatility and climate risks. Sustainable buildings, designed with adaptability and resilience in mind, are better positioned to withstand economic uncertainties and climate-related challenges. This section delves into how sustainable construction practices contribute to economic resilience by mitigating risks associated with resource scarcity, energy price fluctuations, and the physical impacts of climate change.

14.3.6 Job creation and local economic impact

The economic benefits of sustainable construction extend beyond the building itself to the communities in which they are erected. The adoption of sustainable practices often requires skilled labor, contributing to job creation and fostering economic growth. Local sourcing of materials, as encouraged by sustainable construction principles, further bolsters the economic impact on surrounding communities. This section explores how sustainable construction can be a catalyst for regional economic development, creating a ripple effect that goes beyond the immediate construction site.

In conclusion, the economic perspectives on sustainable construction underscore the multi-faceted advantages that extend far beyond the realm of environmental stewardship. From initial costs to operational efficiency, increased property value, and resilience to market dynamics, sustainable construction practices contribute to a robust economic foundation. As the global emphasis on sustainability intensifies, the economic viability of green buildings becomes increasingly evident. The subsequent sections will delve into case studies, emerging trends, and specific strategies that illuminate the transformative potential of adopting an economic lens when approaching sustainable construction projects.

14.4 ENVIRONMENTAL CONSIDERATIONS IN SUSTAINABLE BUILDING DESIGN

Sustainable building design represents a profound commitment to environmental stewardship, with a focus on minimizing negative impacts and fostering ecological resilience (Johan Colding & Stephan Barthel, 2013). This section delves into the key environmental considerations that underpin sustainable building design, encompassing energy efficiency, material selection, water conservation, site planning, and biodiversity preservation. As we navigate the complexities of sustainable design, understanding these environmental considerations becomes essential for creating built environments that coexist harmoniously with the natural world.

14.4.1 Energy efficiency and renewable energy

At the core of sustainable building design is a relentless pursuit of energy efficiency. This involves minimizing energy consumption through passive design strategies, efficient building envelopes, and the integration of renewable energy sources. The section explores how thoughtful orientation, proper insulation, and the incorporation of technologies like solar panels and wind turbines contribute to a reduced carbon footprint. By prioritizing energy efficiency, sustainable buildings play a crucial role in mitigating the environmental impact associated with energy consumption (M. Chitt & M. Khashehchi, 2024).

14.4.2 Material selection and life cycle analysis

The materials chosen for construction exert a substantial influence on a building's environmental footprint. Sustainable building design emphasizes the use of eco-friendly, recycled, and locally sourced materials to reduce embodied energy and minimize resource extraction. Life-cycle analysis, a comprehensive evaluation of a material's environmental impact from extraction to disposal, is a critical tool in material selection. This section delves into how sustainable building design navigates the complex landscape of material choices to create structures that are not only aesthetically pleasing but also environmentally responsible.

14.4.3 Water conservation and management

Water, a finite resource, is a key consideration in sustainable building design (Peter O. Akadiri et al., 2012). The section explores strategies for water conservation, including the use of low-flow fixtures, rainwater harvesting systems, and innovative landscaping practices. Sustainable buildings aim to minimize water consumption, treat, and reuse wastewater, and integrate water-sensitive design principles to reduce the strain on local water supplies. By addressing water scarcity challenges, sustainable building design contributes to the overall resilience of communities in the face of changing climate patterns.

14.4.4 Site planning and urban ecology

The environmental impact of a building extends beyond its immediate structure to the surrounding site. Sustainable building design incorporates thoughtful site planning that preserves natural features, promotes biodiversity, and minimizes disruption to ecosystems. The section explores how principles of urban ecology, green infrastructure, and permeable surfaces contribute to sustainable site development. By harmonizing with the natural environment, sustainable buildings become integral components of resilient and ecologically sensitive urban landscapes.

14.4.5 Biodiversity preservation and green spaces

Sustainable building design seeks not only to minimize harm to the natural world but also to actively contribute to biodiversity preservation. The inclusion of green spaces, native vegetation, and wildlife habitats in and around buildings fosters ecological diversity. This section delves into how sustainable design practices support urban biodiversity, providing refuge for flora and fauna. By creating environments that mimic natural ecosystems, sustainable buildings play a role in maintaining the delicate balance of local ecosystems.

14.4.6 Waste reduction and circular economy

The concept of a circular economy is central to sustainable building design, emphasizing the reduction, reuse, and recycling of materials to minimize waste. The section explores how sustainable buildings incorporate waste reduction strategies during construction and operation, including construction waste management, material repurposing, and the promotion of a cradle-to-cradle approach. By closing the loop on material life cycles, sustainable building design aligns with principles of resource efficiency and waste minimization (Mayara Regina Munaro et al., 2020).

In conclusion, the environmental considerations in sustainable building design are a testament to the profound commitment to ecological integrity and responsible resource use. By prioritizing energy efficiency, selecting sustainable materials, conserving water, embracing thoughtful site planning, preserving biodiversity, and championing a circular economy, sustainable building design becomes a holistic endeavor. As we continue to grapple with global environmental challenges, the principles explored in this section provide a blueprint for creating built environments that contribute to, rather than detract from, the health and resilience of the planet. The subsequent sections will delve into case studies, emerging trends, and specific strategies that further illuminate the transformative potential of environmental considerations in sustainable building design

14.5 INTEGRATING RESEARCH AND TEACHING

The symbiotic relationship between research and teaching forms the cornerstone of academic excellence. This section explores the multifaceted dimensions of integrating research and teaching within academic environments, emphasizing the reciprocal benefits that arise when these two pillars of higher education converge. From enriching student learning experiences to advancing institutional research agendas, the integration of research and teaching fosters a dynamic and intellectually vibrant academic ecosystem.

14.5.1 Enriching student learning experiences

At the heart of the integration of research and teaching is the profound impact on student learning experiences. Engaging students in research activities provides a unique opportunity for them to apply theoretical knowledge to real-world problems. This section delves into how direct research experiences, mentorship from faculty involved in cutting-edge research, and participation in research projects contribute to a more dynamic and enriched learning environment. As students actively contribute to the creation of new knowledge, they develop critical thinking skills, problem-solving abilities, and a deep appreciation for the research process.

14.5.2 Fostering a culture of inquiry

Integrating research and teaching fosters a culture of inquiry and intellectual curiosity within academic institutions. Faculty engaged in research bring the latest developments and discoveries directly into the classroom, creating an environment where curiosity is not only encouraged but also actively cultivated. This section explores how a culture of inquiry stimulates a passion for lifelong learning among students and sets the stage for the development of the next generation of researchers and scholars.

14.5.3 Advancing institutional research agendas

The integration of research and teaching is not only beneficial for individual students but also contributes to the advancement of institutional research agendas. Academic institutions play a pivotal role in shaping the trajectory of knowledge in various disciplines. By aligning teaching activities with ongoing research initiatives, institutions can create a cohesive and synergistic environment where teaching informs research and vice versa. This section delves into strategies for aligning teaching objectives with institutional research priorities, fostering collaboration among faculty members, and leveraging academic resources to advance overarching research agendas.

14.5.4 Mentorship and professional development

The integration of research and teaching creates fertile ground for mentorship and professional development opportunities. Faculty members actively involved in research serve as mentors, guiding students through the intricacies of the research process and providing valuable insights into their respective fields. This section explores the mentor-student relationship, highlighting how it contributes not only to academic growth but also to the development of crucial professional skills. For faculty, engaging in research enhances their expertise, keeps them abreast of the latest advancements in their disciplines, and opens avenues for collaborative scholarship.

14.5.5 Community engagement and knowledge transfer

Integrating research and teaching extends beyond the boundaries of the academic institution, reaching into the broader community. This section explores how research activities undertaken within academic environments can be leveraged for community engagement and knowledge transfer. Collaborative projects, outreach programs, and partnerships with local industries create opportunities for translating research findings into practical solutions. By actively contributing to the betterment of society, academic institutions fulfill their role as engines of knowledge creation and dissemination.

14.5.6 Challenges and strategies for integration

While the benefits of integrating research and teaching are clear, challenges may arise in implementing and sustaining this integration. This section discusses shared challenges such as time constraints, resource limitations, and the need for faculty development. It also explores strategies for overcoming these challenges, including flexible scheduling, interdisciplinary collaborations, and the establishment of supportive institutional policies.

In conclusion, the integration of research and teaching in academic environments is a transformative approach that redefines the landscape of higher education. By intertwining the processes of knowledge creation and knowledge dissemination, academic institutions create a holistic learning ecosystem that nurtures intellectual curiosity, advances research agendas, and prepares students for active participation in their chosen fields. As we navigate the evolving landscape of higher education, the integration of research and teaching stands as a beacon for fostering a culture of inquiry, mentorship, and community engagement within academic institutions. The subsequent sections will delve into case studies, emerging trends, and specific strategies that further illuminate the transformative potential of seamlessly integrating research and teaching.

14.6 FUTURE DIRECTIONS AND CHALLENGES

The seamless integration of research and teaching is a dynamic and ever-evolving endeavor that holds the promise of transforming academic landscapes. As we gaze into the future, this section explores the anticipated directions and potential challenges in the continued pursuit of harmonizing these essential components of higher education. From emerging trends that redefine pedagogy to the persistent challenges that demand innovative solutions, the future of research and teaching integration is shaped by a commitment to excellence, adaptability, and an integrated approach to knowledge creation and dissemination.

14.6.1 Embracing interdisciplinary collaboration

The future of research and teaching integration is poised to witness an increased emphasis on interdisciplinary collaboration. As societal challenges become more complex and interconnected, academic institutions are recognizing the need for collaborative efforts that transcend disciplinary boundaries. This section delves into how future initiatives may prioritize the creation of interdisciplinary research and training teams, fostering an environment where diverse perspectives converge to address multifaceted problems. The integration of knowledge across disciplines is anticipated to enrich both research outcomes and student learning experiences.

14.6.2 Leveraging technology for enhanced learning

Advancements in technology are poised to play a pivotal role in shaping the future of research and teaching integration. Virtual labs, online collaboration platforms, and immersive learning experiences are likely to become integral components of academic environments. This section explores how the integration of innovative technologies can transcend physical barriers, offering students and researchers the opportunity to engage in collaborative endeavors regardless of geographical location. The future holds the potential for technology to augment both the research process and teaching methodologies, creating a more inclusive and accessible learning ecosystem.

14.6.3 Promoting experiential learning opportunities

Future directions in research and teaching integration are expected to place a heightened emphasis on experiential learning opportunities. Internships, research placements, and industry collaborations will become more deeply embedded in academic curricula. This section examines how hands-on experiences provide students with practical insights into the application of theoretical knowledge and research methodologies. By bridging the gap between academic theory and real-world applications, experiential learning opportunities contribute to a more holistic and impactful educational experience.

14.6.4 Enhancing diversity, equity, and inclusion

The future of research and teaching integration must be guided by a commitment to diversity, equity, and inclusion. Efforts to amplify underrepresented voices, cultivate inclusive learning environments, and address systemic barriers are anticipated to gain prominence. This section explores strategies for ensuring that the benefits of integrated research and teaching extend to a diverse student and faculty population. By actively fostering a culture of inclusivity, academic institutions can tap into a wealth of diverse perspectives that enrich both the research landscape and the learning experience.

14.6.5 Addressing global challenges

The pressing challenges facing the global community—climate change, public health crises, socio-economic disparities—call for an academic response that transcends borders. Future directions in research and teaching integration are likely to prioritize collaborative efforts aimed at addressing these global challenges. This section delves into how academic institutions can play a pivotal role in global problem-solving by fostering international collaborations, engaging in cross-cultural research initiatives, and preparing students to be global citizens equipped to tackle the complexities of our interconnected world.

14.6.6 Navigating ethical considerations

As research and teaching integration advances, ethical considerations must be central to future initiatives. This section examines the ethical challenges that may arise, from issues related to research integrity to the responsible use of emerging technologies. Anticipating and proactively addressing ethical considerations ensures that the integration of research and teaching aligns with the highest standards of academic and ethical conduct.

14.6.7 Overcoming resource constraints

A persistent challenge in the integration of research and teaching is resource constraints. The future will demand innovative solutions for overcoming limitations in funding, time, and infrastructure. This section explores strategies for optimizing resource allocation, fostering partnerships with external stakeholders, and creatively leveraging existing resources to maximize the impact of integrated research and teaching initiatives.

In conclusion, the future directions, and challenges in the integration of research and teaching reflect a commitment to innovation, inclusivity, and responsiveness to the evolving needs of the academic landscape. By embracing interdisciplinary collaboration, leveraging technology, promoting experiential learning, enhancing diversity and equity, addressing global challenges, navigating ethical considerations, and overcoming resource constraints, academic institutions can chart a course toward a future where research and teaching are seamlessly woven into the fabric of higher education. As we embark on this journey, the collective efforts of educators, researchers, and institutions will shape a transformative and sustainable future for the integration of research and teaching in academia.

14.7 CONCLUSION

The integration of research and teaching stands as a beacon illuminating the horizons of higher education, transforming traditional paradigms, and shaping a future where academia is a dynamic ecosystem of knowledge creation and dissemination. As we reflect on the multifaceted dimensions explored in this chapter, the conclusions drawn encapsulate the transformative potential, persistent challenges, and the imperative of continual evolution within the integration of research and teaching.

14.7.1 Transformative potential

The journey through the integration of research and teaching has revealed its transformative potential across diverse realms. From enriching student learning experiences through direct research activities to fostering a culture of inquiry and curiosity, the transformative impact is evident. The integration creates a holistic learning environment that prepares students not only with theoretical knowledge but also with critical thinking skills, adaptability, and a passion for lifelong learning. Faculty members engaged in research become mentors, guiding students through the complexities of research methodologies, and contributing to the development of the next generation of scholars.

The transformative potential extends beyond the classroom to the broader community, as research outcomes are translated into practical solutions through community engagement and knowledge transfer. Sustainable practices, interdisciplinary collaboration, and the embrace of technology redefine the landscape of academic research and teaching. The future beckons with the promise of increased diversity, global collaboration, and ethical considerations that further amplify the transformative impact of research and teaching integration.

14.7.2 Persistent challenges

However, the path toward seamless integration is not without challenges. Persistent issues, such as resource constraints, time limitations, and the need for faculty development, present hurdles that demand innovative solutions. The ethical considerations of research and teaching integration require vigilant navigation to ensure the highest standards of academic integrity. As academic institutions strive to create inclusive environments, challenges related to diversity, equity, and inclusion necessitate ongoing efforts and a commitment to dismantling systemic barriers.

The integration of technology, while holding immense promise, requires careful consideration of accessibility and equitable access to ensure that no students are left behind. The interdisciplinary nature of collaborative research and teaching endeavors may encounter institutional barriers that need to be dismantled to fully harness the potential of diverse perspectives.

Addressing these challenges requires a collective and concerted effort from educators, administrators, and policymakers to build a resilient and adaptive framework for research and teaching integration.

14.7.3 Imperative of continual evolution

As we draw conclusions from the exploration of research and teaching integration, the imperative of continual evolution becomes evident. The landscape of higher education is dynamic, influenced by emerging technologies, societal changes, and global challenges. To remain relevant and impactful, academic institutions must embrace a culture of continual evolution, adapting their approaches to meet the evolving needs of students, faculty, and society at large.

The integration of research and teaching must not be viewed as a static achievement but as an ongoing process of innovation and adaptation. The horizons we have glimpsed in this chapter are invitations to explore further, to delve into emerging trends, and to proactively address challenges. The future demands agility, resilience, and a commitment to staying at the forefront of educational excellence.

14.7.4 Call to action

In conclusion, the integration of research and teaching is not merely a model but a call to action—a call to reshape the future of higher education. It is an invitation to educators, researchers, administrators, and policymakers to collaborate in creating environments where the boundaries between research and teaching blur, giving rise to a dynamic and interconnected academic ecosystem. It is a call to empower students with the skills and knowledge needed to navigate a rapidly changing world, to inspire faculty to be catalysts for both discovery and learning, and to foster a commitment to ethical, inclusive, and impactful practices.

As we conclude this chapter, the horizons of research and teaching integration expand into an open landscape of possibilities. The transformative potential, challenges, and imperative of continual evolution beckon us to embark on a journey of exploration and innovation. The integration of research and teaching is not just an academic pursuit; it is a collective endeavor to shape a future where education is a catalyst for both personal and societal transformation.

REFERENCES

F. Grey, 2019, "Space-mate: A framework to harmonize occupant well-being and building sustainability", Thesis for: Doctor of PhilosophyAdvisor: Renate Fruchter, Kincho Law, Sumit Bhargava, Martin Fischer.

J. Colding & S. Barthel, 2013, “The potential of ‘Urban Green Commons’ in the resilience building of cities”, *Ecological Economics*, Volume 86, 156–166.

J. Winslow & L. Coenen, 2023, “Sustainability transitions to circular cities: Experimentation between urban vitalism and mechanism”, *Cities*, Volume 142, 104531.

M. R. Munaro, S. F. Tavares & L. Bragança, 2020, “Towards circular and more sustainable buildings: A systematic literature review on the circular economy in the built environment”, *Journal of Cleaner Production*, Volume 260, 121134. https://doi.org/10.1016/j.jclepro.2020.121134.

M. Ahmadizadeh, M. Heidari, S. Thangavel, E. A. Naamani, M. Khashehchi, V. Verma & A. Kumar, 2024, “Technological advancements in sustainable and renewable solar energy systems”, in *Highly Efficient Thermal Renewable Energy Systems: Design, Design, Optimization and Applications*; V. Verma, S. Thangavel, N. Dutt, A. Kumar, R. Weerasinghe Editors; CRC Press: Boca Raton, FL; Chapter 02; ISBN 9781032595641.

M. Chitt, S. Thangavel, V. Verma & A. Kumar, 2024, “Green hydrogen productions: Methods, designs and smart applications”, in *Highly Efficient Thermal Renewable Energy Systems: Design, Design, Optimization and Applications*; V. Verma, S. Thangavel, N. Dutt, A. Kumar, R. Weerasinghe Editors; CRC Press: Boca Raton, FL; Chapter 16; ISBN 9781032595641.

M. Khashehchi, S. Thangavel, P. Rahmanivahid, M. Heidari, T. Moazzeni, V. Verma & A. Kumar, 2024, “Solar desalination techniques: Challenges and opportunities”, in *Highly Efficient Thermal Renewable Energy Systems: Design, Design, Optimization and Applications*; V. Verma, S. Thangavel, N. Dutt, A. Kumar, R. Weerasinghe Editors; CRC Press: Boca Raton, FL; Chapter 19; ISBN 9781032595641.

P. O. Akadiri, E. A. Chinyio & P. O. Olomolaiye, 2012, “Design of a sustainable building: A conceptual framework for implementing sustainability in the building sector”, *Buildings*, Volume 2, Issue 2, 126–152.

P. K. D. Pramanik, B. Mukherjee, S. Pal, S. Pal & S. P. Singh, 2019, “Green smart building: Requisites, architecture, challenges, and use cases”, in *Green Building Management and Smart Automation*, Chapter: one, IGI Global.

V. Whittem, A. Roetzel, A.-M. Sadick & A. N. Kidd, 2022, “How comprehensive is post-occupancy feedback on school buildings for architects? A conceptual review based upon Integral Sustainable Design principles”, *Building and Environment*, Volume 218, 109109.

V. Verma, S. Thangavel, N. Dutt, A. Kumar & R. Weerasinghe Editors, 2024, *Highly Efficient Thermal Renewable Energy Systems: Design, Design, Optimization and Applications*; CRC Press: Boca Raton, FL; ISBN 9781032595641.

V. Verma, C. S. Meena, S. Thangavel, A. Kumar, T. Choudhary & G. Dwivedi, 2023, “Ground and solar assisted heat pump systems for space heating and cooling applications in the northern region of India – A study on energy and CO2 saving potential”, *Sustainable Energy Technologies and Assessments*, Volume 59, 103405, ISSN 2213-1388, https://doi.org/10.1016/j.seta.2023.103405.

Z. Razaviyn, M. Heidari, V. Verma, S. Thangavel, A. Kumar, K. K Saxena & G. Dwivedi, 2024, “Numerical simulation of a marine energy convertor based on vortex induced vibrations”, *Proceedings of the Institution of Mechanical Engineers, Part E: Journal of Process Mechanical Engineering* (In Press).

Chapter 15

Emerging trends in smart green building technologies

Milad Heidari, Sivasakthivel Thangavel, Eman Al Naamani, and Morteza Khashehchi

15.1 INTRODUCTION

Sustainable development is a worldwide endeavor aimed at addressing current demands without jeopardizing future generations' ability to meet their own. Smart green building technologies have developed as critical components in promoting environmental conservation, resource efficiency, and human well-being within this framework. Smart green buildings are planned and built to have the least amount of environmental effect while maximizing resource efficiency. They combine cutting-edge technology, modern materials, and intelligent systems to minimize energy consumption, optimize water use, and improve indoor environmental quality. These structures function as living ecosystems, interacting with their environment to generate long-lasting, high-performance structures.

The amalgamation of smart technologies with the fundamental principles of green buildings marks a pivotal milestone in sustainable architecture and construction practices. This integration represents a strategic constructive collaboration that capitalizes on the eco-conscious ethos of green building principles and the intelligence of modern technological innovations, yielding a transformative impact on sustainable development.

The necessity of integrating smart technologies with green building principles is underscored by its multifaceted advantages:

Enhanced efficiency and sustainability: Green building principles inherently champion resource conservation, energy efficiency, and environmental stewardship. Integration with smart technologies amplifies these objectives by introducing intelligent systems that continuously monitor, analyze, and optimize resource utilization in real-time, thus elevating the overall operational efficiency and sustainability of buildings

Real-time monitoring and adaptive controls: Smart technologies offer sophisticated monitoring capabilities and adaptive control systems. These technologies facilitate continuous surveillance of a building's performance metrics, enabling instant corrective actions to maintain

DOI: 10.1201/9781003496656-15

or improve sustainability parameters. Adaptive controls adjust various systems, such as lighting, heating, ventilation, and air conditioning (HVAC), and water usage, responding dynamically to environmental changes and occupant behavior.

Optimized occupant experience: Integrating smart technologies not only augments the environmental performance of buildings but also enhances occupant comfort, health, and productivity. Personalized control options, coupled with intelligent sensors regulating indoor environmental quality, contribute to creating healthier and more conducive indoor environments, fostering enhanced occupant well-being.

Smart green building technologies are significant because of their multiple contributions to sustainable development:

Environmental conservation: These technologies significantly reduce the ecological footprint of buildings by employing renewable energy sources, energy-efficient systems, and sustainable construction materials. Through practices like passive design, green roofs, and integrated solar panels, they mitigate greenhouse gas emissions and promote biodiversity.

Resource efficiency: Smart green buildings optimize resource utilization by incorporating systems that monitor and control energy, water, and waste. Through intelligent sensors, IoT-driven management, and adaptive controls, these buildings minimize resource wastage and promote circular economy principles.

Human health and well-being: Smart technology integration helps to create healthier indoor environments. Improved air quality, correct lighting, and thermal comfort all have a favorable influence on occupants' health, productivity, and general well-being, improving their quality of life.

Resilience and future-readiness: Smart green buildings are more resistant to environmental threats such as climate change and natural catastrophes. Advanced monitoring systems and resilient design strategies enable these structures to adapt and withstand adverse conditions, ensuring long-term viability.

Integrating smart technologies with green building principles represents a pivotal step toward fostering sustainable and efficient built environments. By merging these two paradigms, buildings evolve into dynamic ecosystems that optimize resource use, enhance occupant comfort, and contribute significantly to global sustainability goals. The synergy between smart technologies and green building principles redefines traditional construction practices, emphasizing responsiveness, adaptability, and long-term environmental stewardship. This integration aligns technological innovation with eco-conscious design, driving a change in thinking toward intelligent, energy-efficient, and environmentally sensitive buildings.

At the core of this integration lies the transformative potential to revolutionize how buildings operate. Smart technologies offer real-time data monitoring and management, enabling precise control over energy consumption, lighting, HVAC systems, and water usage. This data-driven approach, when aligned with green building principles, empowers structures to optimize resource utilization, reduce waste, and lower carbon footprints. Furthermore, integration promotes adaptive design solutions, which allow buildings to respond dynamically to changing environmental conditions, occupant demands, and external variables. This flexibility promotes healthier and more productive settings inside these sustainable structures by increasing energy efficiency, improving indoor air quality (IAQ), and increasing occupant comfort levels.

15.2 FOUNDATIONS OF SMART GREEN BUILDING TECHNOLOGIES

The foundations of smart green building technologies rest upon a multifaceted approach that converges sustainability, innovation, and technology. At its core, these foundations emphasize the integration of eco-conscious design principles with cutting-edge technological advancements. The fundamental pillars include energy efficiency, resource optimization, and environmental sensitivity. This approach extends beyond traditional construction norms, aiming to create structures that not only minimize their environmental impact but actively contribute to a more sustainable future. Smart green building technologies rely on a combination of renewable energy sources, advanced materials, IoT-driven systems, and data analytics to forge buildings that are responsive, efficient, and resilient in the face of evolving environmental challenges. This foundational framework serves as the blueprint for constructing buildings that harmonize with their surroundings, prioritize resource conservation, and elevate the quality of life for occupants while significantly reducing their carbon footprint. Table 15.1 shows cutting-edge energy technologies used in smart green buildings.

15.2.1 Key components of smart green buildings

Green building technologies refer to innovative methods and practices applied throughout the building lifecycle—design, construction, operation, and demolition—to reduce energy consumption, conserve natural resources, improve indoor air quality, and promote environmental sustainability. Understanding the foundational elements of smart green buildings elucidates their multifaceted nature and the interconnected systems at play.

Table 15.1 Cutting-edge energy technologies in smart green buildings

Energy Technologies	*Key Points*
Smart Lighting Systems	These systems adjust brightness based on occupancy and natural daylight, reducing energy wastage. They can also include energy-efficient lighting such as LED lights and solar-powered lighting [1].
Smart Heating and Cooling Systems	These systems improve energy efficiency by monitoring and modifying heating, ventilation, and air-conditioning (HVAC) systems in real time [2].
Building Energy Management Systems (BEMS)	BEMS uses artificial intelligence (AI) to optimize energy efficiency in green buildings. They collect data on energy usage and adjust systems accordingly to minimize energy consumption [3].
Smart Grid Solutions	These technologies enable efficient energy transmission and distribution, as well as demand response programs, revolutionizing renewable energy integration in metropolitan settings.
Smart Irrigation Systems	These systems optimize water usage in green buildings by monitoring soil moisture and weather conditions, reducing water wastage, and ensuring efficient irrigation [4].
Solar-Powered Streetlights	These lights use solar energy to generate electricity, making them a sustainable and energy-efficient solution for illuminating urban areas.
Wind Energy	Wind turbines may be put in cities to generate electricity or mechanical power, helping to create a greener and more sustainable built environment.

15.2.1.1 Energy efficiency

To minimize reliance on nonrenewable resources, energy efficiency includes improved insulation, passive design tactics, and the incorporation of renewable energy sources such as solar panels and wind turbines. Energy efficiency in smart buildings encompasses strategies and technologies aimed at minimizing energy consumption while maximizing performance and comfort. It involves a comprehensive approach that integrates design, materials, and renewable energy sources to reduce reliance on non-renewable resources. Energy efficiency is fundamental in mitigating climate change, reducing operational costs, enhancing building performance, and ensuring sustainability. It aligns with global initiatives to curb greenhouse gas emissions and promote environmental stewardship.

Investigating the important components of optimized insulation and passive design solutions sheds light on their function in lowering energy consumption. Effective insulation is critical for limiting heat transmission and the requirement for heating and cooling. Advanced insulation materials and techniques, such as high-performance windows, insulation panels, and thermal mass, aid in the maintenance of stable indoor temperatures, lowering HVAC demands. To manage indoor temperatures, passive design makes use of natural components such as sunshine, ventilation, and shade. Techniques such as orientation, natural ventilation, shading devices, and high thermal mass building materials maximize comfort while reducing reliance on mechanical systems.

Investigating the integration of renewable energy sources, notably solar panels and wind turbines, elucidates their role in reducing reliance on non-renewable resources. Solar panels, commonly known as photovoltaic (PV) panels, use the photovoltaic effect to directly convert sunlight into electricity. The panels are made up of numerous solar cells constructed from semiconductor materials such as silicon. When sunlight strikes these cells, it excites electrons, resulting in an electric current. This direct current (DC) is subsequently transformed into useable alternating current (AC) by inverters for consumption or grid integration. They transform solar energy into useable electricity using rooftop solar panels or integrated building facades, offsetting traditional grid power and lowering reliance on fossil fuels.

Wind energy is one of the most promising renewable energy sources since it harnesses the strength of the wind to create electricity. Wind energy is the conversion of kinetic energy from wind into electrical power using wind turbines. These turbines collect the wind's power and transform it into rotational energy, which is then used to create electricity. Wind energy is critical to the worldwide transition to sustainable and renewable energy sources. It provides a renewable alternative to fossil fuels, lowering greenhouse gas emissions and contributing to climate change mitigation.

Integrating wind energy into smart buildings involves thoughtful planning, technology selection, and design considerations to maximize its efficiency and contribution to the building's energy needs. There are a few steps to implement wind energy in smart buildings: The first step is to evaluate the site's wind resources using historical data or on-site measurements to determine the potential for wind energy generation. You need to assess wind speed, direction, and consistency to identify optimal locations for turbines. The second step is to consider the building's design, height, and surroundings to optimize wind turbine placement for maximum exposure to wind. For urban settings or smaller structures, consider vertical-axis wind turbines (VAWTs) that can capture wind from different directions and have a smaller footprint. The third step is to select appropriate wind turbines. Choose wind turbine models based on the building's energy needs, available space, and wind conditions. Consider factors like turbine size, efficiency, noise levels, and visual impact on the building's surroundings. In the next step, you need to integrate the wind turbines with the building's energy management system or smart grid infrastructure for seamless monitoring and control. Utilize sensors and IoT devices to collect real-time data on energy generation, allowing for optimization and predictive maintenance.

In the fifth step, wind energy output must be combined with energy storage options such as batteries to store excess energy for use during low-wind or peak demand periods. Implement smart monitoring systems to track energy production, turbine performance, and building energy consumption. Use automation and control systems to optimize the balance between wind energy generation and other energy sources within the building. In the sixth step, you are required to develop a maintenance schedule for regular inspection and upkeep of the wind turbines to ensure optimal performance and safety. Implement safety measures to prevent accidents and protect against potential hazards associated with wind turbine operations. Finally, educate building occupants about the benefits of wind energy integration, its role in sustainability, and how they can contribute to efficient energy use.

15.2.1.2 IoT and automation

Smart building technologies, at their heart, are a confluence of cutting-edge systems, IoT devices, sensors, automation, and data analytics inside the architecture of a building. Smart green building technologies are growing quickly, with an emphasis on sustainability, energy efficiency, and occupant health. Integration of Internet of Things (IoT) devices and systems, a focus on energy efficiency and sustainability, and improved occupant health and welfare are some of the rising trends in this subject. These technologies allow for real-time monitoring, data collecting, and adaptive management of building components such as HVAC systems, lighting, and security, assuring optimal performance and resource utilization. These technologies provide complete data gathering, analysis, and utilization to improve operational efficiency, increase sustainability, assure occupant comfort, and make data-informed decisions. The Internet of Things (IoT) technological tool allows for the seamless integration of numerous devices and systems, optimizing energy management, monitoring equipment performance, and improving occupant comfort [5]. The rising emphasis on decreasing energy use, minimizing environmental effects, and enhancing indoor air quality is driving these developments. Furthermore, as green technology advances, the use of ecologically friendly materials in buildings has grown more feasible [6].

The Internet of Things (IoT) enables numerous building systems, such as heating, ventilation, air conditioning (HVAC), lighting, security, and energy management systems, to connect and interact with one another. This interconnection allows for more efficient and automatic control of many building processes, resulting in lower energy consumption, better resource utilization, and greater sustainability. Smart HVAC systems may be used to optimize building heating/cooling based on room occupancy, which is sensed using sensors in doors and ceilings [7]. Green construction is also achieving benefits such as increased worker productivity, greater space utilization, and energy savings, all of which lead to a more sustainable and cost-effective built environment [8, 9, 10]. IoT-enabled sensors, for example, may monitor environmental factors such as temperature, humidity, and occupancy levels within a structure. This data may be analyzed in real-time to make changes to HVAC systems, maintaining occupant comfort while minimizing energy waste. Furthermore, IoT may help with predictive maintenance by gathering data from building equipment and identifying faults before they become serious ones, minimizing downtime and wasteful energy use.

15.2.1.3 Material innovations

Material innovation is at the forefront of revolutionizing the construction industry, particularly in the context of smart buildings. This introduction lays the groundwork for understanding the significance and multifaceted nature of material advancements. Material innovations encompass the development and implementation of novel materials, composites, and technologies in construction, aimed at enhancing building performance, sustainability, and functionality. The evolution of material innovations in smart buildings represents a change in thinking, offering solutions to traditional construction challenges. These innovations prioritize sustainability, energy efficiency, occupant comfort, and integration with smart technologies. Advanced and sustainable materials play a crucial role. From recycled materials to those with low environmental impact, innovations in construction materials contribute significantly to the green credentials of smart buildings.

Understanding the many sorts of material innovations explains their various uses and advantages in the construction of smart buildings. Aerogels, phase-change materials, and vacuum-insulated panels are examples of insulation material innovations that increase thermal performance, reduce energy consumption, and improve interior comfort. Energy efficiency, comfort, and sustainability are all improved by advanced insulating materials. Superior R-values are commonly found in advanced insulation materials, demonstrating their capacity to resist heat flow. Heat transmission through walls, ceilings, and floors is reduced, reducing the demand for heating and cooling. By effectively insulating a building, advanced materials help maintain stable indoor temperatures, reducing reliance on HVAC systems and lowering energy consumption for heating and cooling. Superior insulation ensures more consistent indoor temperatures, eliminating hot or cold spots within the building. Some advanced insulation materials also offer soundproofing properties, reducing external noise infiltration for a quieter indoor environment.

Many advanced insulation materials are eco-friendly and contribute to reducing a building's carbon footprint by lowering energy demands. Some insulation materials are made from recycled or bio-based sources, contributing to resource conservation and sustainable practices in construction.

Advanced insulation materials can be integrated with sensors and smart systems to monitor temperature, moisture levels, and energy efficiency in real-time. Smart insulation systems can facilitate predictive maintenance by detecting potential issues early, optimizing performance, and reducing energy losses.

Smart windows represent a transformative technology in modern building design, offering dynamic control over light, heat, and privacy. This introduction sets the stage for understanding their significance and versatility. Smart windows are technologically advanced glazing systems capable of changing their properties, such as transparency, tint, or light transmission, in response to external stimuli like light intensity, temperature, or user commands. These windows play an important role in boosting energy efficiency, occupant comfort, and overall building performance in the context of smart buildings. Their dynamic nature matches sustainability and responsive design ideals. Smart glass and windows in smart buildings require strategic design, technological integration, and maximizing their capabilities to improve energy efficiency, comfort, and overall building performance.

Exploring the various types and underlying technologies of smart windows elucidates their diversity and functionality in smart building applications. Electrochromic smart windows use an electrical charge to change the tint or transparency of the glass, allowing for adjustable light and heat control. Photochromic and thermochromic windows alter their tint or opacity in response to UV light or temperature changes, providing passive control over daylight and heat. Suspended particle devices (SPDs) and liquid crystal windows allow windows to switch between transparent and opaque states using electric voltage, offering instant control over privacy and light transmission.

Examining the practical applications and associated benefits of smart windows demonstrates their versatility and impact on smart building designs. Smart windows optimize natural daylight utilization, reducing the need for artificial lighting while managing solar heat gain, thereby lowering HVAC energy consumption. Dynamic control over glare, heat, and privacy enhances indoor comfort, promoting productivity and well-being among building occupants. Furthermore, smart windows facilitate responsive building design, adjusting to external conditions or user preferences, thus contributing to a more adaptable and sustainable built environment.

15.2.1.4 The benefits of smart green building technologies

The benefits of using smart green building technologies are multifaceted and include improvements in energy efficiency, cost savings, environmental impact, and occupant health and well-being. Table 15.2 shows some key benefits.

Table 15.2 Benefits of smart green building technologies

Benefits	*Details*
Energy Efficiency and Cost Savings	Smart building solutions increase energy efficiency, resulting in considerable long-term energy savings. They offer real-time monitoring and management of energy use, maximizing building performance and lowering running expenses [11, 12].
Environmental Impact	These technologies help to reduce greenhouse gas emissions by improving energy monitoring and controlling carbon emissions from building materials. They also aid in the achievement of carbon neutrality and the reduction of building environmental effects [13].
Occupant Health and Well-Being	By monitoring and regulating heating, ventilation, and air-conditioning systems in real-time, smart green buildings may enhance indoor air quality and occupant health. This has been shown to minimize the incidence of respiratory disease and chronic illnesses [14, 15].
Safety and Security	Through real-time monitoring and control systems, smart building technologies improve safety and security, contributing to a safer and more secure built environment.
Improved Worker Productivity	These technologies have been found to increase worker productivity and tenant happiness while also improving space utilization and overall cost-effectiveness [16, 17].

15.3 INTERSECTION AND INTEGRATION OF SMART TECHNOLOGY WITH GREEN BUILDING CONCEPTS

The intersection and integration of smart technology with green building concepts represent a groundbreaking synergy that reshapes the landscape of sustainable construction. At this convergence point, the ethos of green building, focused on minimizing environmental impact and optimizing resource use, intertwines with the transformative potential of smart technology. Smart systems leverage data-driven insights, automation, and connectivity to enhance the ecological efficiency and operational performance of buildings. These technologies permeate every facet of sustainable design, from energy-efficient lighting and HVAC systems to sophisticated sensors monitoring water consumption, indoor air quality, and waste management. The integration of these systems not only optimizes resource utilization but also fosters intelligent, adaptive buildings capable of responding dynamically to environmental changes and user needs.

This integration promotes a change in thinking in the way buildings are conceived, built, and operated. Smart technologies provide real-time monitoring and control, allowing for more exact management of energy use, water consumption, and waste creation. BIM and IoT sensors provide extensive data collection, enabling predictive analytics that optimize building performance and maintenance schedules. Furthermore, incorporating renewable energy sources such as solar panels or wind turbines into smart grids within buildings increases self-sufficiency while decreasing dependency on non-renewable resources and grid-supplied electricity. This comprehensive approach to sustainability uses technology as a catalyst, propelling green building concepts ahead by minimizing environmental impact while maximizing efficiency and usefulness.

The marriage of smart technology and green building extends beyond operational efficiency to revolutionize occupant experiences. Intelligent building systems, integrated with smart controls and user-centric interfaces, enhance occupant comfort, health, and productivity. Adaptive lighting, responsive climate control, and indoor air quality management systems ensure optimal environmental conditions. Furthermore, smart technology enables personalized, data-driven insights, empowering occupants to engage actively in energy conservation practices and make informed decisions about resource usage. The focus on user-centric design not only improves the quality of life within these buildings but also fosters a culture of environmental awareness and responsibility among occupants, amplifying the impact of sustainable practices beyond the building's physical footprint.

Beyond the immediate benefits, this integration yields far-reaching environmental and economic advantages. By leveraging smart technologies to align with green building principles, buildings achieve higher levels of energy efficiency, reducing operational costs and greenhouse gas emissions. This alignment also positions buildings for compliance with stringent sustainability certifications, enhancing their market value and attracting environmentally conscious investors and tenants. The economic viability and long-term sustainability of these structures foster a ripple effect, influencing industry standards, shaping policies, and driving innovation toward a future where buildings serve as beacons of sustainability, efficiency, and technological advancement.

15.4 SUSTAINABLE MATERIALS AND CONSTRUCTION TECHNIQUES

Sustainable materials and construction techniques epitomize a pivotal shift in the construction industry, steering it toward environmentally conscious practices and resource-efficient methodologies. At the heart of this ethos lies the deliberate selection and utilization of materials that minimize environmental impact across their lifecycle. From eco-friendly sourcing and manufacturing processes to their durability and recyclability, these materials embody sustainability. Additionally, construction techniques rooted in sustainable practices emphasize reduced waste generation, energy efficiency, and innovative approaches that mitigate environmental harm. Embracing concepts like prefabrication, modular construction, and green building certifications, these techniques seek to optimize resource use, minimize carbon footprints, and foster resilient, adaptable structures capable of harmonizing with their surroundings while significantly reducing the ecological burden of the built environment. This concerted focus on sustainable materials and construction techniques underscores a transformative journey toward creating structures that not only endure but also thrive within the fabric of a greener, more sustainable world.

15.4.1 Use of innovative and sustainable materials in construction

The integration of innovative and sustainable materials in construction, particularly within the realm of smart buildings, signifies a transformative leap toward environmentally conscious and technologically advanced structures. These materials include a wide range of advancements, including recycled and bio-based components, sophisticated composites, and nanotechnology-infused materials. Their selection and implementation are subject to strict criteria that consider issues such as durability, energy efficiency, carbon footprint, and life-cycle analysis. These materials are critical in smart buildings for improving energy performance, optimizing thermal management, and lowering environmental impact. From self-healing concrete to energy-efficient insulations and smart glass, these innovations epitomize a fusion of technology and sustainability, reshaping the architectural landscape toward an eco-friendly and responsive built environment.

The adoption of these materials in smart buildings not only amplifies their sustainability quotient but also integrates seamlessly with intelligent systems and design strategies. Embedded with sensors, these materials offer real-time data on structural integrity, energy consumption, and environmental conditions, allowing for predictive maintenance and responsive functionalities. Moreover, their incorporation aligns with the ethos of green building certifications, fostering a confluence of technological innovation and environmental stewardship. Their application in smart buildings extends beyond energy efficiency to bolster occupant comfort, promote healthier indoor environments, and align with the principles of the circular economy by emphasizing material recyclability and reusability. This amalgamation of sustainability and cutting-edge materials within smart buildings represents a paradigm shift in construction, pioneering structures that not only function intelligently but also stand as exemplars of environmental responsibility and long-term resilience.

Furthermore, the utilization of these innovative and sustainable materials underscores a commitment to pushing the boundaries of construction methodologies. The pursuit of advancements in material science, coupled with the integration of smart technologies, drives an evolution toward buildings that adapt, respond, and contribute positively to their surroundings. These materials serve as catalysts for change, sparking an industry-wide shift toward more efficient and ecologically mindful construction practices. Their incorporation in smart buildings marks a pivotal step toward achieving a more sustainable and technologically adept future, where structures not only serve their functional purpose but also harmonize harmoniously with the environment, reflecting a profound commitment to sustainability and innovation.

15.4.2 Advances in construction techniques that promote sustainability and energy efficiency

Advances in construction techniques have undergone a change in thinking, prioritizing sustainability and energy efficiency to meet the evolving demands of the built environment. These advancements encompass a spectrum of methodologies, from modular and prefabricated construction to advanced robotics and 3D printing. Prefabrication, for instance, minimizes material waste, accelerates construction timelines, and allows for precision engineering, reducing on-site labor and energy consumption. Similarly, 3D printing technologies offer opportunities to construct complex structures using sustainable materials, optimize material use, and streamline construction processes. Robotics and AI-driven systems contribute to enhanced precision, safety, and efficiency in construction, reducing errors and optimizing resource utilization. These techniques not only expedite construction timelines but also significantly reduce the environmental footprint of the building process, aligning with sustainability goals and fostering more energy-efficient structures.

Furthermore, innovations in construction processes coexist with the use of renewable energy sources in structures. Solar panels, wind turbines, and geothermal systems are currently used in smart buildings to power operations, optimize energy use, and minimize dependency on non-renewable energy networks. Building-integrated photovoltaics (BIPV), for example, integrates solar panels into building features such as facades or roofing, creating renewable energy while also functioning as structural components. Furthermore, passive design tactics like appropriate orientation, natural ventilation, and daylighting work in conjunction with these building approaches to lower the demand for artificial heating, cooling, and lighting, thus improving energy efficiency and sustainability. These advancements foster a symbiotic link between construction processes and energy solutions, revolutionizing how buildings are conceived, built, and operated in the direction of a more sustainable and energy-efficient future.

Furthermore, the evolution of sustainable construction techniques extends beyond the building process to encompass the entire life cycle of structures. Concepts like deconstruction and adaptive reuse champion circular economy principles by repurposing materials from existing buildings, minimizing waste, and reducing the need for raw materials in new construction. Building Information Modeling (BIM) and digital twins facilitate efficient building operations, enabling predictive maintenance and optimizing energy consumption throughout the building's lifespan. Life cycle assessment tools aid in evaluating environmental impacts, allowing for informed decision-making regarding materials and construction methodologies, and emphasizing long-term sustainability over the entire lifespan of structures. These holistic approaches redefine construction practices, emphasizing efficiency, durability, and environmental responsibility, thereby shaping a more sustainable trajectory for the construction industry and the built environment.

15.5 ENERGY MANAGEMENT AND EFFICIENCY

Building energy management is critical for optimizing resource use, lowering operational costs, and decreasing environmental impact. Implementing energy-efficient technology, such as smart lighting, HVAC controls, and occupancy sensors, enables accurate energy monitoring and adjustment. Building management systems (BMSs) or energy management systems (EMSs) provide real-time data collecting, allowing for informed decision-making and the optimization of energy usage patterns. Furthermore, by incorporating smart technologies such as IoT devices and machine learning algorithms, buildings may dynamically alter energy use based on occupancy, external weather conditions, and energy demand, assuring maximum performance while minimizing waste.

Investing in energy-efficient building envelopes, insulation, and windows also helps to reduce heat loss, optimize thermal comfort, and reduce dependency on heating and cooling systems. Renewable energy sources, such as solar panels or wind turbines, improve energy management in buildings by augmenting grid electricity and supporting sustainability. Furthermore, energy audits and performance evaluations help to discover areas for improvement, optimize equipment efficiency, and increase overall energy performance. Buildings may achieve improved energy efficiency, minimize environmental impact, and pave the path for a more sustainable future by using a comprehensive strategy that combines technical breakthroughs, strategic design, and continual monitoring.

15.5.1 Implementation of energy-efficient systems in smart buildings

Implementing energy-efficient systems in smart buildings necessitates a strategic strategy that combines cutting-edge technologies with principles of sustainable design. Smart buildings optimize energy use by utilizing technological technology such as building management system (BMS), IoT sensors, and artificial intelligence (AI). These systems gather real-time data on occupancy, weather trends, and energy use, enabling intelligent management and regulation of lighting, HVAC, and other energy-consuming devices. Furthermore, machine learning algorithms analyze previous data to forecast use trends, allowing for proactive energy-saving modifications. Smart buildings that integrate these technologies respond dynamically to changing conditions, minimizing waste and maximizing energy savings without sacrificing occupant comfort or operational efficiency.

Implementing energy-efficient systems in smart buildings entails more than just individual components; it entails a full, integrated ecosystem. Smart lighting systems that use occupancy sensors and daylight harvesting alter lighting settings based on occupancy and natural light availability, saving energy. Similarly, sophisticated HVAC systems with predictive maintenance and zoned controls optimize heating and cooling requirements, guaranteeing exact temperature management in various building sections. Furthermore, including renewable energy sources such as solar panels or geothermal systems improves energy efficiency by augmenting grid electricity and decreasing dependency on nonrenewable sources. This comprehensive integration of energy-efficient technology into smart buildings offers an intelligent framework, boosting sustainability, lowering operational costs, and establishing these structures as models of efficient energy consumption and environmental responsibility.

Furthermore, the successful implementation of energy-efficient systems in smart buildings necessitates ongoing monitoring, analysis, and adaptation. Continuous data collection and analysis enable building managers to identify areas for improvement, fine-tune system settings, and implement energy-saving strategies. Collaborative efforts among architects, engineers, facility managers, and occupants are vital for fostering a culture of energy conservation and ensuring the effective operation of these systems. Regular maintenance, upgrades, and technological advancements also play a pivotal role in enhancing the efficiency and longevity of energy-efficient systems. By embracing an iterative approach that integrates technological innovation, occupant engagement, and continual improvement, smart buildings can achieve and sustain high levels of energy efficiency while contributing significantly to environmental conservation.

15.5.2 Role of IoT in energy management

The Internet of Things (IoT) is transforming energy management by enabling real-time monitoring, control, and optimization of energy consumption across several industries. IoT devices and sensors provide extensive data collection on energy consumption patterns, ambient conditions, and equipment performance in the sphere of buildings and infrastructure. When connected to energy management systems, these devices provide detailed insights into energy consumption, exposing inefficiencies and potential for optimization. Furthermore, IoT-driven analytics and machine learning algorithms digest this data, resulting in actionable insights that enable proactive decision-making. This technology allows for predictive maintenance, load balancing, and the fine-tuning of energy-consuming systems, ensuring optimal operational efficiency while minimizing waste. By fostering connectivity and intelligence among devices, IoT transforms energy management into a dynamic, adaptive process, enabling businesses and industries to achieve greater energy efficiency and sustainability goals.

Furthermore, IoT's impact on energy management extends beyond individual buildings or sectors, contributing to the larger landscape of smart grids and energy ecosystems. Smart grids leverage IoT-enabled devices to facilitate bidirectional communication between energy providers and consumers, enabling demand-response mechanisms and dynamic pricing models. IoT sensors embedded in smart meters or appliances enable real-time energy consumption tracking, allowing consumers to make informed decisions about energy usage based on pricing fluctuations or peak demand periods. Additionally, IoT-driven systems support the integration of renewable energy sources, such as solar or wind, by optimizing their generation and distribution within the grid. This interconnected infrastructure, powered by IoT technologies, not only enhances energy management efficiency but also fosters a more resilient, responsive energy ecosystem capable of meeting evolving demands while promoting sustainability and cost-effectiveness.

15.6 THE INTEGRATION OF RENEWABLE ENERGY SYSTEMS IN SMART GREEN BUILDINGS

The integration of renewable energy systems in smart green buildings has various benefits, including increased renewable energy reliability and stability, lower power costs, and climate change mitigation. These systems cut energy usage while simultaneously contributing to a cleaner and more sustainable future by capturing natural resources such as sunshine, wind, and water [18]. Green buildings use sophisticated insulation techniques, energy-efficient equipment, and renewable energy sources such as solar panels to reduce energy use significantly. They also help to reduce the urban heat island effect, save water, and contribute to a more pleasant and ecologically aware urban environment [19]. Smart building technologies, including renewable energy integration, are expected to rise fast, providing benefits in terms of sustainability, safety, energy savings, and increased worker productivity [20]. Overall, renewable energy integration in smart green buildings is critical for attaining energy efficiency, economic savings, and environmental sustainability [21]. Renewable energy sources can be integrated into the design of a building in various ways (Table 15.3).

Architects and designers must consider optimizing the architecture of the building to maximize available space for renewable energy systems. Implementing renewable energy systems in buildings necessitates careful planning, coordination, and awareness to address problems such as restricted space, high upfront cost, and energy supply dependability and stability [25].

Table 15.3 Renewable energy integrated into the design of a building

Source	*Key Points*
Solar Energy	Solar panels can be mounted on the building's roof or walls to generate energy or heat water. The direction and shading of the structure should be addressed to maximize the quantity of sunshine received [22].
Wind Energy	Wind turbines can be installed on the roof or walls of a building to generate electricity or mechanical power. The building's location and wind patterns should be considered to maximize the amount of wind energy harnessed [23].
Geothermal Heat Pump	Geothermal heat pumps can be installed to heat and cool the building by utilizing the constant temperature of the earth. The building's foundation and soil conditions should be considered to maximize the efficiency of the system [24].
Biomass	Biomass boilers, which create heat by burning organic resources such as wood, agricultural waste, and municipal solid waste, can be installed. To maximize system efficiency, the location of the building and the availability of biomass fuel should be considered.

However, there are obstacles connected with the use of renewable energy in these buildings, such as significant upfront costs, installation complexity, energy supply dependability, integration with existing structures and regulations, and restricted installation space [26]. Despite these obstacles, the use of renewable energy in smart green buildings is critical for creating a greener, more sustainable built environment, as well as lowering reliance on fossil fuels and mitigating climate change [27, 28]. The dependability of the energy supply in smart greenhouses is an important issue influencing their overall performance and sustainability. Renewable energy sources, such as solar panels and wind turbines, provide environmental advantages; however, they are intermittent and changeable owing to variables like as weather and availability [29]. This fluctuation can make keeping a continuous energy supply in smart greenhouses difficult.

Smart buildings can use tactics such as energy storage systems, demand response programs, and advanced control systems to optimize energy use and provide a consistent power supply to meet these difficulties [30]. Smart buildings may maximize the use of renewable resources and minimize their reliance on fossil fuels by combining energy-efficient equipment and renewable energy systems. Furthermore, using smart IoT-based materials in greenhouses can increase energy supply dependability by monitoring and modifying energy use depending on occupancy, weather conditions, and other variables [31].

15.7 INDOOR ENVIRONMENTAL QUALITY AND HEALTH

The significance of indoor air quality (IAQ) and occupant health within smart green buildings underscores a holistic approach to sustainable and human-centric design. IAQ stands as a pivotal determinant of occupant health, productivity, and overall well-being. Smart green buildings prioritize IAQ by employing advanced ventilation systems, air quality sensors, and filtration technologies. These systems continuously monitor and regulate indoor air, minimizing pollutants, volatile organic compounds (VOCs), and allergens. By maintaining optimal air quality levels, these buildings create environments that mitigate health risks associated with poor IAQ, such as respiratory issues, allergies, and fatigue, thereby fostering healthier and more productive spaces for occupants.

Moreover, the integration of smart technologies within green buildings extends beyond passive IAQ management to proactive and adaptive solutions. IoT-driven sensors monitor IAQ parameters in real-time, providing insights into pollutant levels, humidity, and temperature variations. This data informs intelligent HVAC systems that adjust ventilation and air circulation to maintain optimal IAQ levels, responding dynamically to fluctuating conditions or occupancy patterns. Additionally, smart buildings utilize natural ventilation strategies, integrating operable windows or automated systems that facilitate fresh air circulation while minimizing energy consumption. Prioritizing IAQ within smart green buildings not only safeguards occupant health but also underscores a commitment to creating environments that foster comfort, productivity, and overall well-being.

Furthermore, the emphasis on IAQ and occupant health within smart green buildings aligns with global standards and certifications, reinforcing the buildings' commitment to environmental responsibility and occupant-centric design. Certifications like WELL Building Standard or RESET prioritize IAQ, establishing stringent criteria for air quality, ventilation, and toxin control within buildings. Compliance with these standards not only validates a building's dedication to health-focused design but also enhances its market value, attracting environmentally conscious tenants or investors. Additionally, promoting occupant health and well-being aligns with corporate social responsibility initiatives, demonstrating a commitment to creating healthier communities and reducing the environmental footprint of buildings. As smart green buildings continue to evolve, the integration of IAQ management strategies remains central, ensuring that these structures not only function efficiently but also prioritize the health and comfort of their occupants, thus promoting a sustainable and people-centric approach to building design.

15.7.1 Technologies addressing ventilation, air purification, and thermal comfort

In the realm of smart buildings, technologies addressing ventilation, air purification, and thermal comfort play a pivotal role in creating healthier, more sustainable indoor environments. Advanced ventilation systems within smart buildings leverage IoT sensors and real-time data analytics to regulate airflow and maintain optimal indoor air quality (IAQ). These systems adjust ventilation rates based on occupancy, outdoor air quality, and pollutant levels, ensuring a constant supply of fresh air while minimizing energy consumption. Additionally, smart buildings integrate natural ventilation strategies, incorporating operable windows or automated systems that facilitate airflow, allowing for passive cooling and improved IAQ without relying solely on mechanical systems. This integration of smart ventilation technologies optimizes air circulation, enhancing comfort and health for occupants while reducing energy demands.

Air purification technologies embedded within smart buildings further elevate IAQ by removing pollutants and contaminants from indoor air. High-efficiency particulate air (HEPA) filters activated carbon filtration, and UV-C disinfection systems are among the advanced technologies employed. These systems trap airborne particles, allergens, and volatile organic compounds (VOCs), and inactivate microbes, ensuring cleaner, healthier air for building occupants. Moreover, smart purification systems equipped with sensors and IoT connectivity provide real-time monitoring of air quality parameters, allowing for immediate adjustments to maintain optimal IAQ levels. By integrating these purification technologies, smart buildings create environments that prioritize health and comfort, fostering productivity and well-being among occupants.

Additionally, technologies addressing thermal comfort within smart buildings utilize innovative solutions to optimize indoor temperature and humidity levels. Advanced HVAC systems leverage machine learning algorithms and predictive analytics to regulate climate control, adjusting temperatures based on occupancy patterns and external weather conditions. Zoning systems within smart buildings enable personalized comfort settings in different areas, ensuring tailored climate control while conserving energy. Furthermore, smart windows and shading systems employ electrochromic or thermochromics technologies that dynamically adjust transparency and solar heat gain, optimizing natural light while controlling indoor temperatures. These technologies not only enhance thermal comfort but also contribute to energy efficiency, reducing reliance on mechanical heating and cooling systems and promoting sustainable building practices.

In essence, the integration of smart technologies addressing ventilation, air purification, and thermal comfort within buildings epitomizes a comprehensive approach to creating healthier, more comfortable, and energy-efficient indoor environments. These advancements underscore a commitment to occupant well-being while advancing sustainability goals, shaping smart buildings that not only function intelligently but also prioritize the health, comfort, and productivity of their occupants.

15.7.2 Strategies for promoting a healthy indoor environment

Promoting a healthy indoor environment encompasses a multifaceted approach that integrates various strategies aimed at optimizing air quality, enhancing comfort, and fostering well-being for occupants. Ventilation strategies stand as a cornerstone, emphasizing the importance of adequate airflow and fresh air circulation. Implementing natural ventilation through operable windows, passive design strategies, and utilizing smart ventilation systems ensures a constant supply of fresh air while expelling indoor pollutants. Furthermore, air quality monitoring, coupled with advanced filtration technologies like HEPA filters and UV-C disinfection systems, targets the removal of allergens, pollutants, and microbes, maintaining cleaner and healthier indoor air. These strategies, combined with smart building technologies and real-time monitoring, ensure an environment conducive to occupant health and productivity.

Moreover, optimizing thermal comfort is pivotal in creating a healthy indoor environment. Implementing temperature and humidity control systems tailored to occupants' preferences fosters comfort while minimizing the risk of health issues related to extreme temperatures. Innovative solutions like smart windows, shading systems, and energy-efficient HVAC systems offer precise climate control, balancing thermal comfort with energy efficiency. Additionally, employing materials with thermal properties that regulate temperatures and enhance insulation contributes to maintaining consistent comfort levels while reducing reliance on mechanical heating and cooling systems. By integrating these strategies, buildings create environments that prioritize occupant well-being, ensuring comfort and health while minimizing energy consumption.

Additionally, promoting a healthy indoor environment involves strategies that focus on holistic well-being beyond air quality and thermal comfort. Encouraging access to natural light through well-designed windows, skylights, and interior layouts that maximize daylight exposure improves mental health and boosts productivity among occupants. Providing spaces for physical activity, relaxation, and connection with nature, such as green spaces or biophilic design elements, further enhances occupants' mental and emotional well-being. Furthermore, fostering a culture of wellness through educational programs, health-focused initiatives, and ergonomic design features promotes healthy behaviors and a sense of community within the building. By embracing an integrated approach that integrates multiple strategies addressing air quality, thermal comfort, and overall well-being, buildings evolve into spaces that nurture the health and vitality of their occupants.

15.8 CHALLENGES AND FUTURE DIRECTIONS

Implementing smart green building technologies presents a change in basic assumptions in the construction industry, yet it also confronts several challenges that impede its widespread adoption. One prominent challenge is the initial high cost associated with integrating advanced technologies and sustainable features into building designs. The upfront investment required for incorporating smart systems, energy-efficient materials, and renewable energy sources often deters stakeholders, despite the long-term benefits. Moreover, the lack of standardized guidelines or regulatory frameworks specific to smart green buildings can create uncertainty and complexity during the design and construction phases, further complicating the implementation process.

Interoperability and compatibility issues among different technologies pose another significant challenge. The integration of various smart systems, IoT devices, and energy management platforms from different vendors may encounter compatibility issues, hindering seamless communication and optimal functionality. This fragmentation complicates the integration process, requiring thorough planning and coordination to ensure interoperability and effective system operation. Additionally, the rapid pace of technological advancements poses a challenge in selecting future-proof technologies and systems that remain relevant and adaptable over the building's lifecycle. This constant evolution necessitates careful planning and flexibility to accommodate future upgrades without rendering current systems obsolete.

Furthermore, overcoming the knowledge gap and skill shortage in the workforce presents a notable challenge in implementing smart green building technologies. The successful execution of these technologies requires specialized expertise in areas like IoT, data analytics, sustainable design, and energy management. However, the shortage of professionals with a comprehensive understanding of these technologies and sustainable practices poses a barrier to the effective implementation and maintenance of smart green buildings. Bridging this skill gap through education, training, and capacity-building initiatives becomes crucial for ensuring the successful adoption and operation of these advanced systems within the construction industry.

Furthermore, in the construction industry, resistance to change and a conservative attitude can stymie the adoption of innovative technology and sustainable practices. Traditional construction processes and established standards can create resistance to adopting new approaches, causing the adoption of smart green building technology to be delayed. To overcome this inertia, stakeholders must promote a culture of innovation, information sharing, and creative partnerships to accelerate acceptance and implementation of these transformational technologies. To traverse the intricacies and promote the wider application of smart green building technologies toward a more sustainable built environment, industry players, governments, and educational institutions must work together to address these difficulties.

15.8.1 Future trends and potential advancements in the field

The future of smart green building technologies holds promising advancements poised to redefine construction practices and sustainability standards. One notable trend on the horizon is the integration of artificial intelligence (AI) and machine learning algorithms into building management systems. AI-driven platforms will enable buildings to autonomously optimize energy consumption, predictive maintenance, and occupant comfort. These systems will continuously learn from data collected through IoT sensors, allowing for more sophisticated decision-making in real-time, thus enhancing operational efficiency and reducing energy waste. Additionally, advancements in smart materials, such as self-healing concrete or adaptive insulation, are anticipated to become more prevalent. These materials will offer enhanced durability, energy efficiency, and resilience, contributing significantly to the longevity and sustainability of structures while minimizing maintenance needs.

Moreover, the evolution of net zero and positive-energy buildings stands as a significant trajectory in the field of smart green buildings. Future advancements aim to push the boundaries of energy efficiency, aiming for buildings that produce more energy than they consume. Integrating innovative energy storage solutions, like advanced batteries or hydrogen-based systems, will play a crucial role in balancing energy demands and surpluses within these structures. Additionally, the rise of biophilic design principles and living architecture signifies a shift toward creating buildings that mimic natural ecosystems, fostering connections with nature, and promoting occupant well-being. Green roofs, vertical gardens, and integrated urban agriculture will not only enhance aesthetics but also contribute to improved air quality, biodiversity, and overall sustainability. Embracing these future trends and advancements reflects a commitment to pushing the boundaries of sustainability, innovation, and occupant-centric design in the realm of smart green building technologies.

15.9 CONCLUSION

This chapter illuminates the powerful constructive interaction between smart technologies and green building principles, ushering in a new era of sustainable development. This integration transcends mere structures; it signifies a fundamental shift in how we perceive construction and design. At its core, it's a commitment to a future where buildings coexist harmoniously with their surroundings, evolving into living entities that embody environmental consciousness. This amalgamation does not just optimize resources; it births a new breed of intelligent, adaptable structures. Beyond efficiency metrics lie transformative cultural impacts, educating occupants and fostering collective responsibility for sustainability. These buildings become trailblazers, influencing industry norms, inspiring innovation, and setting global benchmarks for a future where sustainability is intrinsic to progress. This integration is not just a solution; it is a guiding compass steering us toward a future where technology and unwavering environmental commitment coalesce for generations to thrive sustainably.

REFERENCES

1. Zou, H., Zhou, Y., Jiang, H., Chien, S.C., Xie, L. and Spanos, C.J., 2018. WinLight: A WiFi-based occupancy-driven lighting control system for smart building. *Energy and Buildings*, 158, pp.924–938.
2. Merabet, G.H., Essaaidi, M., Haddou, M.B., Qolomany, B., Qadir, J., Anan, M., Al-Fuqaha, A., Abid, M.R. and Benhaddou, D., 2021. Intelligent building control systems for thermal comfort and energy-efficiency: A systematic review of artificial intelligence-assisted techniques. *Renewable and Sustainable Energy Reviews*, 144, p.110969.
3. Aguilar, J., Garces-Jimenez, A., R-moreno, M.D. and García, R., 2021. A systematic literature review on the use of artificial intelligence in energy self-management in smart buildings. *Renewable and Sustainable Energy Reviews*, 151, p.111530.
4. Obaideen, K., Yousef, B.A., AlMallahi, M.N., Tan, Y.C., Mahmoud, M., Jaber, H. and Ramadan, M., 2022. An overview of smart irrigation systems using IoT. *Energy Nexus*, 7, p.100124. https://doi.org/10.1016/j.nexus.2022.100124
5. Jia, M., Komeily, A., Wang, Y. and Srinivasan, R.S., 2019. Adopting Internet of Things for the development of smart buildings: A review of enabling technologies and applications. *Automation in Construction*, 101, pp.111–126.
6. Patil, M., Boraste, S. and Minde, P., 2022.A comprehensive review on emerging trends in smart green building technologies and sustainable materials. *Materials Today: Proceedings*, 65, 1813–1822. https://doi.org/10.1016/j.matpr.2022.04.866
7. Bae, Y., Bhattacharya, S., Cui, B., Lee, S., Li, Y., Zhang, L., Im, P., Adetola, V., Vrabie, D., Leach, M. and Kuruganti, T., 2021. Sensor impacts on building and HVAC controls: A critical review for building energy performance. *Advances in Applied Energy*, 4, p.100068.
8. Kats, G., 2003. *Green building costs and financial benefits* (pp. 2–8). Boston, MA: Massachusetts Technology Collaborative.
9. Miller, N., Pogue, D., Gough, Q. and Davis, S., 2009. Green buildings and productivity. *Journal of Sustainable Real Estate*, 1(1), pp.65–89.
10. Kibert, C.J., 2016. *Sustainable construction: Green building design and delivery*. John Wiley & Sons.
11. Ejidike, C.C. and Mewomo, M.C., 2023. Benefits of adopting smart building technologies in building construction of developing countries: Review of literature. *SN Applied Sciences*, 5, 52. https://doi.org/10.1007/s42452-022-05262-y]
12. King, J. and Perry, C., 2017. *Smart buildings: Using smart technology to save energy in existing buildings* (pp. 1–46). Washington, DC: American Council for an Energy-Efficient Economy.
13. Lu, W., Tam, V.W., Chen, H. and Du, L., 2020. A holistic review of research on carbon emissions of green building construction industry. *Engineering, Construction and Architectural Management*, 27(5), pp.1065–1092.
14. Dong, B., Prakash, V., Feng, F. and O'Neill, Z., 2019. A review of smart building sensing system for better indoor environment control. *Energy and Buildings*, 199, pp.29–46.

15. Che, W.W., Tso, C.Y., Sun, L., Ip, D.Y., Lee, H., Chao, C.Y. and Lau, A.K., 2019. Energy consumption, indoor thermal comfort and air quality in a commercial office with retrofitted heat, ventilation and air conditioning (HVAC) system. *Energy and Buildings*, 201, pp.202–215.
16. Nurick, S. and Thatcher, A., 2021. The relationship of green office buildings to occupant productivity and organizational performance: A literature review. *Journal of Real Estate Literature*, 29(1), pp.18–42.
17. Kwong, Q.J., 2020. Light level, visual comfort and lighting energy savings potential in a green-certified high-rise building. *Journal of Building Engineering*, 29, p.101198.
18. Ellabban, O., Abu-Rub, H. and Blaabjerg, F., 2014. Renewable energy resources: Current status, future prospects and their enabling technology. *Renewable and Sustainable Energy Reviews*, 39, pp.748–764.
19. Gago, E.J., Roldan, J., Pacheco-Torres, R. and Ordóñez, J., 2013. The city and urban heat islands: A review of strategies to mitigate adverse effects. *Renewable and Sustainable Energy Reviews*, 25, pp.749–758.
20. Nižetić, S., Djilali, N., Papadopoulos, A. and Rodrigues, J.J., 2019. Smart technologies for promotion of energy efficiency, utilization of sustainable resources and waste management. *Journal of Cleaner Production*, 231, pp.565–591.
21. Hafez, F.S., Sa'di, B., Safa-Gamal, M., Taufiq-Yap, Y.H., Alrifaey, M., Seyedmahmoudian, M., Stojcevski, A., Horan, B. and Mekhilef, S., 2023. Energy efficiency in sustainable buildings: A systematic review with taxonomy, challenges, motivations, methodological aspects, recommendations, and pathways for future research. *Energy Strategy Reviews*, 45, p.101013.
22. Valladares-Rendón, L.G., Schmid, G. and Lo, S.L., 2017. Review on energy savings by solar control techniques and optimal building orientation for the strategic placement of façade shading systems. *Energy and Buildings*, 140, pp.458–479.
23. Zhou, H., Lu, Y., Liu, X., Chang, R. and Wang, B., 2017. Harvesting wind energy in low-rise residential buildings: Design and optimization of building forms. *Journal of Cleaner Production*, 167, pp.306–316.
24. Omer, A.M., 2016. Experimental investigation of the performance of a ground source heat pump system for buildings heating and cooling. *International Journal of Innovative Mathematics, Statistics and Energy Policies*, 4(1), pp.10–44.
25. Kistelegdi, I., Horváth, K.R., Storcz, T. and Ercsey, Z., 2022. Building geometry as a variable in energy, comfort, and environmental design optimization—A review from the perspective of architects. *Buildings*, 12(1), p.69.
26. Curtius, H.C., 2018. The adoption of building-integrated photovoltaics: Barriers and facilitators. *Renewable Energy*, 126, pp.783–790.
27. Bungau, C.C., Bungau, T., Prada, I.F. and Prada, M.F., 2022. Green buildings as a necessity for sustainable environment development: Dilemmas and challenges. *Sustainability*, 14(20), p.13121.
28. Andrić, I., Koc, M. and Al-Ghamdi, S.G., 2019. A review of climate change implications for built environment: Impacts, mitigation measures and associated challenges in developed and developing countries. *Journal of Cleaner Production*, 211, pp.83–102.

29. Li, J., Chen, S., Wu, Y., Wang, Q., Liu, X., Qi, L., Lu, X. and Gao, L., 2021. How to make better use of intermittent and variable energy? A review of wind and photovoltaic power consumption in China. *Renewable and Sustainable Energy Reviews*, 137, p.110626.
30. Samad, T., Koch, E. and Stluka, P., 2016. Automated demand response for smart buildings and microgrids: The state of the practice and research challenges. *Proceedings of the IEEE*, 104(4), pp.726–744.
31. Maraveas, C., Piromalis, D., Arvanitis, K.G., Bartzanas, T. and Loukatos, D., 2022. Applications of IoT for optimized greenhouse environment and resources management. *Computers and Electronics in Agriculture*, 198, p.106993.

Chapter 16

Implementation of renewable technologies in energy-efficient urban infrastructure

Development to promote sustainable socio-economic growth

Sumanta Bhattacharya and Bhavneet Kaur Sachdev

16.1 INTRODUCTION

Sustainable socio-economic growth is greatly aided by renewable technologies, which are essential in promoting the development of energy-efficient urban infrastructure. Incorporating renewable energy sources is crucial for building resilient and environmentally conscious urban settings as cities worldwide face the triple threat of population increase, industrialization, and climate change. Distributed solar power systems are an important part of renewable energy integration. It is possible to create clean power from sunshine by properly placing rooftop and ground-mounted solar panels in metropolitan settings. In addition to lowering emissions of greenhouse gases, this lessens our need for traditional fossil fuels, which helps alleviate environmental problems. Solar power is becoming more appealing to city planners as a result of technological developments that have made it both efficient and affordable [1].

One of the most important renewable energy sources for city infrastructure is wind power. Installing wind turbines on land or in the water allows for the collection of wind energy, which can then be used to produce energy. The advantages of wind farms are enormous, but their integration into urban development needs meticulous attention to regional circumstances. Wind power, when added to a diverse energy portfolio, helps provide a steady supply of electricity while reducing pollution levels. Using renewable energy sources like solar and wind, together with biomass and geothermal power, makes city infrastructure far more environmentally friendly. Heat, power, and biofuels may all be made from biomass, which is essentially any organic material [2]. A steady and dependable source of electricity, geothermal energy is extracted from the Earth's core. Integrating these renewable energy sources into city power grids makes the grid more robust and distributed, making it less susceptible to interruptions in the supply chain. Renewable energy sources may be more seamlessly integrated into city planning with the help of smart grid technology. Smart grids improve grid resilience, optimize resource utilization, and enable efficient energy distribution. A steady and dependable energy supply for cities is ensured

DOI: 10.1201/9781003496656-16

by smart grids, which integrate modern metering systems, energy storage technologies, and real-time monitoring to make renewable energy sources work seamlessly [1].

Not only does it help the environment, but it also helps the economy and creates jobs when renewable technologies are used to build urban infrastructure. Manufacturing, installation, maintenance, and R&D are all areas where qualified individuals might find work in the renewable energy business. Global economic growth and competitiveness may be further stimulated by encouraging innovation in renewable technology, which in turn can give rise to new sectors. A multi-pronged strategy that promotes sustainable socio-economic progress is the integration of renewable technology into energy-efficient urban infrastructure. A combination of smart grid technology and renewable energy sources including solar, wind, biomass, and geothermal heat can help cities create sustainable and resilient cityscapes. By promoting economic development and reducing the negative impacts of climate change, this transition helps bring about a balance between human activities and the planet's ecological well-being [3].

16.2 ENVIRONMENT CONSIDERATION IN ENERGY-EFFICIENT URBAN INFRASTRUCTURE

In order to address environmental concerns and promote sustainable socio-economic progress, it is crucial to include renewable technology in energy-efficient urban infrastructure construction. Cutting down on emissions of greenhouse gases is one of the main advantages to the environment. Cities may help the fight against climate change by switching from traditional energy sources that rely on fossil fuels to renewables like solar, wind, biomass, and geothermal power. This would drastically reduce their carbon footprint. In addition, switching to renewable energy sources reduces the negative environmental impacts of conventional power generation on both land and water. To improve air quality and protect water resources, renewable energy sources are preferable than fossil fuels since they do not contribute to acid rain or discharge hazardous pollutants. Because of the correlation between improved air and water quality and a decrease in the incidence of respiratory and water-related illnesses, this has direct beneficial effects on the health and welfare of urban populations [4].

Another important factor in protecting natural ecosystems is the use of renewable energy sources. Habitat loss and environmental deterioration are common outcomes of fossil fuel extraction and combustion. Sustainable energy sources, on the other hand, can be seamlessly incorporated into cityscapes without wreaking havoc on native flora and fauna. Urban areas may now satisfy their energy demands in a way that is good for the environment and biodiversity. Using sustainable energy sources also contributes to waste reduction. Sustainable energy from renewable sources is virtually endless, unlike fossil fuels, which have finite stocks (Figure 16.1). By doing so, we lessen the strain on our limited resources and lessen the societal and environmental consequences of extracting, transporting, and processing those resources [5]. Decentralized and dispersed energy-generating alternatives are frequently provided by renewable technologies, which greatly improve energy efficiency. A more effective supply of energy is achieved through this decentralization, which aids in lowering transmission and distribution losses. Improving the integration of renewable sources into the urban energy infrastructure, optimizing energy consumption, and minimizing wastage are all ways in which smart grid technologies increase the overall efficiency of energy systems. Sustainable urban planning and design are also part of the environmental considerations that go beyond energy generation. Green areas, energy-efficient buildings, and environmentally friendly transit alternatives are all part of well-planned urban development

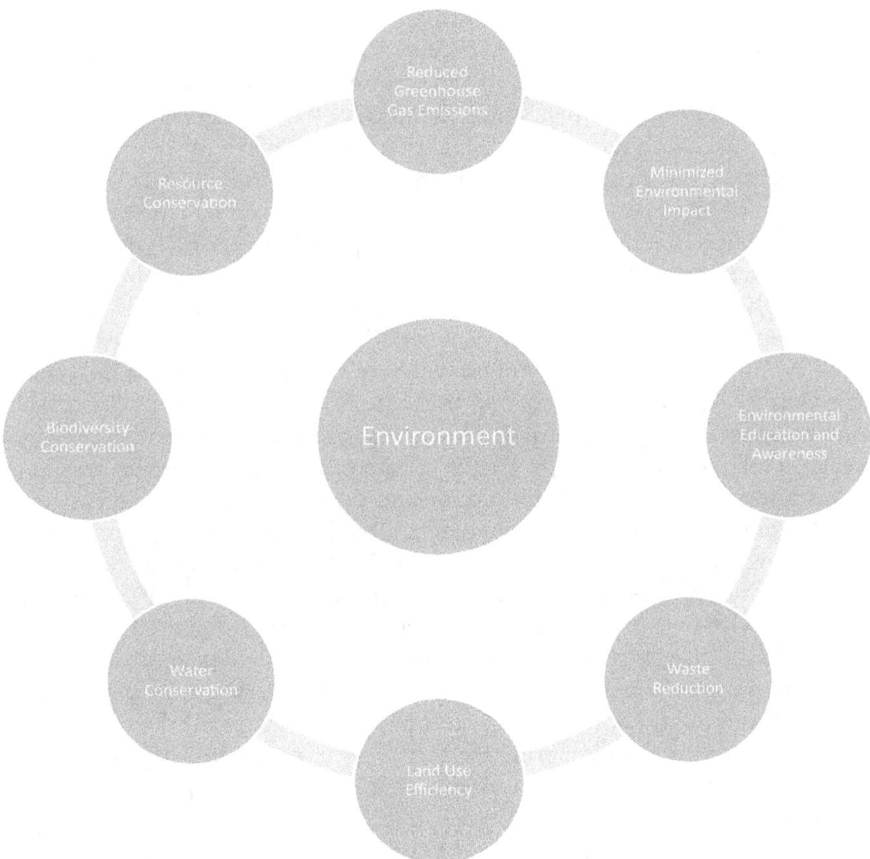

Figure 16.1 Impact on environment

initiatives that may help integrate renewable technology. To better withstand and adapt to the effects of climate change, this all-encompassing strategy encourages a balanced interaction between human-made structures and their natural surroundings. Sustainable socio-economic progress and a core commitment to environmental stewardship are both facilitated by the integration of renewable technology into energy-efficient urban infrastructure [6]. A future where economic development and environmental preservation coexist may be achieved via the implementation of eco-friendly urban design methods and the prioritization of renewable energy sources. Cities have the power to shape this future.

16.3 ECOLOGICAL CONSIDERATION

In order to foster economic growth that is both environmentally sensitive and socially responsible, it is very necessary to incorporate renewable energy sources into the design of urban infrastructure that is responsible for energy efficiency. It is very necessary to make use of renewable energy sources in order to mitigate the adverse consequences of climate change and the rising energy requirements brought about by urbanization.

To begin, it is abundantly evident that solar electricity is a significant component in the process of fostering environmental sustainability. The utilization of solar panels to gather energy from the sun has the potential to assist cities in reducing their consumption of fossil fuels, which in turn leads to a reduction in pollution and greenhouse gas emissions. Switching to solar power, which lowers the environmental harm caused by existing energy extraction procedures, has other benefits, like cleaner air and water [7]. The use of wind energy in urban areas, which is yet another significant renewable source, has positive effects on the state of the environment. Electricity may be generated by wind turbines without causing any harm to the environment or squandering any of the limited resources that are available. Wind power, which is both abundant and environmentally friendly, has a smaller impact on the environment than traditional power plants, which are powered by coal and natural gas. In the event that they are appropriately positioned, wind farms have the potential to fit in with cityscapes and offer an environmentally beneficial alternative that reduces the load on natural regions. Energy sources that are sustainable, such as geothermal heat and biomass, contribute to the maintenance of urban infrastructure that is in harmony with Mother Earth. Biomass, which is derived from organic waste, provides not only a solution to the issue of waste management but also a source of energy that is environmentally friendly [8]. Instead of putting organic waste in landfills, cities might reduce their influence on the environment by converting it into energy instead of dumping it somewhere else. When compared to more conventional methods of power generation, geothermal energy, which is based on the heat that is naturally present on

Earth, provides a reliable and ecologically beneficial supply of electricity that does less damage to the ecosystems that are located in the surrounding area. Concerns about the environment are exacerbated when smart grid technologies are included in the electrical infrastructure of metropolitan regions. Smart grids make it feasible to distribute renewable energy supplies in an effective manner by reducing the amount of transmission losses that occur and making the most of the resources that are available. Smart networks support a more careful and effective use of energy in urban environments by minimizing energy waste, which in turn helps with environmental sustainability and resource conservation. Smart grids are becoming increasingly popular across the world. In addition to the generation of energy, urban planning and design also have the potential to reduce environmental impacts [9]. Green roofs, good insulation, and sustainable materials are some of the characteristics that may be included in the design of buildings in order to make them more environmentally friendly and energy efficient. The incorporation of urban forests and green areas into city design not only improves air quality but also lessens the influence of the urban heat island and boosts biodiversity. All of these factors contribute to a more sustainable and healthy urban ecosystem. Renewable technologies, in addition to being beneficial to the environment, bring about economic growth and the creation of new job opportunities. Cities are more likely to invest in renewable energy projects, which in turn leads to the creation of job opportunities in areas like research and development, manufacturing, installation, and maintenance [10]. In addition to being in accordance with ecological principles, this economic growth is created by the industry of renewable energy, which is to the overall advantage of communities. It is essential to include renewable technology in the design of energy-efficient urban infrastructure in order to support sustainable socio-economic growth while also taking into consideration environmental concerns. It is possible for cities to attain a sustainable energy balance with the assistance of smart grid technologies, renewable energy sources (including solar, wind, biomass, and geothermal heat), and higher utilization of these sources. This transition will result in cityscapes that are more resilient and ecologically conscious, which will help maintain the planet livable for decades to come. In addition to fixing the environmental problems that are generated by urbanization, this shift will also result in the improvement of the environment [11].

16.4 ETHICAL IMPACT

There are significant ethical considerations involved in encouraging sustainable socio-economic growth through the use of renewable technology in the building of energy-efficient urban infrastructure (Figure 16.2). Stewardship of the environment, social justice, and the welfare of present and future generations are all ethical concerns. Renewable energy sources

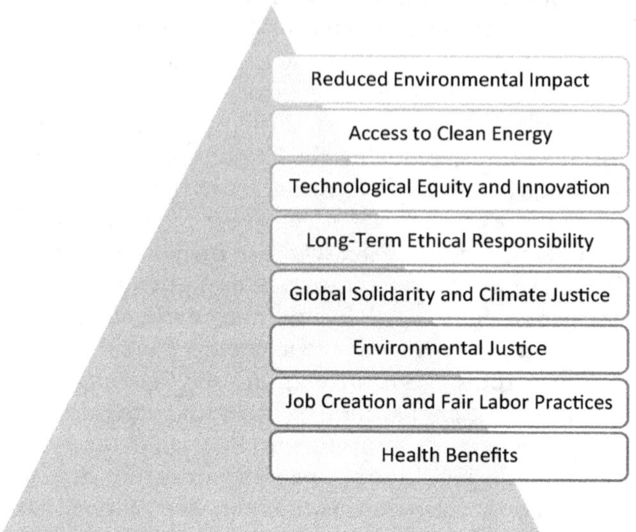

Figure 16.2 Ethical implication

should be considered an ethical need rather than a just practical option for cities dealing with the twin crises of climate change and urbanization. Lowering carbon emissions and slowing the rate of climate change are two ways in which renewable technology adoption might help with environmental ethics. Disposing of fossil fuels shows that cities care about the environment because they do a lot of damage and contribute to climate change [12]. A responsibility to preserve ecosystems, biodiversity, and Earth's general health for the benefit of all living things is at the heart of this ethical position. One of the most important moral considerations when using renewable energy sources is social justice. The equitable distribution of the advantages of urban growth may be achieved through the deployment of renewable sources, which also provide access to clean and sustainable energy. Cities may help create a more equitable and inclusive society by reducing socio-economic gaps through the provision of inexpensive and dependable electricity, especially in neglected and disadvantaged areas. The moral implications go into the welfare of city dwellers [13]. Using renewable energy sources helps reduce air pollution, which improves air quality and has a direct impact on public health. Promoting the preservation of clean air and water is morally important since it improves the standard of living for locals and shows that we care about their health and safety. In addition, the use of renewable energy sources is consistent with ethical ideals that promote equity across the generations. It is the civic duty of cities to build sustainable infrastructure so that the earth is habitable and prosperous for centuries to come. In light of this moral imperative, we must act now to ensure that future generations have access to healthy ecosystems, abundant natural resources, and a predictable climate. An important part of integrating renewable energy sources is smart grid technology, which raises ethical questions about data security and privacy [14]. Data gathering and use problems must be addressed in order to optimize energy distribution and consumption. In order to protect citizens' privacy and prevent data breaches, cities should have open and honest policies for handling smart grid data. Creating jobs and boosting the economy are two other ways in which renewable technologies have an ethical influence. More manufacturing, installation, maintenance, and R&D jobs will be created as a result of investments in renewable energy. Fostering innovation in sustainable technologies while also providing communities with meaningful jobs and economic stability is emphasized by this ethical dimension. Ultimately, there are many ethical questions that go beyond technical ones when it comes to developing energy-efficient urban infrastructure with renewable technology. Urban areas may ethically face the issues of climate change and urbanization by focusing on environmental stewardship, social fairness, public health, intergenerational justice, and responsible data management. Together, these ethical principles pave the way for a better, more equitable future in urban areas and serve as a compass for decision-making [15].

16.5 ECONOMIC IMPLICATION AND THE ROLE OF INDUSTRIAL SECTOR IN RENEWABLE ENERGY

The inclusion of renewable technology into the development of energy-efficient urban infrastructure, which has far-reaching financial effects, is a significant factor that contributes significantly to the preservation of socio-economic growth that is sustainable (Figure 16.3). Both residents of the city and enterprises in the city stand to gain commercially, professionally, and monetarily from the switch to renewable energy sources. One of the ways that investments in renewable technologies contribute to economic growth is through the establishment of new businesses and employment opportunities. Jobs in manufacture, installation, and maintenance are created by renewable energy technologies such as solar panels, wind turbines, and other technologies that are employed in the creation of renewable energy [16]. This helps foster a workforce that is trained, which is beneficial for the economy as a whole, in addition to lowering the number of people who are considered unemployed. One of the advantages of using renewable energy sources in the construction of new city infrastructure rather than depending on fossil fuels imported from other countries is an increase in energy security. This shift toward renewable energy sources has the potential to promote economic stability in the region by lowering the region's dependency on petroleum, which is a resource that is notoriously difficult to anticipate. In addition, the diversity of energy sources reduces reliance on

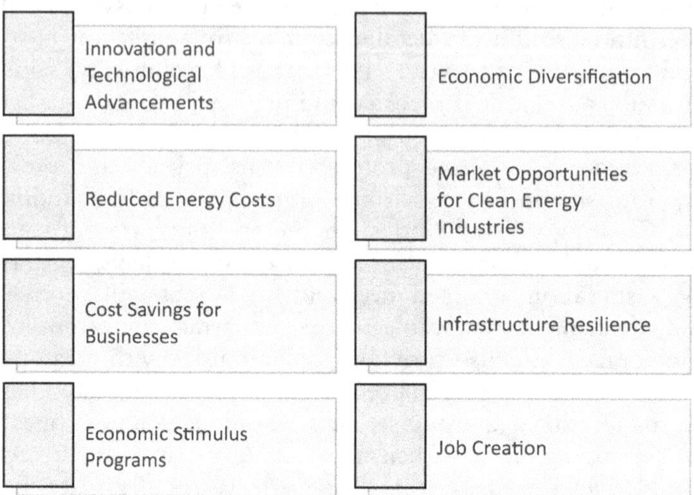

Figure 16.3 Renewable energy contribution to economic growth and development

a small number of critical markets that are vulnerable to geopolitical instability. The use of renewable technologies also has the long-term benefit of reducing financial expenditures [17]. When compared to conventional fossil fuel infrastructure, renewable energy systems often offer lower operational and maintenance costs. This is despite the fact that they may need a higher initial investment. Eventually, communities may be able to reap the benefits of decreased energy expenditures, which may be transferred to other essential services and infrastructure projects, finally leading to an improvement in the economic well-being of the community.

Local economies frequently receive a boost in capital through the implementation of renewable energy projects, which in turn attract corporate investments and public-private partnerships. This infusion of finance helps not only to strengthen the development of renewable projects but also to promote an environment that is suitable for the expansion of businesses and the generation of new ideas. When municipalities establish themselves as leaders in the implementation of renewable technology, they have the potential to kickstart a positive cycle of economic expansion. It is also true that sustainable development initiatives increase the appeal of a city to both individuals and businesses. Those businesses who are concerned about the environment and want to lessen their influence on the environment could be interested in a site that is committed to energy efficiency and renewable energy sources. This, in turn, can lead to an increase in property prices as well as tax revenues, which provides municipalities with additional funds to invest in growth [18].

In the process of the emergence of a green economy, which is characterized by activities that are responsible for the environment, the adoption of renewable technology is an essential component of this trend. At the same time as a new economic paradigm is being created, green employment is also being born. Works in areas such as energy efficiency, renewable energy, and sustainable infrastructure are included in this category. Through the integration of green economic principles and urban development, cities have the potential to position themselves for long-term prosperity while simultaneously addressing social and environmental problems [19]. The transition to renewable energy sources also helps to slow down the rate of global warming because it reduces the amount of greenhouse gases that are released into the atmosphere. It is possible that climate change will have far-reaching economic effects, in addition to the costs that are connected with responding to weather events that are occurring more frequently and with greater severity. Cities should make investments in renewable technologies right now in order to lessen the chance of having to deal with these dangers in the future. This will help cities' economies become more robust in the face of climate change. The creation of urban infrastructure that is energy-efficient through the application of renewable technologies is not only important for the environment, but it also has the potential to push long-term economic and social advancement. Through the promotion of job creation, the reduction of energy prices, the attraction of investments, and the advancement of a green economy, cities have the potential to take the initiative in the development of a future that is more sustainable and economically robust.

16.6 ROLE OF INDUSTRIAL SECTOR

Sustainable socio-economic growth is aided by industries' pivotal role in promoting the use of renewable technology and their incorporation into energy-efficient urban infrastructure. The path of urban development toward a future that is more sustainable and ecologically sensitive is being shaped by these industries, which cover a wide range from production and installation to research and development. The renewable technology industry relies on the industrial sector as its foundation. Businesses that make renewable energy components, such as solar panels, wind turbines, and energy-efficient appliances, play a crucial role in satisfying the increasing demand for such solutions. Renewable technologies become increasingly feasible for extensive urban application as a result of technology breakthroughs and economies of scale in these industries, which make them more accessible and affordable [20].

Renewable energy technologies also provide substantial job possibilities during installation and maintenance. Solar panels, wind farms, and other infrastructure components need the assistance of skilled professionals for installation, monitoring, and maintenance. Opportunities for employment arise as a result of cities' growing investment in renewable initiatives, which in turn helps local people build their skills and create more jobs. This leads to a more fair distribution of economic gains and improves socio-economic well-being. Innovation and efficiency in renewable technology can only be driven by research and development (R&D). Research and development industries are always working to make renewable technology better and cheaper [21]. If we want renewable energy solutions to be competitive with existing energy sources and integrated into urban infrastructure faster, we need breakthroughs in energy storage, efficiency improvements, and novel materials (Figure 16.4).

Key players also include sectors involved in the manufacturing of smart grid technology and energy-efficient home products. Urban energy efficiency is enhanced by HVAC, lighting, and other systems that use less energy. In a similar vein, smart grid technologies allow for more precise control of power distribution, which in turn minimizes waste and maximizes efficiency. These sectors lay the groundwork for an urban energy system that is both smarter and more responsive. They provide distributed and decentralized energy solutions, and businesses in the renewable technology sector greatly aid in making cities more resilient. The decentralization of power systems makes cities less dependent on centralized power grids, which improves energy security and makes backup power sources more

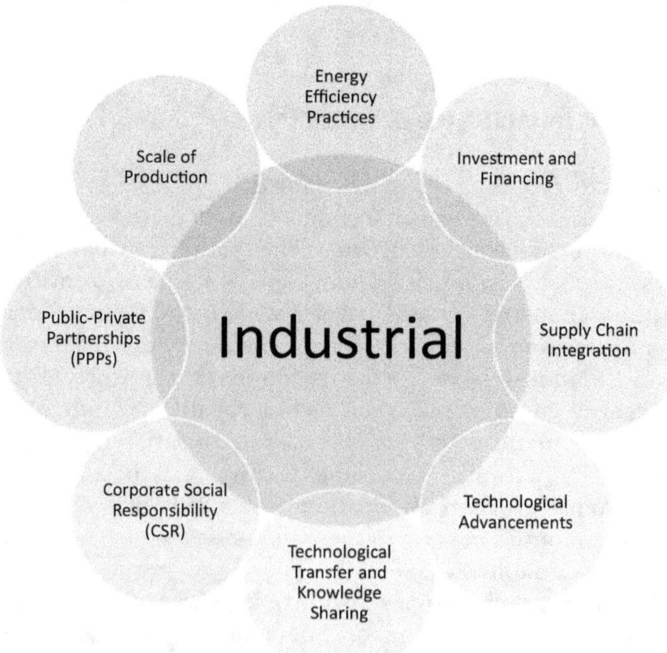

Figure 16.4 Role of industrial sector in energy efficiency

dependable in case of catastrophes. Building resilient urban infrastructure is a shared responsibility of the industries working on microgrids and other decentralized energy solutions. Renewable technology-related businesses also help keep the planet habitable. These sectors are important in the fight against climate change, reduction of greenhouse gas emissions, and the negative effects of urbanization on the environment. In addition to fostering a model of sustainable and environmentally responsible urban development, this is in line with worldwide initiatives to shift towards an economy that uses less carbon. An additional role for the renewable technology sector is to encourage more global commerce and collaboration. Exporting knowledge and goods from countries with developed renewable technology sectors might encourage international cooperation. This helps with diplomatic relations and also makes it easier to share information and technologies, which is great for building sustainable cities all across the world [22].

The development of energy-efficient urban infrastructure is a key component of sustainable socio-economic progress, and the industries linked to renewable technologies play an important part in this process. Jobs, economic growth, and environmental sustainability are all boosted by these industries, which include manufacturing, installation, and research and development. Their combined efforts are fueling the shift toward a future where cities are more resilient, egalitarian, and eco-conscious.

16.7 ROLE OF STAKEHOLDERS AND NGOS

One of the most important factors that contributes to sustainable socio-economic advancement is the incorporation of renewable technology into the building of energy-efficient urban infrastructure. Non-governmental organizations (NGOs) and stakeholders play an essential part in this process (Figure 16.5). In order to expedite the transition toward urban energy systems that are more environmentally friendly and sustainable, these groups play a crucial role in the creation of laws, the dissemination of information, and the implementation of projects [1].

There are several layers of government who have a strong interest in seeing the promotion of renewable technologies. Establishing legislative frameworks, providing incentives, and drafting rules are all things that are being done in order to encourage the adoption of renewable energy sources into city infrastructure systems. By offering tax breaks, subsidies, and many other financial incentives to businesses and developers, governments have the potential to support the development of projects that utilize renewable energy sources. The flourishing of these initiatives and their contribution to the socio-economic growth of metropolitan regions would be facilitated as a result of this [23].

In addition, local communities, which are essential stakeholders, are involved in the process of adopting sustainable technology. By actively

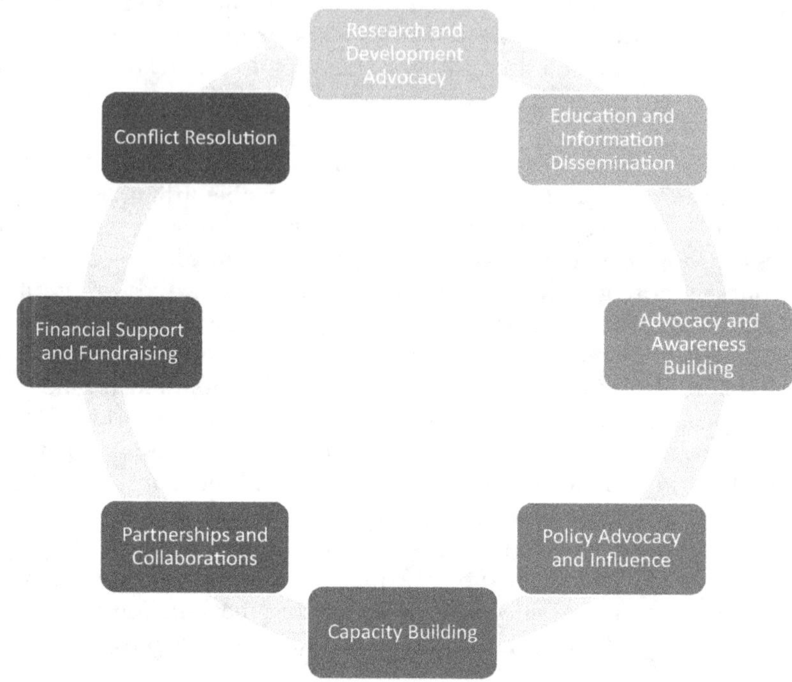

Figure 16.5 Role of NGO in renewable energy in sustainable development

participating in and providing support for urban energy efficiency projects, it is possible that these programs may be successful. The promotion of renewable energy, the resolution of community concerns, and the participation of individuals in decision-making are all common goals of non-governmental organizations (NGOs), which commonly collaborate with local communities. The local communities are given the opportunity to play an active role in the process of encouraging the sustainable growth of their city as a result of their actions. When it comes to the development of renewable energy sources, businesses and other forms of private sector participation are absolutely necessary [24]. Through investments in and participation in energy programs, businesses have the potential to reduce their negative influence on the environment and stimulate economic growth. In order for businesses to satisfy the growing demand from customers for activities that are environmentally responsible, they should increase their corporate social responsibility (CSR) initiatives by embracing renewable technologies. As a result of their emphasis on community engagement and advocacy, non-governmental organizations (NGOs) are bringing about a change in the landscape of renewable energy. They contribute to the dissemination of information on the significance of renewable energy, provide key stakeholders with information regarding the benefits of this type of energy, and monitor the development of energy to ensure that it adheres to sustainable standards. It is common practice for non-governmental organizations (NGOs) to collaborate with public and commercial organizations, as well as with local communities, in order to fulfill their objective of promoting sustainable urban development that takes into consideration social and environmental concerns [25].

In addition, international groups and financial institutions exert a significant amount of influence over the development of innovative renewable technologies. In particular, they provide financial assistance, knowledge, and other resources to cities that are expanding in order to support huge renewable energy efforts. These institutions contribute to the promotion of long-term social and economic development by making it simpler for local governments or businesses to embrace renewable energy solutions that they may not be able to purchase. Academic institutions and research organizations are a significant driving force behind the development of new technologies and innovations in the field of renewable technology. Their research has led to the development of technology that is superior and more effective, which is excellent news for the initiatives that are being undertaken to improve municipal infrastructure. It is possible that the implementation of contemporary renewable energy systems in city centers might be accelerated by collaborative efforts between academics, corporations, and governments [26].

Utility firms and energy providers are also considered stakeholders since they are involved in the process of effectively integrating renewable technology into the existing energy infrastructure. This collaboration is essential for the successful integration of renewable technology. Through their active engagement in the improvement of infrastructure and the creation of smart grid technologies, they provide a smooth transition to a city that is more environmentally friendly and energy efficient. The media has a significant impact on both the decisions that are made about public policy and the opinions of the general population. Through the promotion of sustainable practices and the attraction of more investors, the media may be able to emphasize the good elements of renewable technology and successful urban development efforts [27].

Banks and investment organizations, among other entities, play an essential part in the process of funding projects related to renewable energy. Because to their assistance, we will be able to guarantee that renewable technologies in urban infrastructure will receive the financing that is necessary for their development, installation, and maintenance operations. It is necessary for a large number of interested parties to work together in order to successfully incorporate renewable technology into the process of developing energy-efficient city infrastructure. These parties include national and regional governments, businesses, non-governmental organizations, universities, media outlets, and banks. They are able to establish a complementary approach by working jointly, which ensures the lifespan and health of urban ecosystems while simultaneously supporting socio-economic advancement over the long term.

16.8 ROLE OF POLICIES AND GOVERNANCE

Sustainable socio-economic growth is greatly influenced by policies and governance, which have a pivotal role in determining the prevalence of renewable technology in energy-efficient urban infrastructure development. The incorporation of renewable energy sources into urban planning may be facilitated by well-crafted policies, which are then put into action and enforced by competent leaders. Renewable energy adoption objectives, incentives, and regulatory frameworks are created by national governments through policymaking. With these rules in place, communities can build their infrastructure in a way that will lead to a low-carbon future. By setting lofty goals for the deployment of renewable energy, we can boost economic growth and reduce the negative effects of urbanization on the environment [28].

When it comes to implementing national policy in particular metropolitan regions, local governments play a crucial role. Renewable technology integration into city infrastructure may be encouraged through the application of zoning restrictions, construction requirements, and land-use policies. Sustainable transportation systems, energy-efficient buildings, and solar panel installations are all promoted by enabling environments created by local governments, which in turn lead to socio-economic progress.

Policies that effectively stimulate the adoption of renewable technology often include incentive programs, which offer financial and regulatory assistance. Renewable energy projects are made financially feasible for people and businesses through feed-in tariffs, tax credits, and subsidies, which encourage investment in these areas. The renewable energy sector may benefit from well-planned incentive programs because they encourage private sector involvement, which in turn drives innovation and helps to create jobs. To guarantee that renewable technology deployments are safe, environmentally friendly, and of high quality, regulatory frameworks are put in place. Regulatory agencies ensure that renewable energy is properly integrated into urban infrastructure by keeping an eye on compliance, enforcing standards, and handling any problems that may arise. Investors and developers benefit from clear laws because they lessen the risks involved in implementing projects. One collaborative governance strategy that has the potential to hasten the spread of renewable energy sources is public-private partnerships (PPPs). [29] Procurement partnerships (PPs) allow for the successful execution of large-scale, significant projects by bringing together public and private organizations. When it comes to removing logistical and financial obstacles to sustainable urban growth, this collaborative governance paradigm shines. Achieving a successful implementation requires decision-making procedures that are both transparent and participative. Communities and non-governmental organizations are only two examples of the types of stakeholders who should be allowed to weigh in on renewable technology policymaking through inclusive governance. The public's confidence, understanding, and support for sustainable urban projects are all bolstered by this method. In order to measure the efficacy of programs and guarantee that government is accountable, monitoring and evaluation systems are essential. Policies may be fine-tuned, tactics can be adjusted, and new difficulties can be tackled with the help of regular evaluations. Public trust and dedication to sustainable urban development are bolstered by open reporting on the status of renewable energy projects. A worldwide effort is underway to promote renewable technology, and this endeavor relies on international collaboration and policy alignment [30]. By working together, countries can tackle climate change and promote long-term economic and social prosperity by exchanging information and resources. By building a more conducive global environment, international agreements and collaborations increase the effectiveness of particular initiatives. Renewable technology governance plans must include capacity building and education programs. Governments may help facilitate the shift toward sustainable urban development by funding training programs and spreading information on the advantages of renewable energy. This equips professionals, lawmakers, and the broader public to engage actively in this movement. Overall, there are many moving parts when it comes to regulations and governance and how they affect the incorporation of renewable technology into efficient urban infrastructure. Sustainable socio-economic growth is built upon successful policies and governance frameworks, which include national objectives, local rules, incentives, and collaborative governance models. If we want to build low-carbon cities that can withstand natural disasters and provide a better life for generations to come, we need policies with vision and strong governance systems to work together.

16.9 CHALLENGES

The widespread adoption of renewable technologies is hampered by a variety of difficulties, which is unfortunate because these technologies would be of tremendous assistance in the development of energy-efficient urban infrastructure and in the promotion of long-term social and economic success.

One of the most significant challenges presents itself in the form of the inherent intermittent nature of renewable energy sources such as wind and solar. In order to provide a consistent and uninterrupted supply of energy for urban infrastructure, it is essential to have energy storage systems that are both efficient and effective. This is because power generation irregularities can lead to unreliability within the system. It is vital to create energy storage systems that are both effective and inexpensive in order to combat the intermittent nature of generation from renewable sources of electricity. Due of their short lifespan, high environmental impact, and low capacity, present storage technologies such as batteries make it impossible to accomplish unreliable energy storage on a big enough scale to fulfill the demands of cities [31].

The process of retrofitting older buildings and systems is especially challenging when one is attempting to integrate renewable technology into urban infrastructure that already exists. Improving infrastructure so that it can take renewable sources may be a challenging and expensive procedure, and there are presently no defined rules that can make the process any simpler. In terms of aesthetics and land use, large-scale renewable energy installations such as solar farms, wind turbines, and other similar establishments may need substantial area utilization. In densely populated metropolitan regions, it becomes increasingly challenging to conserve natural areas, agricultural land, and urban aesthetics while simultaneously satisfying the need for renewable energy without compromising the quality of life [32].

Although the costs of renewable technology have decreased, it is still difficult to get started with it. In spite of the fact that renewable energy projects may offer large long-term benefits, the high initial expenses associated with them may prohibit them from being commercially feasible. It will be necessary for governments and businesses to overcome these financial constraints in order to ensure widespread adoption. It is possible that legislative and regulatory frameworks that are inconsistent or imprecise would act as a barrier to the development and implementation of initiatives involving renewable energy. Supporting policies, such as feed-in tariffs and renewable portfolio standards, are two examples of policies that are important for the growth of the renewable energy sector [1]. The continual research and development of renewable technologies is necessary in order to address technological limitations and optimize these technologies for use in urban settings. There is also a lack of research in this particular field. It is necessary for there to be quick innovation in the transmission, storage, and collection of energy; nevertheless, there may be a delay in advancement owing to a lack of research and unknowns When it comes to the broad adoption of renewable energy sources, one of the most significant problems that might be encountered is a lack of awareness and support from the general people. Destroying misconceptions, educating the general public about the benefits of renewable energy, and finding solutions to problems relating to noise and aesthetics are all vital steps to take in order to garner widespread support. The availability of renewable resources varies from place to place across the planet. There is a possibility that certain renewable energy sources are not feasible in regions that have rare or excessive sunshine or wind patterns that are unpredictable. Additional challenges are brought about by the requirement to maximize the utilization of resources and choose suitable locations for renewable energy infrastructure.

The production of renewable technologies and the extraction of raw materials are both components of the global supply chain for these technologies, which brings with it a number of dependencies and dangers. The availability of renewable technology and the cost of that technology are vulnerable to disruptions in supply chains, geopolitical conflicts, and fluctuations in the price of resources. In order to effectively address these challenges, it is imperative that local communities, regional governments, enterprises, and research institutions all work together in a manner that is both comprehensive and collaborative. If these challenges are to be conquered, it is imperative that the full potential of renewable technology to foster sustainable socio-economic growth in urban environments be fulfilled [33].

16.10 FUTURE PERSPECTIVE

Sustainable socio-economic growth may be fostered through the use of renewable technology in the future, particularly in the context of developing energy-efficient urban infrastructure. Renewable energy in urban settings is predicted to be shaped by a number of trends and breakthroughs, which will ultimately lead to cities that are more resilient and ecologically friendly. Technological progress smart grids, energy storage systems, solar photovoltaics, wind turbines, and other renewable technologies are expected to continue seeing significant breakthroughs in the near future. Renewable energy systems will become more efficient and cost-effective as a result of innovations like next-generation solar panels, high-capacity batteries, and more effective wind turbines.

In order to counteract the intermittent nature of renewable power, innovations in energy storage are essential. New energy storage technologies, such as improved batteries and unconventional storage media, will allow cities to harness extra power at production peaks and release it when demand is high. There will probably be a change to decentralized energy systems in the future. Cities will be better prepared to handle catastrophes and disruptions if they use distributed energy resources like rooftop solar panels and local energy storage. This will decrease their dependence on centralized power plants and strengthen urban energy systems. A major factor will be the incorporation of renewable energy sources into smart city programs and the Internet of Things (IoT). Sustainable social and economic development will be aided by smart grids and intelligent energy management systems, which will optimize energy use, enhance efficiency, and allow for the control and monitoring of urban infrastructure in real-time.

Reusing and recycling materials is a key component of the circular economy, which is a set of practices that future urban development is likely to adopt. This method is in line with the objective of making cities that are more sustainable and resilient while also minimizing waste and the negative effects of urban growth on the environment. The use of green building certifications and standards is going to grow in popularity. A combination of eco-friendly materials, green roofs, and energy-efficient designs is becoming more commonplace in urban infrastructure, which means less energy use and better environmental sustainability. More and more, renewable energy systems will be hybrid, meaning they combine several types of renewable energy. To overcome the unpredictability of individual renewable sources, hybrid systems that combine wind, solar, and energy storage provide a more stable and continuous supply of electricity. Through community-based energy programs, local communities will be able to generate more renewable energy. Community empowerment and localized sustainability may be achieved through shared solar projects, local energy cooperatives, and joint initiatives to build and maintain renewable infrastructure.

Governments are expected to offer more robust policy backing for renewable energy, with more incentives, subsidies, and regulatory frameworks in place. This will be accompanied by more international cooperation. There will also be a significant role for international collaboration, with joint attempts to combat climate change and provide access to renewable energy sources. Equal and inclusive access advancements in renewable energy technology going forward will put an emphasis on making clean energy accessible to everybody. Social and economic progress, including the creation of more resilient and sustainable communities, will result from efforts to close the energy access gap in underprivileged metropolitan areas and emerging regions. Finally, technical advancement, decentralized systems, smart city integration, and a dedication to sustainability define the future of renewable technologies in energy-efficient urban infrastructure development. Cleaner and more resilient energy solutions are becoming more accessible as these trends develop, providing cities throughout the world with a chance to boost their economies and reduce their environmental impact simultaneously.

16.11 CONCLUSION

One game-changing step toward long-term social and economic prosperity is the incorporation of renewable energy sources into the design of energy-efficient city infrastructure. Renewable energy sources provide a dynamic option to build resilient urban settings, which is crucial as cities face the difficulties of population growth, climate change, and limited resources. Smart grid technology, sustainable design techniques, and renewable energy sources like solar, wind, biomass, and geothermal heat may lessen cities' environmental impact while simultaneously fostering economic growth, new employment possibilities, and technological advancement. To ensure a sustainable and fair quality of life for generations to come, communities, stakeholders, and politicians may work together to prioritize renewable technology in urban development. This will help cities prosper in harmony with the environment. Understanding the interdependence of ecological, social, and economic aspects is crucial as we negotiate the complexity of urbanization. Sustainable urban infrastructure development with renewable technology requires teamwork, well-informed policymaking, and a common goal of reducing environmental impact. In addition to solving their current energy problems, cities may set themselves up for a more equitable, resilient, and ecologically responsible future by tapping into the power of renewable energy sources. The goal of achieving sustainable socio-economic growth in future cities is no longer a pipe dream thanks to the coming together of technology, policy, and community participation.

REFERENCES

1. Kumar, J.C.R., Majid, M.A. (2020). Renewable energy for sustainable development in India: Current status, future prospects, challenges, employment, and investment opportunities. *Energy, Sustainability and Society*, 10, 2. https://doi.org/10.1186/s13705-019-0232-1
2. Nazir, M.S., Wang, Y., Bilal, M., Abdalla, A.N. (2022). Wind energy, its application, challenges, and potential environmental impact. In: Lackner, M., Sajjadi, B., Chen, W.Y. (eds.), *Handbook of Climate Change Mitigation and Adaptation*. Springer, New York, NY. https://doi.org/10.1007/978-1-4614-6431-0_108-2
3. Hasna, Z., Jaumotte, F., Pienknagura, S. (2023). How green innovation can stimulate economies and curb emissions. IMF Blog. https://www.imf.org/en/Blogs/Articles/2023/11/06/how-green-innovation-can-stimulate-economies-and-curb-emissions
4. Osman, A.I., Chen, L., Yang, M. et al. (2023). Cost, environmental impact, and resilience of renewable energy under a changing climate: A review. *Environmental Chemistry Letters*, 21, 741–764. https://doi.org/10.1007/s10311-022-01532-8
5. Bertrand, S. (2021). Fact sheet | Climate, environmental, and health impacts of fossil fuels (2021). Environment and Energy Studies Institute. https://www.eesi.org/papers/view/fact-sheet-climate-environmental-and-health-impacts-of-fossil-fuels-2021
6. Kabeyi, M.J.B., Olanrewaju, O. (2023). Smart grid technologies and application in the sustainable energy transition: A review. *International Journal of Sustainable Energy*, 42(1), 685–758. https://doi.org/10.1080/14786451.2023.2222298
7. Strielkowski, W., Civín, L., Tarkhanova, E., Tvaronavičienė, M., Petrenko, Y. (2021). Renewable energy in the sustainable development of electrical power sector: A review. *Energies*, 14, 8240. https://doi.org/10.3390/en14248240
8. Charabi, Y., Abdul-Wahab, S. (2020). Wind turbine performance analysis for energy cost minimization. *Renewables*, 7, 5. https://doi.org/10.1186/s40807-020-00062-7
9. Bethem, J., Frigo, G., Biswas, S. et al. (2020). Energy decisions within an applied ethics framework: An analysis of five recent controversies. *Energy,*

Sustainability and Society, 10, 29. https://doi.org/10.1186/s13705-020-00261-6

10. Perivoliotis, D., Arvanitis, I., Tzavali, A., Papakostas, V., Kappou, S., Andreakos, G., Fotiadi, A., Paravantis, J.A., Souliotis, M., Mihalakakou, G. (2023). Sustainable urban environment through green roofs: A literature review with case studies. *Sustainability*, 15, 15976. https://doi.org/10.3390/su152215976
11. Chen, L., Hu, Y., Wang, R. et al. (2023). Green building practices to integrate renewable energy in the construction sector: A review. *Environmental Chemistry Letters*. https://doi.org/10.1007/s10311-023-01675-2
12. Fallah Shayan, N., Mohabbati-Kalejahi, N., Alavi, S., Zahed, M.A. (2022). Sustainable Development Goals (SDGs) as a framework for Corporate Social Responsibility (CSR). *Sustainability*, 14, 1222. https://doi.org/10.3390/su14031222
13. Kumar, M. (2020). *Social, Economic, and Environmental Impacts of Renewable Energy Resources*. IntechOpen. https://doi.org/10.5772/intechopen.89494
14. Bansard, J., Schröder, M. (2021). The sustainable use of natural resources: The governance challenge. IISD. https://www.iisd.org/system/files/2021-04/still-one-earth-natural-resources.pdf
15. Martínez-Peláez, R., Ochoa-Brust, A., Rivera, S., Félix, V.G., Ostos, R., Brito, H., Félix, R.A., Mena, L.J. (2023). Role of digital transformation for achieving sustainability: Mediated role of stakeholders, key capabilities, and technology. *Sustainability*, 15, 11221. https://doi.org/10.3390/su151411221
16. Almihat, M.G.M., Kahn, M.T.E., Aboalez, K., Almaktoof, A.M. (2022). Energy and sustainable development in smart cities: An overview. *Smart Cities*, 5, 1389–1408. https://doi.org/10.3390/smartcities5040071
17. Owusu, P.A., Asumadu-Sarkodie, S., Dubey, S. (Reviewing Editor). (2016). A review of renewable energy sources, sustainability issues and climate change mitigation. *Cogent Engineering*, 3, 1. https://doi.org/10.1080/23311916.2016.1167990
18. Yasir, N., Babar, M., Mehmood, H.S., Xie, R., Guo, G. (2023). The environmental values play a role in the development of green entrepreneurship to achieve sustainable entrepreneurial intention. *Sustainability*, 15, 6451. https://doi.org/10.3390/su15086451
19. Söderholm, P. (2020). The green economy transition: The challenges of technological change for sustainability. *Sustain Earth*, 3, 6. https://doi.org/10.1186/s42055-020-00029-y
20. Bradu, P., Biswas, A., Nair, C. et al. (2023). Recent advances in green technology and Industrial Revolution 4.0 for a sustainable future. *Environmental Science and Pollution Research*, 30, 124488–124519. https://doi.org/10.1007/s11356-022-20024-4
21. ILO. (2022). Renewable energy jobs hit 12.7 million globally. https://www.ilo.org/global/about-the-ilo/newsroom/news/WCMS_856515/lang--en/index.htm
22. Farzaneh, H., Malehmirchegini, L., Bejan, A., Afolabi, T., Mulumba, A., Daka, P.P. (2021). Artificial intelligence evolution in smart buildings for energy efficiency. *Applied Sciences*, 11, 763. https://doi.org/10.3390/app11020763

23. Tryndina, N., An, J., Varyash, I., Litvishko, O., Khomyakova, L., Barykin, S., Kalinina, O. (2022). Renewable energy incentives on the road to sustainable development during climate change: A review. *Frontiers in Environmental Science*, 10, 1016803. https://doi.org/10.3389/fenvs.2022.1016803
24. Zebra, E.I.C., van der Windt, H.J., Nhumaio, G., Faaij, A.P.C. (2021). A review of hybrid renewable energy systems in mini-grids for off-grid electrification in developing countries. *Renewable and Sustainable Energy Reviews*. https://doi.org/10.1016/j.rser.2021.111036.
25. Tilt, C.A. (2016). Corporate social responsibility research: The importance of context. *International Journal of Corporate Social Responsibility*, 1, 2. https://doi.org/10.1186/s40991-016-0003-7
26. OECD. (2019). Innovation and business/market opportunities associated with energy transitions and a cleaner global environment. https://www.oecd.org/g20/summits/osaka/OECD-G20-Paper-Innovation-and-Green-Transition.pdf
27. Kabeyi, M.J.B., Olanrewaju, O.A. (2022). Sustainable energy transition for renewable and low carbon grid electricity generation and supply. *Frontiers in Energy Research*, 9, 743114. https://doi.org/10.3389/fenrg.2021.743114
28. Lu, Y., Khan, Z.A., Alvarez-Alvarado, M.S., Zhang, Y., Huang, Z., Imran, M. (2020). A critical review of sustainable energy policies for the promotion of renewable energy sources. *Sustainability*, 12, 5078. https://doi.org/10.3390/su12125078
29. Enerdatics. (2023). Unveiling renewable energy policies and incentives. https://enerdatics.com/blog/unveiling-renewable-energy-policies-and-incentives/
30. Sarangi, G.K. (2018). Green energy finance in India: Challenges and solutions. ADBI Working Paper 863. Tokyo: Asian Development Bank Institute. https://www.adb.org/publications/green-energy-finance-india-challenges-and-solutions
31. Worku, M.Y. (2022). Recent advances in energy storage systems for renewable source grid integration: A comprehensive review. *Sustainability*, 14, 5985. https://doi.org/10.3390/su14105985
32. Amoah, C., Smith, J. (2022). Barriers to the green retrofitting of existing residential buildings. *Journal of Facilities Management*, ahead-of-print. https://doi.org/10.1108/JFM-12-2021-0155
33. McKinsey & Company. (2023). Renewable-energy development in a net-zero world: Disrupted supply chains. https://www.mckinsey.com/industries/electric-power-and-natural-gas/our-insights/renewable-energy-development-in-a-net-zero-world-disrupted-supply-chains

Chapter 17

A comprehensive review on pipe layout and life cycle energy analysis of earth-air heat exchanger

Sustainable solution for energy-efficient building

Veena Chaudhary, Ashish Bilatiya, Aditya Kumar, and Chandan Swaroop Meena

17.1 INTRODUCTION

Energy consumption due to conventional heating and cooling systems is a significant contributor to the overall energy usage in buildings [1]. These systems, often powered by fossil fuels or electricity generated from fossil fuels, account for a substantial portion of the energy demand in the building sector. In many regions, space heating and cooling needs constitute nearly one-third of the total energy consumption in buildings [2]. Unfortunately, the use of conventional heating and cooling systems leads to increased carbon dioxide (CO_2) emissions and contributes to global warming, exacerbating the greenhouse effect [3]. As concerns over environmental impacts and energy efficiency grow, there is a growing interest in exploring and adopting more sustainable alternatives, such as renewable energy-based heating and cooling solutions and energy-efficient technologies, to reduce the environmental footprint of buildings.

The scientific and technical community has worked very hard in recent years to investigate energy-efficient options for space heating and cooling that rely on renewable energy sources (RES). These strategies seek to preserve energy while also preserving the ecosystem [4–6]. Passive heating and cooling methods and algorithms stand out as particularly noteworthy. They have the ability to significantly lower energy use and improve the microclimate in cities [7].

The main features of this chapter may be summarised as follows:

- This chapter explores a diverse range of earth-air heat exchanger (EAHE) systems, classified based on air circulation patterns and pipe orientations, as demonstrated in Figure 17.1. This systematic study of various EAHE configurations provides valuable insights into

 DOI: 10.1201/9781003496656-17

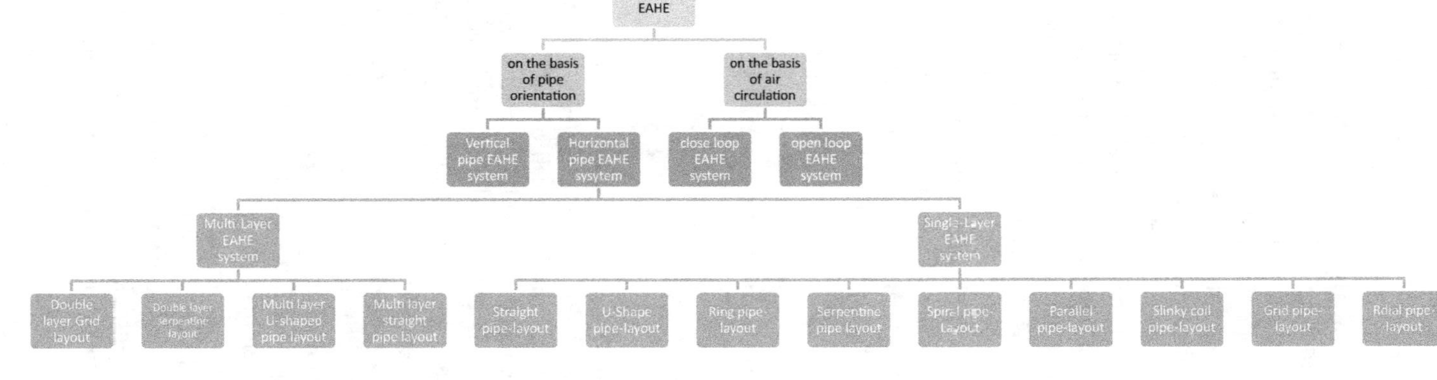

Figure 17.1 Classification of earth-air heat exchanger system

identifying the most practical design that is adapted to a given landmass's characteristics and heating and cooling requirements.
- It covers a various modeling methods mainly divided into three categories, i.e., numerical, analytical, and data-driven methods [8]. Numerical methods are predominantly based on differential equations, which describe the heat transfer processes involving conduction in the soil and convection inside the tubes during EAHE operation, while analytical methods involve solving the heat and mass transfer equation analytically [8] and data-driven methods involve the testing and training of historical data to predict the result using advanced techniques such as machine learning, artificial neural network, etc.
- The chapter extensively covers several studies conducted on life cycle energy analysis, providing valuable insights into the total energy consumption of EAHE systems from the cradle to the grave. By considering the entire lifecycle, including manufacturing, installation, operation, and eventual decommissioning, researchers gain a comprehensive understanding of the system's overall energy footprint. This information proves crucial in making informed decisions regarding the most sustainable and environmentally friendly options for EAHE system implementation.

17.1.1 Types of earth-air heat exchanger

- *On the basis of air circulation*

Earth-air heat exchanger (EAHE) systems can be categorized into two main types based on air circulation: open-loop and closed-loop systems. In the open-loop system, ambient fresh air is drawn from the surroundings through the subterranean EAHE pipes and directly introduced into the building. During this process, the air undergoes heat exchange with the surrounding earth, providing a heating or cooling effect, before being discharged back into the atmosphere. This system maintains a continuous flow of fresh outdoor air, promoting natural ventilation and ensuring a constant supply of fresh air for occupants. In contrast, the closed-loop EAHE system involves the recirculation of air within the system. The closed-loop configuration aims to maintain a stable indoor temperature and reduce the amount of heat transfer between the air and the ground. Although both open-loop and closed-loop systems have their merits, open-loop systems are generally preferred due to their ability to provide a continuous supply of fresh outdoor air. Therefore, open-loop systems offer a more desirable combination of energy-efficient heat exchange and enhanced indoor air quality, contributing to improved occupant comfort and overall building performance [9–12].

- *On the basis of pipe orientation*

Previous studies show that on the basis of pipe orientation, EAHE systems are divided into two categories: horizontal EAHE system and vertical EAHE system.

- *Horizontal EAHE system:*

The horizontal pipe arrangement is commonly employed for the EAHE system since it is less expensive to install than the vertical pipe system. Single-layer and multi-layer pipe arrangement are the two basic categories in horizontal EAHE systems.

- *Single-layer-pipe EAHE system:*

Numerous studies have extensively evaluated the performance of single-layer-pipe EAHE systems under varying environmental conditions and installation configurations. In this section, different types of pipe layouts are discussed for single-layer-pipe EAHE system.

- *Straight pipe layout:*

The straight pipe configuration is widely preferred for earth-air heat exchanger (EAHE) systems due to its ease of installation and minimal pressure loss. Bansal et al. [13, 14] investigated the heating and cooling performance of an earth-air heat exchanger in Ajmer, India. In a study conducted by Khabbaz et al. [15] in Marrakech, a hot semi-arid climate, an innovative cooling approach was employed for a residential building.

- *Grid pipe layout*

In this layout, a network of parallel pipes is installed in the ground, forming a grid-like pattern. The atmospheric air is drawn into the system through a common intake header pipe and then circulated through the underground pipes for heat exchange. The conditioned air is subsequently delivered to the indoor space via an output header pipe. In a study by Ahmed et al. [16], an earth-air heat exchanger (EAHE) system with a grid pipe arrangement buried at a depth of 2 meters was used. A significant 4.11°C drop in air temperature was accomplished by the device. Grosso et al. [17] conducted a study in Imola, Italy, where they implemented a grid-layout EAHE system for a school. The system incorporated a wind tower at the inlet section for air supply. Remarkably, the EAHE system was capable of meeting 64.6% and 76.9% of the total heating and cooling load for the building, respectively.

- *Serpentine pipe layout:*

The serpentine pipe layout, characterized by its intricate and meandering configuration, presents a highly effective solution tailored for medium-sized earth-air heat exchanger (EAHE) systems. Rangarajan et al. [18] conducted an experiment in Pune, India, where they installed an earth-air heat exchanger (EAHE) within a 252 m² area, positioning it at a shallower depth. The EAHE utilized a serpentine pipe configuration. They developed and simulated a numerical model using MATLAB, taking into account inputs such as net radiation, ambient air temperature, relative humidity, material thermal conductivity, EAHE dimensions, and buried depth. Serageldin et al. [19] investigated a serpentine pipe layout in Egypt utilizing 5.5-m-long pipes (0.05m in diameter) at 2 m depth.

- *U-shaped pipe layout:*

This design distinctively positions both the inlet and outlet segments of the pipe at the same end of the trench, forming a distinctive U shape. The EAHE system with U-shaped pipes is also employed for low heating and cooling load requirements. Yusof et al. [21] conducted a laboratory-scale experimental setup using a U-shaped PVC pipe. The study achieved a significant maximum air temperature drop of 9.62°C with 88% effectiveness. Misra et al. [22] used a U-shaped PVC pipe with a length of 60 meters and a diameter of 0.1 meters at a depth of 3.7 meters to cool an air conditioning unit's condenser coil. The results showed that the air conditioner with the earth-air heat exchanger system used 16.11% less energy than the air conditioner without EAHE.

- *Spiral pipe layout:*

Straight and U-shaped EAHE pipes entail a lengthy trench with a high aspect ratio (depth-to-length ratio). Mathur et al. [23] compared the thermal performance of a spiral pipe earth-air heat exchanger (EAHE) system with a trench aspect ratio of 1 to that of a U-shaped pipe EAHE system with a trench aspect ratio of 22. The results demonstrated that the spiral pipe layout performed nearly as well as the U-shape pipe layout.

- *Parallel pipe layout*

Parallel pipe layout is commonly used when the cooling/heating load is significant and a high airflow rate is necessary. Shojaee et al. [24] conducted a study on the energy-saving capacity of an earth-pipe-air heat exchanger (EPAHE) system consisting of four-parallel pipes in Iranian climatic conditions. The research findings indicated significant monthly energy savings in July, with the four-piped EPAHE system achieving 396 kWh and 438.7 kWh for central and lateral pipes, respectively. This study highlights the effectiveness of the EPAHE system in reducing energy consumption for cooling purposes, making it a promising solution for enhancing energy efficiency and reducing environmental impact in hot climate regions like Iran.

- *Ring pipe layout:*

The ring-pattern earth-air heat exchanger arrangement is a highly cost-effective solution since it uses the building's existing foundation trench, minimizing the need for new excavation. Grosso et al. [17] built a 48-meter-long ring-pattern PVC pipe with a diameter of 0.25 meters and a depth of 3.2 meters around a single-family house in Rubiana, Italy. For pipe lengths of 14 and 48 meters, respectively, the study found system efficacy of 0.2 and 0.45. These findings highlight the effectiveness of the ring-pattern EAHE layout in delivering good heat exchange and improving building thermal performance, making it an appealing and cost-effective option for sustainable cooling and heating solutions.

- *Radial pipe layout:*

In earth-air heat exchanger systems, the pipes are connected radially to a central collection tank, avoiding the need for manifold pipes. This arrangement is useful since it requires less land area than the parallel pipe style. A notable disadvantage is the difficulty in excavating near the collection tank. Despite this difficulty, the radial pipe layout is still a feasible alternative for EAHE systems, providing a cost-effective approach that maximizes heat exchange efficiency while minimizing the need for extra components. The position of the collection tank and excavation planning must be carefully considered to ensure the effective deployment and optimal performance of the radial pipe layout EAHE system.

- *Slinky coil pipe layout:*

This pipe layout requires less land area for installation. Kumar et al. [25] conducted a laboratory experiment on a slinky coil earth-air heat exchanger (EAHE) using sand-bentonite as backfill material. Results indicated a temperature drop of 18.70°C with a 30-m-long pipe, achieving 14.20°C drop with dry sand-bentonite at 30 m and the same drop at 23.4 m with wet sand-bentonite. Chong et al. [26] investigated a slinky-coil loop for ground source heat pump systems using numerical modeling. They adjusted its thermal performance by taking heat transfer, pipe material, and excavation cost into account. The results showed that decreasing the loop pitch and loop diameter improved the performance of the GSHP system slightly.

- ***Multi-layer EAHE system:***

As single-layer pipe plans require a significant land area for installation, researchers used multi-layer pipe layouts to maximize the thermal potential of the EAHE system (Figure 17.2). Jesus et al. (2013) studied and compared the utilization of multi-layer pipe layouts to the standard single-layer pipe configuration. The investigation unveiled that the inclusion of two or three layers of heat exchangers resulted in a significant reduction in the thermal load imposed by these units, with a decrease of 3–6% observed in comparison to a singular heat exchanger configuration.

Tarnawski [28] et al. used a double-layer serpentine pipe configuration in series, with one layer inserted at a depth of 0.5 meters and the other at a depth of 1.0 meters, with a horizontal spacing of 0.5 meters, in their investigation. The purpose of the study was to examine the ground surface area required for the installation of a 300-meter-long pipe in single-layer and double-layer designs. Fujii [29] et al. developed a numerical model to compare the performance of single- and double-layer slinky-coil EAHE over

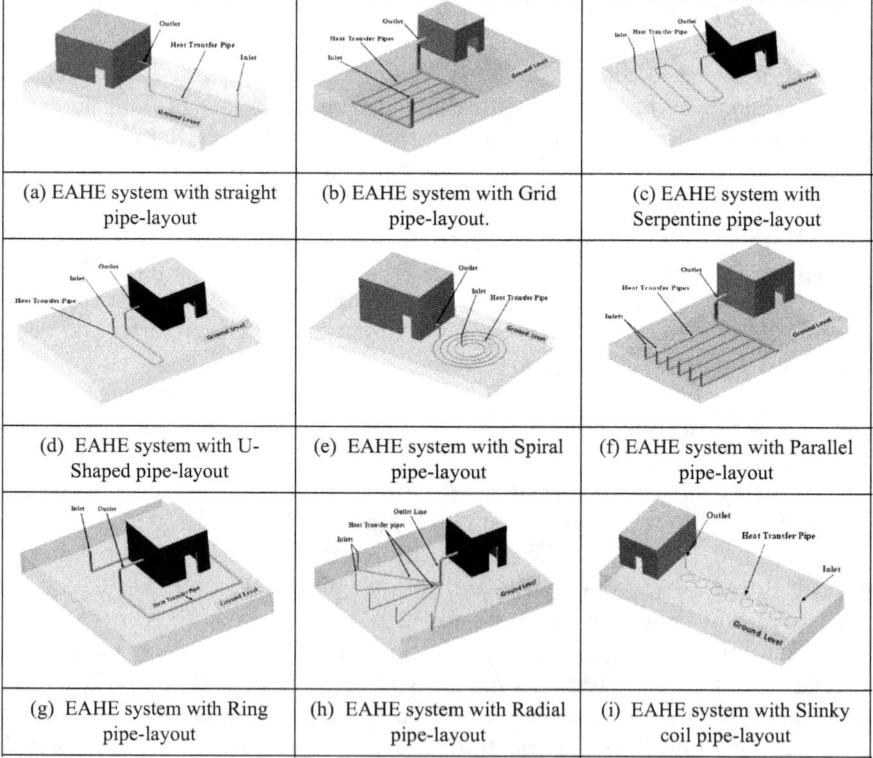

Figure 17.2 Different pipe layouts for single-layer-pipe EAHE system [27]

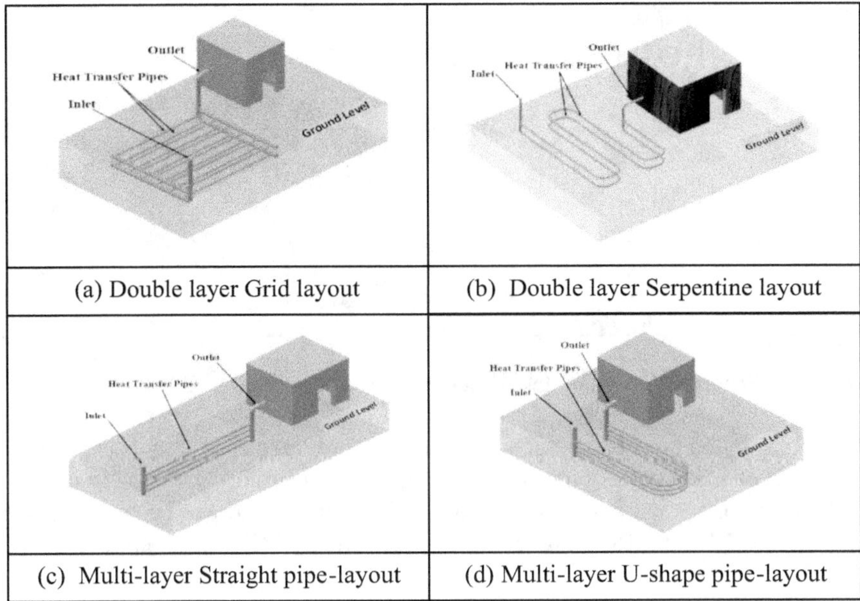

(a) Double layer Grid layout | (b) Double layer Serpentine layout

(c) Multi-layer Straight pipe-layout | (d) Multi-layer U-shape pipe-layout

Figure 17.3 EAHE system includes multi-layer pipes [27]

time. The simulation findings showed that the heat exchange rates per unit land area for the double-layer EAHE system were higher than for the single-layer version. Furthermore, the study discovered that the double-layer heat exchanger greatly lowered the necessary land area, making it a more feasible and efficient option for broader EAHE system implementation.

- ***Vertical EAHE system (VEAHE):***

According to Liu et al. [30], vertical earth-air heat exchanger (EAHE) systems require less land area and have a greater geothermal energy usage efficiency than horizontal EAHE systems (Figures 17.3 and 17.4). The biggest disadvantage of the vertical EAHE system is its much higher installation cost. As a result, vertical pipe layout is rather unusual in EAHE systems, although it is widely used in ground source heat pump (GSHP) systems. It was also seen that at lower velocity, outlet temperature reduced in summer and raised in winter. For this, VEAHE energy payback time of 8.2 years was estimated with air velocity of 1 m/s as compared to conventional air conditioners. Also 17.5 years was estimated as monetary payback period.

Patel et al. [31] carried out a comparison of vertical and horizontal earth-air heat exchanger (EAHE) systems. They erected an RCC pipe with a 0.104-meter inner diameter and a length of 25 meters at a depth of 3 meters for the horizontal EAHE system. They employed four PVC pipes with a

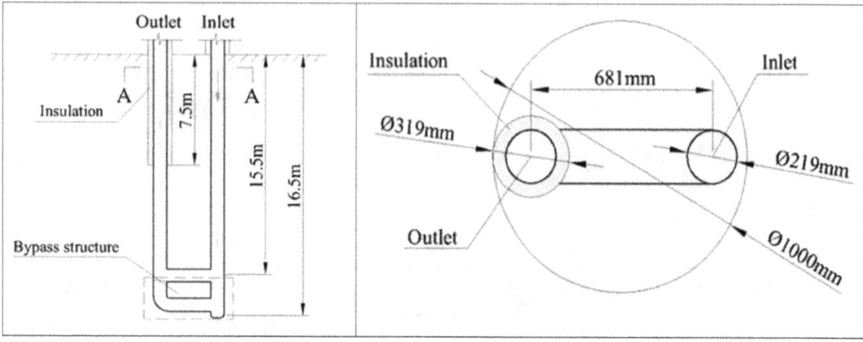

Figure 17.4 Schematic diagram of vertical earth-air heat exchanger [30]

diameter of 0.11 meters, put at 8 meters underground in a bore with a diameter of 0.61 meters, for the vertical EAHE system. The results showed that the horizontal buried tube heat exchanger caused an air temperature decrease of 14°C to 18°C, whereas the vertical EAHE system caused an air temperature drop of 5°C to 10°C.

17.2 MODELING OF EAHE

The scientific literature contains various models for predicting the thermal performance of earth-air heat exchanger (EAHE) systems, as shown in Figure 17.5, which can be categorized into three main types: numerical, analytical, and data-driven. Numerical models offer complex numerical solutions. Analytical models solve heat and mass transfer equations analytically.

Data-driven models use historical data to train and test a designed network for modeling and prediction purposes. Many of these models have successfully predicted the thermal behavior of EAHE devices and have been experimentally validated [20].

17.2.1 Numerical methods

Over the past few decades, numerous numerical models have emerged for simulating and predicting the thermal performance of earth-air heat exchanger (EAHE) systems in various applications, such as heating/cooling buildings or agricultural greenhouses. These models are predominantly based on differential equations, which describe the heat transfer processes involving conduction in the soil and convection inside the tubes during EAHE operation. This section includes various numerical methods used by researchers over the years, to investigate the thermal performance of an earth-air heat exchanger (EAHE) system for space heating/cooling in three towns in Mexico. Xaman et al. [32] proposed a two-dimensional

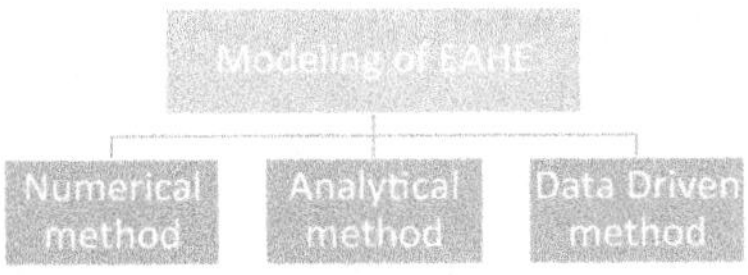

Figure 17.5 Methods for modeling of EAHE

pseudo-transient numerical model. The model was built on discretized and solved energy, mass, and momentum conservation equations with particular boundary conditions at the ground surface, pipe, and substantial depth.

Four Reynolds values (100, 500, 1000, and 1500) were used in the calculations, and the outcomes showed the system's astounding potential for efficient heating and cooling applications; similarly, a two-dimensional transient numerical model was developed by Niu et al. [33] considering both conductional and convectional heat transfer processes for an earth-air heat exchanger (EAHE) system. The model utilized a control volume formulation for discretization, dividing the soil into control volumes, and setting boundary conditions at the soil surface, undisturbed ground, and the tube.

Puri [34] published a transient model for a single earth-air heat exchanger (EAHE) system, which was based on a pair of differential equations. These equations explain how moisture and heat are transferred simultaneously in the soil around the EAHE. The model took into account a one-dimensional axial geometry and made the assumption that the soil's initial moisture and temperature were constant. The distribution of ground temperatures and moisture content was unaffected by the tube's existence at large axial distances from it. The model was then used to express the fundamental heat and mass transfer equations (1) and (2) [34]. This one-dimensional transient model contributes to the improvement and understanding of linked heat and moisture transmission mechanisms in the soil surrounding the EAHE.

$$\rho_d C_s \frac{\partial T}{\partial t} = \frac{1}{r}\frac{\partial}{\partial r}\left(rk_s \frac{\partial T}{\partial r}\right) \tag{1}$$

$$\frac{\partial W}{\partial t} = \frac{1}{r\rho_d}\frac{\partial}{\partial r}\left(\left(r\rho_a D_s \frac{\partial W}{\partial T}\right)\frac{\partial T}{\partial r}\right) + \frac{1}{r}\frac{\partial}{\partial r}\left(rK\frac{\partial W}{\partial r}\right) \tag{2}$$

where

T is soil temperature

W is soil moisture

ρ_d is soil density

ρ_a is air density

C_s is specific heat of wet soil

D_s is soil vapour diffusion coefficient

k_s is overall soil thermal conductivity
K is soil moisture diffusivity
r is polar coordinate
t is time variable

The equations were discretized using the finite elements approach to examine the thermal behavior. The results showed that soil temperature profiles change more quickly than equivalent profiles of moisture content.

To evaluate the thermal performance of an earth-air heat exchanger (EAHE) system used for energy saving in an agricultural greenhouse, researchers created a transient numerical model in reference [35]. Equation 3, a heat transfer differential equation, served as the model's foundation and explained how energy transferred into the soil in three dimensions (x, y, and z) [35]. The model was designed to examine how heat is diffused and transported in the soil around the EAHE system.

$$\boldsymbol{C}\frac{\partial \boldsymbol{T}}{\partial t} = \frac{\partial}{\partial x}\left(\boldsymbol{k}\frac{\partial \boldsymbol{T}}{\partial \boldsymbol{x}}\right) + \frac{\partial}{\partial y}\left(\boldsymbol{k}\frac{\partial \boldsymbol{T}}{\partial \boldsymbol{y}}\right) + \frac{\partial}{\partial z}\left(\boldsymbol{k}\frac{\partial \boldsymbol{T}}{\partial z}\right) + \boldsymbol{S} \tag{3}$$

where C is the volumetric heat capacity, T is the temperature within the soil, t is time variable, k is thermal conductivity, S is the source term, and x, y, z are space coordinates.

Belatrache et al. [36] and Fazlikhani et al. [37] suggested approaches for estimating the thermal performance and efficiency of earth-air heat exchanger (EAHE) systems under particular climatic conditions. Belatrache addressed dry areas in Algeria, Fazlikhani concentrated on hot and cold desert climates in Iran. Both approaches utilized conductive heat transmission in the soil to simulate temperature distribution at varying depths below the surface and were based on the laws of heat transfer. The researchers successfully validated their models against experimental data by numerically resolving the pertinent differential equations. These methods offer useful instruments for precisely evaluating the thermal performance and efficiency.

To forecast the thermal behavior of an earth-air heat exchanger (EAHE) system used for space heating/cooling in the environment of Egypt, Serageldin et al. [19] developed a transient equation. The model's foundation was a one-dimensional energy conservation equation that made a number of assumptions about the soil's qualities, temperature characteristics, and air flow inside the pipe. Equation 4 [19] is a representation of the energy conservation equation that takes both convection and diffusion processes into account. The researchers used this formulation in an effort to shed light on how EAHE systems function in the unique environmental conditions of Egypt.

$$\frac{\partial \acute{A}\boldsymbol{T}}{\partial t} + \underbrace{\mathbf{div}\left(\acute{A}\boldsymbol{T}\boldsymbol{u}\right)}_{\text{Convection term}} = \underbrace{\mathbf{div}\left(\boldsymbol{k}\,\mathbf{grad}\boldsymbol{T}\right)}_{\text{Diffusion term}} + \underbrace{\boldsymbol{Q}_{\text{Air}}^{\text{Soil}}}_{\text{Source term}} \tag{4}$$

The model incorporated a one-dimensional energy conservation equation, considering fluid density (ρ), flowing air temperature (T), turbulence kinetic energy (k), heat flux from/to the subsurface (Q_{Air}^{soil}), and velocity components (u, y, w) in the x, y, and z directions. The air flow and temperature inside the pipe were predicted using a convectional three-dimensional model based on CFD simulation software. The model's output was rigorously tested against experimental data.

17.2.2 Analytical methods

In this section, we explore into the diverse range of analytical models employed by researchers to better understand the dynamics of earth-air heat exchangers. Ajmi et al. [38] introduced a 1-D, steady state, analytical model to predict the outlet air temperature distribution of an earth-air heat exchanger (EAHE) system utilized for cooling a residence in a climate of extreme heat. To calculate the convective heat transfer coefficient, the model incorporated correlations with Reynolds and Nusselt numbers for both laminar and turbulent flow conditions. The researchers successfully validated the model by comparing its results with experimental data and theoretical parametric studies, and according to simulation data, the EAHE might reduce the peak cooling load by 1700 W and lower indoor temperatures by 2.8°C during the summer's greatest demand (mid of July). Similarily, to simulate a closed-loop earth-air heat exchanger (EAHE) system used for cooling residential structures in humid and hot climates, Do and Sung [39] created an steady-state, 1-D, analytical model.

Moncef et al. [40] introduced a simplified, transient analytical model to forecast the thermal performance of earth-air heat exchangers (EAHEs). This model effectively estimated the outlet air temp of EAHE at any given time. It was established on heat transfer equations encompassing conduction and convection phenomena, and it assumed that the system reaches a quasi-steady state operation after a few days of functioning. Model was validated against experimental resluts.

Mihalakakou et al. [41], applied parametric model was devised to estimate the thermal behavior of an earth-air heat exchanger (EAHE) system. The model was constructed using a systematic analytical parametric approach, utilizing regression analysis. The following four parameters, influencing the system's thermal performance, were used for the analysis: (a) burial depth below the ground surface; (b) pipe length; (c) pipe radius; and (d) air velocity inside the tube. The model's accuracy was validated by comparing its results with precise numerical data and experimental findings, demonstrating its reliability and effectiveness.

Deglin et al. [42] introduced a comprehensive three-dimensional model tailored to forecast the performance of subsoil earth-air heat exchanger (EAHE) systems in cooling livestock structures. The model focused primarily on convectional heat transfer processes occurring between the air inside the tubes and the surrounding soil. The model's predictions were checked against experimental data, affirming its accuracy and reliability in simulating the system's thermal behavior.

17.2.3 Data-driven methods

Data-driven models are computing systems that depend on historical data instead of the traditional mathematical and physical equations that are typically utilized in deterministic models. These models are designed to learn patterns and relationships from the data, enabling them to make predictions and other analyses without explicitly incorporating complex mathematical formulations. Machine learning (ML) and artificial neural network (ANN) models belong to the category of data-driven models. Data-driven models offer significant advantages, with their key strength lying in their efficiency and ability to work solely based on existing historical data. This efficiency allows them to effectively model complex and non-linear systems and processes with a satisfactory level of accuracy [43].

A neural network model was presented by Kumar et al. [44] to estimate the air temperature distribution at an EAHE system's outlet. The neural network included 15 tan-sigmoid neurons in the hidden layer and one linear neuron in the output layer, and it had a multi-layered feedforward design based on the back-propagation approach. Six key input parameters—ambient air temperature, ground temperature at burial depth, ground surface temperature, ambient air humidity, air mass flow rate, and pipe length—were used to train and test the network. The neural model's performance was closely compared to that of a numerically verified model [45], and the results showed that it performed very well and was well suited for real-world use.

A thermodynamic model was created by Diaz et al. [46] to simulate the thermal performance of an EAHE system. To improve the system's energy efficiency, they also suggested a fuzzy logic controller as an alternative to the traditional on/off controller. The findings showed that, in contrast to the on-off controller, which always used the maximum air mass flow rate, the fuzzy logic controller effectively identified the desired air temperature and mass flow rate. Due to the fuzzy logic controller's implementation, an astonishing 80% less energy was used to heat and cool a building.

An artificial neural network (ANN) approach was suggested in the study by Mihalakakou [47] to estimate the air temperature at the tube outlet of an EAHE system. The procedure had three steps: creating the neural network architecture, putting it through training, and evaluating how well it worked. The backpropagation technique, which uses gradient descent in order to reduce the mean square error of the weights, is the foundation of the neural network's architecture, which consists of a multi-layered feedforward structure. The network had an output layer with one linear neuron and a hidden layer with 16–21 log–sigmoid neurons. The ANN model was subjected to rigorous training and testing, and its performance was verified by comparing its results with those obtained from a highly accurate deterministic model. The successful validation demonstrated that the ANN approach efficiently predicts the thermal behavior of an EAHE system.

17.3 LIFE CYCLE ENERGY ANALYSIS (LCEA)

A life cycle energy analysis (LCEA) of an earth-air heat exchanger (EAHE) involves assessing the energy consumption and environmental impacts of the system throughout its entire life cycle, from manufacturing and installation to operation and disposal. The analysis takes into account factors such as energy used in materials production, construction, and maintenance, as well as the energy savings achieved during the operational phase, as demonstrated in Figure 17.6. By evaluating the overall energy efficiency and environmental footprint, researchers can determine the sustainability and long-term benefits of using EAHE systems for heating and cooling applications in buildings and other structures. Shazib et al. [48] evaluated the performance and life cycle analysis of earth-air heat exchanger (EAHE).

The study shows that energy consumption during the extraction and transportation stages is relatively minimal when compared to the significant energy demand involved in the manufacturing of heat exchanger materials as shown in Table 17.1. EAHE is used for space cooling of room having volume 20m3. EAHE system consists of a pipe 47ft in length having diameter 1.5in, made of PVC material buried at shallow depth, and a blower, which was installed at inlet of EAHE, was connected with PV panel which make the system self-sustainable. Experimental results showed that there was average rise of 9°C noticed during winter season and average drop of 6°C noticed during summer season. The lifecycle analysis was done for two pipe material, i.e., MS (mild steel) and PVC (polyvinyl-chloride), and it was found that MS consumes less energy than PVC during its life cycle, as represented in Table 17.1. Also, the energy payback time for MS was less than that of PVC.

Table 17.1 Energy consumption by PVC and MS materials during its entire life cycle [48]

Stages of Life Cycle	*PVC (MJ)*	*MS (MJ)*
Extraction of raw material	4.66	2.52×10^{-4}
Manufacturing of component materials	301	96
Transportation	1.679	0.422
Total life cycle energy consumption	307	96.4

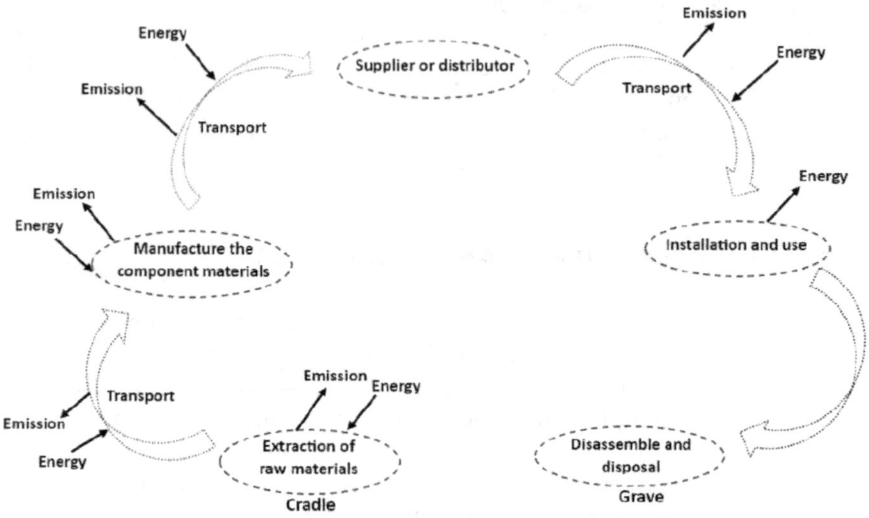

Figure 17.6 Representing several stages in life cycle energy analysis [48]

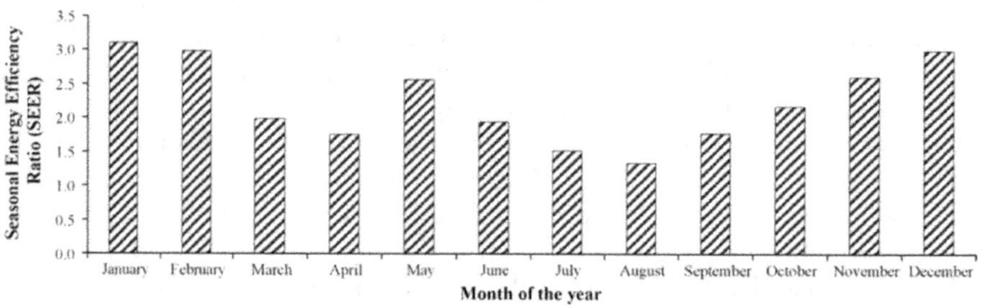

Figure 17.7 Representing the seasonal energy efficiency ratio (SEER) for different months of the year [49]

Chel et al. [56] conducted a study for a vault roof building integrated with earth-air heat exchanger (EAHE) system. The study focused on the energy saving potential, and it was found that the integration of earth-air heat exchanger (EAHE) in the building led to a considerable increase in energy savings. Prior to integration, the building's energy saving potential was 4946 kWh/year, but after incorporating the EAHE system, the energy savings significantly improved to 10321 kWh/year. Additionally, it was determined that employing the earth-air heat exchanger (EAHE), the seasonal average energy efficiency ratios (SEER) for heating and cooling were 1.8 and 2.9, respectively. This suggests that the EAHE technology is more effective at heating than cooling as demonstrated by Figure 17.7.

Energy metrics such as energy payback time (EPBT) and seasonal energy efficiency ratio (SEER) were calculated by Bisoniya et al. [50] based on the simulation results. The embodied energy for the EAHE system was found to be 1663.88 kWh, with maximum heating and cooling potentials calculated as 191.06 kWh and 247.25 kWh, respectively. The yearly energy output for the system was 1290.53 kWh at an air flow velocity of 5 m/s, and the EPBT was estimated to be 1.29 years. The SEER values were computed as 1.34 for the typical summer months and 1.10 for the winter months.

Soni et al. [52] conducted a study on a hybrid earth-air heat exchanger (EAHE) system at Energy Park in Bhopal. The approach involved integrating an EAHE with a 1.5 TR window air conditioner. The experiment comprised three distinct arrangements: arrangement 1 involved solely the window AC in the test room, arrangement 2 incorporated both the EAHE and the AC, and arrangement 3 connected the EAHE's outlet to condenser tubes for cooling. The outcomes revealed that arrangements 2 and 3 consumed 10.32 kWh and 9.84 kWh of energy, respectively, while arrangement 1 consumed 11.04 kWh. These findings demonstrated that after 8 hours of operation, arrangement 1 consumed more energy than both arrangements 2 and 3. Similarly, Mendez et al. [53] used a smart control called PID (proportional, integral and derivative) to save energy in an earth-air heat exchanger (EAHE) system. Their tests showed that using PID reduced energy use by a big 87%. This smart control helps manage airflow and saves energy in the EAHE system.

The study conducted by Nayak and Tiwari [51] at IIT Delhi assesses the system's effectiveness across different climatic conditions in Srinagar, Mumbai, Jodhpur, New Delhi, and Bangalore. The analysis is based on embodied energy and annual energy outputs. The results highlight that Jodhpur experiences the best overall thermal energy, annual electrical energy savings, and annual exergy at 29,156.8 kWh, 1185 kWh, and 1366.4 kWh, respectively, due to higher solar intensity and sunshine hours. Conversely, Srinagar exhibits the lowest values for these metrics. Moreover, Jodhpur boasts the lowest energy payback time at 16.7 years, while Srinagar records the highest at 21.6 years. The highest electricity production factor (EPF) is observed in Jodhpur at 2.04, and Srinagar showcases the highest life cycle conversion efficiency (LCCE). The study also observes an increase in LCCE with an increase in the life cycle (Table 17.2).

17.4 CONCLUSIONS

In conclusion, the review of earth-air heat exchanger (EAHE) technology for sustainable building applications underscores the significant potential and benefits that these systems offer in the pursuit of environmentally friendly and energy-efficient structures. The synthesis of research and technological advancements in EAHE has revealed several key findings and implications for the future of sustainable building technology. Technologies for passive heating and cooling have grown in popularity in recent decades as a practical way to improve interior and outdoor thermal comfort while conserving energy in

Table 17.2 The EPBT and carbon mitigation in a year for different locations

Energy Payback Time (EPBT) (In years)	*Carbon Mitigation (ton/year)*	*Location*	*Setup*	*References*
19.8	–	Bangalore	Photovoltaic and earth-air heat exchanger integrated with greenhouse	[51]
19.3		Jodhpur		
20.7		Mumbai		
21.7		New Delhi		
24.9		Srinagar		
Less than 2	16	New Delhi	Earth-air heat exchanger integrated with adobe house	[49]
1.29	2.026	Bhopal	Earth-air heat exchanger Installed at hot and dry climate	[52]
9 days	2.281 energy payback time for MS was less than	Bangladesh	Earth-air heat exchanger installed for a room with 20m³ in volume	[48]
3 days		Bangladesh	Earth-air heat exchanger installed for a room with 20m³ in volume	

buildings. Among these green technologies, earth-air heat exchangers have attracted a lot of interest and attention for both study and application.

This chapter explores the remarkable progress of EAHE, including study on diverse range of earth-air heat exchanger (EAHE) systems, categorizing them based on air circulation patterns and pipe orientations. The study identifies practical designs suitable for specific landmass characteristics and heating and cooling requirements. Additionally, the chapter extensively covers life cycle energy analysis, aiding in informed decisions for sustainable EAHE system implementation, promoting energy efficiency and eco-friendly building practices. Implementing earth-air heat exchangers aligns with green building standards and certifications, such as LEED (Leadership in Energy and Environmental Design), showcasing a commitment to sustainable building practices.

The utilization of EAHE systems presents a compelling solution to enhance building efficiency, reduce energy consumption, and mitigate the environmental impact associated with traditional HVAC systems. By leveraging the stable temperature of the earth, these heat exchangers contribute to both heating and cooling processes, offering a holistic and sustainable approach to maintaining indoor comfort. From residential structures to commercial complexes, the adaptability and scalability of these systems demonstrate their potential to become integral components in the design and construction of sustainable buildings.

In essence, the review affirms that earth-air heat exchanger technology stands at the forefront of sustainable building solutions, offering a pathway to energy-efficient, eco-friendly structures. The key findings of this study are:

- The ring pipe configuration is the most economical pipe layout since it eliminates the need for excavation by utilizing an existing trench in the building's base.
- For a medium-sized EAHE system, a serpentine pipe layout is the best option because lengthy pipes may be put in relatively short trenches.
- EAHE pipe material like polyvinyl chloride (PVC) consumes more energy than mild steel (MS) during its lifecycle.
- EAHE can enhance indoor air quality by continuously providing fresh air and reducing the reliance on recirculated air. This is particularly beneficial for occupants' health and comfort.
- Earth-air heat exchangers are less susceptible to external climate variations. The stable temperature of the ground provides a consistent heat source or sink, offering resilience to temperature fluctuations and extreme weather conditions.

In summary, earth-air heat exchangers play a crucial role in enhancing building sustainability by promoting energy efficiency, reducing costs, minimizing environmental impact, and improving indoor comfort and air quality.

17.4.1 Future scope

The combination of EAHE systems with renewable energy sources, such as solar and wind power, can further enhance their energy efficiency and make it self sustainable. Green building design and construction will be greatly aided by EAHE technology. The demand for energy-efficient heating and cooling systems will increase as the world adopts more environmentally friendly practices, and EAHEs provide a great substitute for traditional HVAC systems. As urbanization continues, there is potential for the incorporation of earth-air heat exchangers into the planning and design of sustainable urban developments. This could involve the integration of such systems into green buildings and eco-friendly urban infrastructure. The use of artificial intelligence, machine learning, and predictive analytics could optimize the operation of earth-air heat exchangers. Smart systems could adapt to changing environmental conditions, occupancy patterns, and energy demand, maximizing efficiency and comfort. Advances in monitoring technologies may enable real-time performance monitoring of EAHE systems. Predictive maintenance algorithms could help prevent downtime and ensure the long-term reliability of these systems.

REFERENCES

1. R. Singh, R. L. Sawhney, I. J. Lazarus, and V. V. N. Kishore, "Recent advancements in earth air tunnel heat exchanger (EATHE) system for indoor thermal

comfort application: A review," *Renewable and Sustainable Energy Reviews*, vol. 82, pp. 2162–2185, Feb. 2018, doi: 10.1016/j.rser.2017.08.058.

2. K. K. Agrawal, R. Misra, G. D. Agrawal, M. Bhardwaj, and D. K. Jamuwa, "The state of art on the applications, technology integration, and latest research trends of earth-air-heat exchanger system," *Geothermics*, vol. 82, pp. 34–50, Nov. 2019, doi: 10.1016/j.geothermics.2019.05.011.
3. S. Karytsas and H. Theodoropoulou, "Public awareness and willingness to adopt ground source heat pumps for domestic heating and cooling," *Renewable and Sustainable Energy Reviews*, vol. 34, pp. 49–57, Jun. 2014, doi: 10.1016/j.rser.2014.02.008.
4. A. Mustafa Omer, "Ground-source heat pumps systems and applications," *Renewable and Sustainable Energy Reviews*, vol. 12, no. 2, pp. 344–371, Feb. 2008, doi: 10.1016/j.rser.2006.10.003.
5. A. M. Omer, "Energy, environment and sustainable development," *Renewable and Sustainable Energy Reviews*, vol. 12, no. 9, pp. 2265–2300, Dec. 2008, doi: 10.1016/j.rser.2007.05.001.
6. H. Lund, "Renewable energy strategies for sustainable development," *Energy*, vol. 32, no. 6, pp. 912–919, Jun. 2007, doi: 10.1016/j.energy.2006.10.017.
7. M. Santamouris and D. Kolokotsa, "Passive cooling dissipation techniques for buildings and other structures: The state of the art," *Energy and Buildings*, vol. 57, pp. 74–94, Feb. 2013, doi: 10.1016/j.enbuild.2012.11.002.
8. G. Mihalakakou *et al.*, "Applications of earth-to-air heat exchangers: A holistic review," *Renewable and Sustainable Energy Reviews*, vol. 155, p. 111921, Mar. 2022, doi: 10.1016/j.rser.2021.111921.
9. M. Kaushal, "Geothermal cooling/heating using ground heat exchanger for various experimental and analytical studies: Comprehensive review," *Energy and Buildings*, vol. 139, pp. 634–652, Mar. 2017, doi: 10.1016/j.enbuild.2017.01.024.
10. O. Ozgener, L. Ozgener, and D. Y. Goswami, "Seven years energetic and exergetic monitoring for vertical and horizontal EAHE assisted agricultural building heating," *Renewable and Sustainable Energy Reviews*, vol. 80, pp. 175–179, Dec. 2017, doi: 10.1016/j.rser.2017.05.056.
11. A. Yildiz, O. Ozgener, and L. Ozgener, "Exergetic performance assessment of solar photovoltaic cell (PV) assisted earth to air heat exchanger (EAHE) system for solar greenhouse cooling," *Energy and Buildings*, vol. 43, no. 11, pp. 3154–3160, Nov. 2011, doi: 10.1016/j.enbuild.2011.08.013.
12. T. S. Bisoniya, A. Kumar, and P. Baredar, "Experimental and analytical studies of earth–air heat exchanger (EAHE) systems in India: A review," *Renewable and Sustainable Energy Reviews*, vol. 19, pp. 238–246, Mar. 2013, doi: 10.1016/j.rser.2012.11.023.
13. V. Bansal, R. Misra, G. D. Agrawal, and J. Mathur, "Performance analysis of earth–pipe–air heat exchanger for winter heating," *Energy and Buildings*, vol. 41, no. 11, pp. 1151–1154, Nov. 2009, doi: 10.1016/j.enbuild.2009.05.010.
14. V. Bansal, R. Misra, G. D. Agrawal, and J. Mathur, "Performance analysis of earth–pipe–air heat exchanger for summer cooling," *Energy and Buildings*, vol. 42, no. 5, pp. 645–648, May 2010, doi: 10.1016/j.enbuild.2009.11.001.
15. M. Khabbaz, B. Benhamou, K. Limam, P. Hollmuller, H. Hamdi, and A. Bennouna, "Experimental and numerical study of an earth-to-air heat exchanger for air cooling in a residential building in hot semi-arid climate,"

Energy and Buildings, vol. 125, pp. 109–121, Aug. 2016, doi: 10.1016/j.enbuild.2016.04.071.

16. S. F. Ahmed, M. T. O. Amanullah, M. M. K. Khan, M. G. Rasul, and N. M. S. Hassan, "Parametric study on thermal performance of horizontal earth pipe cooling system in summer," *Energy Conversion and Management*, vol. 114, pp. 324–337, Apr. 2016, doi: 10.1016/j.enconman.2016.01.061.
17. M. Grosso and L. Raimondo, "Horizontal air-to-earth heat exchangers in Northern Italy – Testing, design and monitoring," *International Journal of Ventilation*, vol. 7, no. 1, pp. 1–10, Jun. 2008, doi: 10.1080/14733315.2008.11683794.
18. V. Rangarajan, R. Singh, and P. Kaushal, "Model development and performance evaluation of an earth air heat exchanger under a constrained urban environment," *Modeling Earth Systems and Environment*, vol. 5, no. 1, pp. 143–158, Mar. 2019, doi: 10.1007/s40808-018-0524-z.
19. A. A. Serageldin, A. K. Abdelrahman, and S. Ookawara, "Earth-Air heat exchanger thermal performance in Egyptian conditions: Experimental results, mathematical model, and Computational Fluid Dynamics simulation," *Energy Conversion and Management*, vol. 122, pp. 25–38, Aug. 2016, doi: 10.1016/j.enconman.2016.05.053.
20. C.-E. Mehdid *et al.*, "Thermal design of Earth-to-air heat exchanger. Part II a new transient semi-analytical model and experimental validation for estimating air temperature," *Journal of Cleaner Production*, vol. 198, pp. 1536–1544, Oct. 2018, doi: 10.1016/j.jclepro.2018.07.063.
21. T. M. Yusof, H. Ibrahim, W. H. Azmi, and M. R. M. Rejab, "Thermal analysis of earth-to-air heat exchanger using laboratory simulator," *Applied Thermal Engineering*, vol. 134, pp. 130–140, Apr. 2018, doi: 10.1016/j.applthermaleng.2018.01.124.
22. R. Misra, V. Bansal, G. Das Agarwal, J. Mathur, and T. Aseri, "Evaluating thermal performance and energy conservation potential of hybrid earth air tunnel heat exchanger in hot and dry climate—In situ measurement," *Journal of Thermal Science and Engineering Applications*, vol. 5, no. 3, p. 031006, Sep. 2013, doi: 10.1115/1.4023435.
23. A. Mathur, Priyam, S. Mathur, G. D. Agrawal, and J. Mathur, "Comparative study of straight and spiral earth air tunnel heat exchanger system operated in cooling and heating modes," *Renewable Energy*, vol. 108, pp. 474–487, Aug. 2017, doi: 10.1016/j.renene.2017.03.001.
24. S. M. N. Shojaee and K. Malek, "Earth-to-air heat exchangers cooling evaluation for different climates of Iran," *Sustainable Energy Technologies and Assessments*, vol. 23, pp. 111–120, Oct. 2017, doi: 10.1016/j.seta.2017.09.007.
25. K. K. Agrawal, R. Misra, and G. D. Agrawal, "Thermal performance analysis of slinky-coil ground-air heat exchanger system with sand-bentonite as backfilling material," *Energy and Buildings*, vol. 202, p. 109351, Nov. 2019, doi: 10.1016/j.enbuild.2019.109351.
26. C. S. A. Chong, G. Gan, A. Verhoef, R. G. Garcia, and P. L. Vidale, "Simulation of thermal performance of horizontal slinky-loop heat exchangers for ground source heat pumps," *Applied Energy*, vol. 104, pp. 603–610, Apr. 2013, doi: 10.1016/j.apenergy.2012.11.069.
27. K. K. Agrawal, R. Misra, G. D. Agrawal, M. Bhardwaj, and D. K. Jamuwa, "Effect of different design aspects of pipe for earth air tunnel heat exchanger

system: A state of art," *International Journal of Green Energy*, vol. 16, no. 8, pp. 598–614, Jun. 2019, doi: 10.1080/15435075.2019.1601096.
28. V. R. Tarnawski, W. H. Leong, T. Momose, and Y. Hamada, "Analysis of ground source heat pumps with horizontal ground heat exchangers for northern Japan," *Renewable Energy*, vol. 34, no. 1, pp. 127–134, Jan. 2009, doi: 10.1016/j.renene.2008.03.026.
29. H. Fujii, S. Yamasaki, T. Maehara, T. Ishikami, and N. Chou, "Numerical simulation and sensitivity study of double-layer Slinky-coil horizontal ground heat exchangers," *Geothermics*, vol. 47, pp. 61–68, Jul. 2013, doi: 10.1016/j.geothermics.2013.02.006.
30. Z. Liu *et al.*, "Experimental investigation of a vertical earth-to-air heat exchanger system," *Energy Conversion and Management*, vol. 183, pp. 241–251, Mar. 2019, doi: 10.1016/j.enconman.2018.12.100.
31. R. D. Patel and P. V. Ramana, "Experimental analysis of horizontal and vertical buried tube heat exchanger air conditioning system," *Indian Journal of Science and Technology*, vol. 9, no. 35, Sep. 2016, doi: 10.17485/ijst/2016/v9i35/100510.
32. J. Xamán, I. Hernández-López, R. Alvarado-Juárez, I. Hernández-Pérez, G. Álvarez, and Y. Chávez, "Pseudo transient numerical study of an earth-to-air heat exchanger for different climates of México," *Energy and Buildings*, vol. 99, pp. 273–283, Jul. 2015, doi: 10.1016/j.enbuild.2015.04.041.
33. F. Niu, Y. Yu, D. Yu, and H. Li, "Investigation on soil thermal saturation and recovery of an earth to air heat exchanger under different operation strategies," *Applied Thermal Engineering*, vol. 77, pp. 90–100, Feb. 2015, doi: 10.1016/j.applthermaleng.2014.11.069.
34. V. M. Puri, "Heat and mass transfer analysis and modeling in unsaturated ground soils for buried tube systems," *Energy in Agriculture*, vol. 6, no. 3, pp. 179–193, Nov. 1987, doi: 10.1016/0167-5826(87)90001-5.
35. C. Gauthier, M. Lacroix, and H. Bernier, "Numerical simulation of soil heat exchanger-storage systems for greenhouses," *Solar Energy*, vol. 60, no. 6, pp. 333–346, Jun. 1997, doi: 10.1016/S0038-092X(97)00022-4.
36. D. Belatrache, S. Bentouba, and M. Bourouis, "Numerical analysis of earth air heat exchangers at operating conditions in arid climates," *International Journal of Hydrogen Energy*, vol. 42, no. 13, pp. 8898–8904, Mar. 2017, doi: 10.1016/j.ijhydene.2016.08.221.
37. F. Fazlikhani, H. Goudarzi, and E. Solgi, "Numerical analysis of the efficiency of earth to air heat exchange systems in cold and hot-arid climates," *Energy Conversion and Management*, vol. 148, pp. 78–89, Sep. 2017, doi: 10.1016/j.enconman.2017.05.069.
38. F. Al-Ajmi, D. L. Loveday, and V. I. Hanby, "The cooling potential of earth–air heat exchangers for domestic buildings in a desert climate," *Building and Environment*, vol. 41, no. 3, pp. 235–244, Mar. 2006, doi: 10.1016/j.buildenv.2005.01.027.
39. S. L. Do, J.-C. Baltazar, and J. Haberl, "Potential cooling savings from a ground-coupled return-air duct system for residential buildings in hot and humid climates," *Energy and Buildings*, vol. 103, pp. 206–215, Sep. 2015, doi: 10.1016/j.enbuild.2015.05.043.
40. M. Krarti and J. F. Kreider, "Analytical model for heat transfer in an underground air tunnel," *Energy Conversion and Management*, vol. 37, no. 10, pp. 1561–1574, Oct. 1996, doi: 10.1016/0196-8904(95)00208-1.

41. G. Mihalakakou, M. Santamouris, D. Asimakopoulos, and I. Tselepidaki, "Parametric prediction of the buried pipes cooling potential for passive cooling applications," *Solar Energy*, vol. 55, no. 3, pp. 163–173, Sep. 1995, doi: 10.1016/0038-092X(95)00045-S.
42. D. Deglin, L. Van Caenegem, and P. Dehon, "Subsoil heat exchangers for the air conditioning of livestock buildings," *Journal of Agricultural Engineering Research*, vol. 73, no. 2, pp. 179–188, Jun. 1999, doi: 10.1006/jaer.1998.0401.
43. K. Chakraborty, K. Mehrotra, C. K. Mohan, and S. Ranka, "Forecasting the behavior of multivariate time series using neural networks," *Neural Networks*, vol. 5, no. 6, pp. 961–970, Nov. 1992, doi: 10.1016/S0893-6080(05)80092-9.
44. R. Kumar, S. C. Kaushik, and S. N. Garg, "Heating and cooling potential of an earth-to-air heat exchanger using artificial neural network," *Renewable Energy*, vol. 31, no. 8, pp. 1139–1155, Jul. 2006, doi: 10.1016/j.renene.2005.06.007.
45. G. Mihalakakou, M. Santamouris, and D. Asimakopoulos, "Modelling the thermal performance of earth-to-air heat exchangers," *Solar Energy*, vol. 53, no. 3, pp. 301–305, Sep. 1994, doi: 10.1016/0038-092X(94)90636-X.
46. S. E. Diaz, J. M. T. Sierra, and J. A. Herrera, "The use of earth–air heat exchanger and fuzzy logic control can reduce energy consumption and environmental concerns even more," *Energy and Buildings*, vol. 65, pp. 458–463, Oct. 2013, doi: 10.1016/j.enbuild.2013.06.028.
47. G. Mihalakakou, "On the heating potential of a single buried pipe using deterministic and intelligent techniques," *Renewable Energy*, vol. 28, no. 6, pp. 917–927, May 2003, doi: 10.1016/S0960-1481(02)00183-0.
48. Md. S. Uddin, R. Ahmed, and M. Rahman, "Performance evaluation and life cycle analysis of earth to air heat exchanger in a developing country," *Energy and Buildings*, vol. 128, pp. 254–261, Sep. 2016, doi: 10.1016/j.enbuild.2016.06.088.
49. A. Chel and G. N. Tiwari, "Performance evaluation and life cycle cost analysis of earth to air heat exchanger integrated with adobe building for New Delhi composite climate," *Energy and Buildings*, vol. 41, no. 1, pp. 56–66, Jan. 2009, doi: 10.1016/j.enbuild.2008.07.006.
50. T. S. Bisoniya, A. Kumar, and P. Baredar, "Energy metrics of earth–air heat exchanger system for hot and dry climatic conditions of India," *Energy and Buildings*, vol. 86, pp. 214–221, Jan. 2015, doi: 10.1016/j.enbuild.2014.10.012.
51. S. Nayak and G. N. Tiwari, "Energy metrics of photovoltaic/thermal and earth air heat exchanger integrated greenhouse for different climatic conditions of India," *Applied Energy*, vol. 87, no. 10, pp. 2984–2993, Oct. 2010, doi: 10.1016/j.apenergy.2010.04.010.
52. S. K. Soni, M. Pandey, and V. N. Bartaria, "Energy metrics of a hybrid earth air heat exchanger system for summer cooling requirements," *Energy and Buildings*, vol. 129, pp. 1–8, Oct. 2016, doi: 10.1016/j.enbuild.2016.07.063.
53. S. E. Diaz-Mendez, C. Patiño-Carachure, and J. A. Herrera-Castillo, "Reducing the energy consumption of an earth–air heat exchanger with a PID control system," *Energy Conversion and Management*, vol. 77, pp. 1–6, Jan. 2014, doi: 10.1016/j.enconman.2013.09.033.

Chapter 18

Retrofitting of existing buildings to achieve energy-efficient buildings—A case study of residential buildings in India

Lilly Rose Amirtham, K. Parameswari, and Aravamudhan Swaminathan

18.1 INTRODUCTION

In India, the construction sector contributes to 24% of CO_2 emissions and 40% of primary energy use (Kishore & Chouhan, 2014). Cement, burnt clay bricks, and steel are the three major conventional materials that contribute to around 80% of the emissions during their manufacturing processes. In recent years, with significant emphasis on environmental awareness, it has become imperative that the emission of greenhouse gases into the atmosphere needs to be reduced drastically. The natural resources become scarce, and it is therefore important to manage the resources sustainably, and the construction industry must take the primary role through appropriate material selection and techniques for construction. With the rapid increase in the pollution levels caused by the manufacturing of building materials, it has become mandatory for the building materials and products to indicate their chemical compositions along with information on the level of toxicity and potential hazards to the environment. Organizations that promote green buildings in India such as Green Rating for Integrated Habitat Assessment (GRIHA), Indian Green Buildings Council (IGBC), etc. recommend building materials that are manufactured using recycled materials that have a higher potential for recycling. Building materials used for construction consume energy all through their life cycle right from manufacturing to disposal.

In developed countries, there is a growing awareness of the need to reduce energy consumption for heating and cooling purposes. While these countries have the financial resources to meet the high costs of energy demands, this is not the case for the majority of the population in developing countries. As a result, most of the low-income people in these countries are unable to afford the costs associated with the heating and cooling of their homes. Hence, it is crucial to evaluate the thermal efficiency of walls and roofs of naturally ventilated buildings, particularly in developing nations. The building envelope is crucial in maintaining a comfortable indoor environment. A well-designed building envelope can help to minimize heat transfer and reduce the need for active heating and cooling systems, which

 DOI: 10.1201/9781003496656-18

in turn can significantly lower energy consumption (Barrios et al., 2012). Zhou et al. (2011) state that various components of a building's envelope, such as walls, windows, doors, and roofs, can have different impacts on the exchange of heat. The performance of materials in terms of thermal insulation can vary depending on the climate. Designing buildings in warm and humid climates can be particularly challenging. The best approach is to utilize lightweight construction materials for walls and roofs with lower U-values and lower heat capacity, as storing heat is not desirable in such climates (ASHRAE handbook, 2013 & ISO 6946, 2007; Givoni, 1994; Kumar et al., 2020) and reflective or resistive insulation can enhance indoor comfort (Haase & Amato, 2009). It is crucial to choose the appropriate insulation material and calculate the ideal insulation thickness to ensure effective insulation (Yu et al., 2009; Kumar et al., 2014).

Embodied energy (EE) of a material is the energy consumed by the building material all through its life cycle. Similarly, the embodied energy of a building is the energy consumed in all three phases of construction, namely, the preconstruction phase (mining/extraction, processing, manufacturing, and transportation), construction phase (product delivery and assembly), and post-construction phase (maintenance, renovation, demolition, and disposal). It is possible to reduce the use of conventional materials through the use of alternate sustainable materials which results in a considerable amount of reduction in energy consumption and CO_2 emissions. Oka et al., 1993; Buchanan & Honey, 1994; Debnath et al., 1995; Suzki et al., 1995; Reddy & Jagadish, 2003; Demir, 2008; Kofoworola & Gheewala, 2009; Praseeda et al., 2015; Mastrucci & Rao, 2018; Bansal et al., 2019; Mohamed & Nehal, 2022 studied the embodied energy requirements of building materials and their impacts on the environment. Debnath et al., 1995 estimated the energy requirements for residential buildings in India as 3 to 5 GJ/m^2 of built-up area and concluded that the major component of the energy costs was due to the energy consumption of conventional materials such as bricks, cement, and steel. Oka et al., 1993 found the energy consumption for single-family house construction in Japan as 3GJ/m^2. Buchanan & Honey, 1994; Reddy & Jagadish, 2003 revealed that the construction based on wood and the use of soil-based low-energy materials reduced the embodied energy significantly and was found to be between 2.32 and 5.53 GJ/ m^2. Suzki et al., 1995 found that the embodied energy of the houses ranged from 2.7 to 10.4 GJ/ m^2 and found that the environmental impact of a wooden house was significantly lower when compared to a steel and reinforced concrete structure. Morel et al., 2001; Kofoworola, & Gheewala, 2009 suggested optimum usage of vernacular materials/resources in conserving EE. The EE in a renovated adobe passive house was estimated to be 2.3 GJ/ m^2 (Chel et al., 2009). Monahan & Powell, 2011 calculated the embodied energy for three types of low energy houses and found that timber contributed to the lowest EE of the building which ranged from 5.7 to 8.2 GJ/m^2. Kariyawasam and Jayasinghe, 2016 compared the use of vernacular materials (rammed earth stabilized soil blocks, etc.) with brick masonry for walls in urban dwellings and found that the EE of vernacular materials was less than 2.0 GJ/m^2 whereas the EE of brick walls ranged from 2.40 to 2.90 GJ/ m^2. Mohamed and Nehal, 2022 stated the need for a shift in the development and promotion of embodied energy research.

Tiwari and Parikh, 1995 showed that building envelopes in residential buildings had good potential in reducing the EE as they constitute 62% of the materials used for walls (46%) and roofs (16 %). Thormark, 2000, 2001, 2002 studied recycling nationally produced building waste and found that the potential energy savings by recycling was about 50% of the EE. Morel et al., 2001 found that the EE of a house constructed with stone masonry (246%) and the rammed house (240%) performed better than that of a conventional house. Reddy, 2009 conducted a study on stabilized mud blocks (SMB) and found a saving of around 60–70% energy when compared to burnt bricks. Praseeda et al., 2014, 2015 proposed mud and stone for rural housing and carried out EE and operating energy (OE) assessment in various climatic zones in India. The study showed that the EE in the life cycle energy significantly depended on the characteristics of materials used in the building, and at times the EE exceeded the OE. Djamil, 2016 found that the EE of reinforced concrete is 99.5 times higher than the embodied energy of raw earth. The study showed that raw earth stabilized with natural materials was 100% reusable. Samuel et al., 2017 revealed that sun-dried blocks reduced 80% of the carbon emission when compared to fired bricks. Mastrucci & Rao, 2018 examined the life cycle energy and costs of alternative building materials and their impacts on carbon emission and thermal comfort of the occupants in India, and found stabilized earth blocks as a substitute for fire bricks. Bansal et al., 2019 analyzed the embodied energy for components and found that 82% to 85% of the total EE of houses in India was consumed by the walls, roofs, and foundations. The study also found that the EE of walls in single-story houses was 39% and that of roofs was 18%. A study on sustainable construction through advanced bio-materials was conducted by Khitab & Anwar, 2020 and inferred that bio-materials provide similar properties as that of synthetic materials without going through numerous chemical reactions. This reduction in chemical processes resulted in the use of bio-materials with lower EE. The use of natural, bio, and locally sourced materials in building construction has been found to have a significant impact on energy efficiency. Furthermore, these materials have been shown to improve the thermal performance of buildings, leading to more comfortable and sustainable living spaces.

The India Meteorological Department (IMD) has classified the climate of India as composite, hot and dry, warm and humid, temperate, and cold climates. The Energy Conservation Building Code (ECBC) in India classifies building typologies based on their energy consumption patterns into three categories: ECBC, ECBC Plus, and ECBC Super. Research on the EE analysis of sustainable wall and roof assemblies that comply with the Energy

Conservation Building Code (ECBC) can help identify the most appropriate assembly for retrofit options for existing buildings. It can also aid in exploring new sustainable alternative materials. This study aims to analyze the embodied energy of various wall, roof, and floor assemblies for retrofit using natural, bio, and salvaged materials. The study will be conducted in the warm and humid climate of Vijayawada, India.

18.2 MATERIALS AND METHODS

A single-story residential building in Vijayawada, India, has been chosen for the present study. Fifteen alternative wall, roof, and floor assemblies have been analyzed for sustainable retrofit options. Embodied energy of the alternative assemblies has been analyzed to identify the appropriate wall, roof, and floor retrofit options for the selected single-story residence with framed structure. The plan and section of the residence are shown in Figures 18.1 and 18.2. The thickness of the wall and roof assemblies has been considered as 280 mm and 250 mm respectively for all the retrofit options in the study. The material characteristics of the residence are shown in Table 18.1.

In recent years, there has been a growing trend toward using natural, biomaterials, and salvaged materials as alternative building materials due to their eco-friendly and sustainable properties. These materials have unique thermal characteristics that can enhance the energy efficiency of buildings and can be utilized in various building components including walls, roofs, and floors. Sustainable building materials with potential thermal benefits for warm humid climates were selected from literature as retrofit options in the study. Table 18.2 shows the alternative wall, roof, and floor assemblies chosen for the study. Four options of natural materials, six options of biomaterials, and five options of salvaged materials have been chosen for the present study.

18.3 ASSESSMENT OF EMBODIED ENERGY

The embodied energy of a material is calculated as the product of the quantity of materials used in the structure and their embodied energy values. The quantities and specifications of materials used in the single-story residential building were calculated based on the plan and section of the selected building. The information regarding the embodied energy of different building materials has been gathered from relevant literature sources (UNCHS, 1991; Yu et al., 2011; IFC, 2017; CECP, 2020). The EE of the materials is based on the cradle-to-gate approach. The total embodied energy of the building is calculated as the sum of all the EE of materials used in the structure. The EE of the single-story structure using conventional materials (Table 18.3) is considered as the base case. Further, the EE of the retrofit options of natural

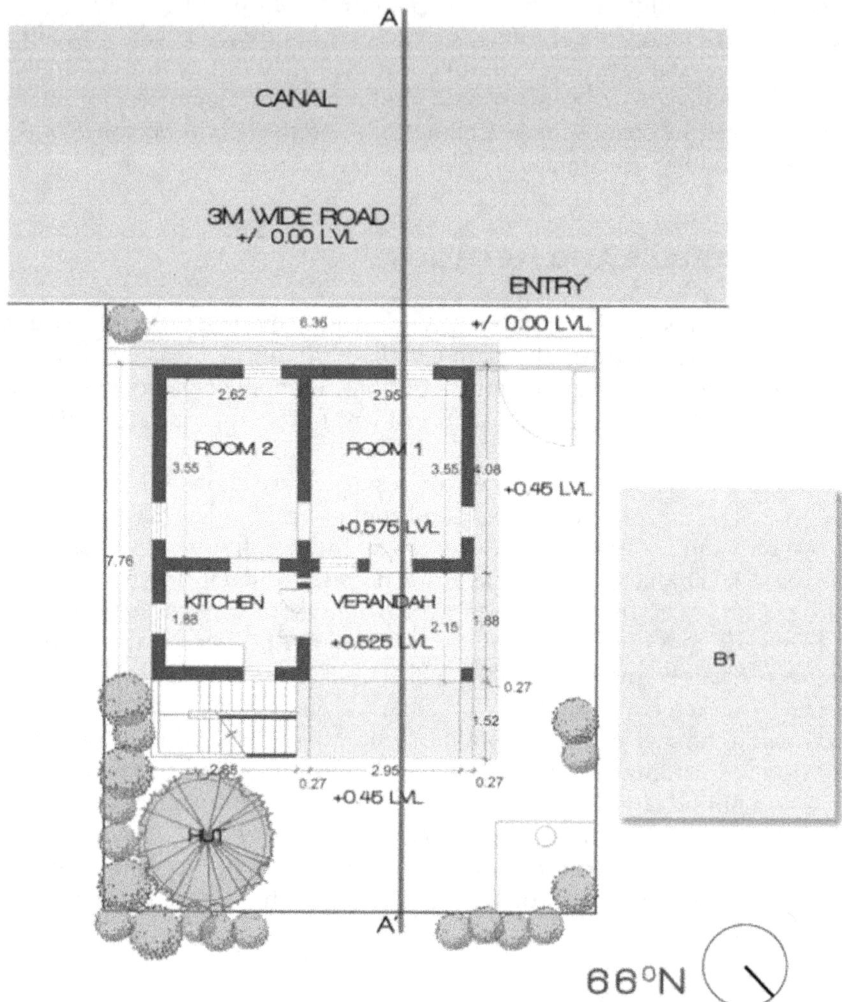

Figure 18.1 Plan of the single storey Residence

material, bio-materials, and salvaged materials are calculated to analyze the savings in the embodied energy compared with the base case.

18.4 RESULTS AND DISCUSSION

18.4.1 Embodied energy of the base case

The EE of the base case using conventional materials of cement, steel, and brick was calculated. EE of alternative materials for wall, roof, and flooring

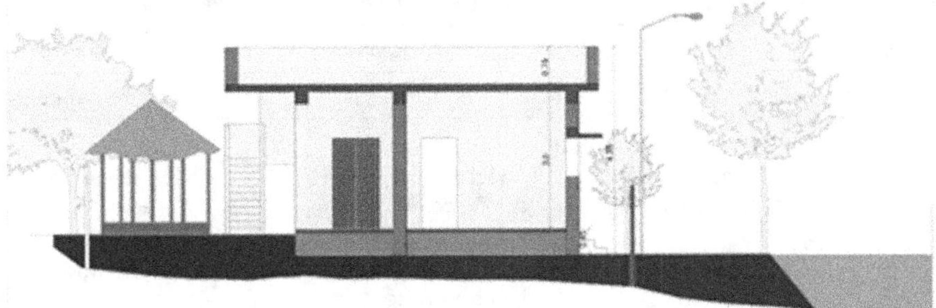

Figure 18.2 Section of the single-storey Residence

Table 18.1 Characteristics of the single-story structure

Total Built-Up Area	31.6 m²
RCC	Beams, columns, slab
PCC	Flooring
Clay fired bricks	Walls

with four options of natural materials, six options of Biomaterials, and five options of Salvaged materials were calculated. The total EE of the base case was found to be 101.89 GJ which is 3.22 GJ/ m^2, similar to the study of Debnath et al. (1995). It was found that 65.07% of the total EE of the building was due to the wall component in the base case, which is predominantly due to the high embodied energy of fired brick and cement. The roof contributed to 22.85% while flooring contributed to 13.97% of the total EE of the structure.

18.4.2 Analysis of EE of wall, roof, and floor retrofits

Using alternative building materials like natural materials, biomaterials and salvaged materials with lesser embodied energy than conventional building materials can aid in reducing the EE of the building significantly.

18.4.3 Analysis of EE of wall retrofits

The EE of 13 retrofits using natural, bio, and salvaged materials were assessed and compared with the conventional brick wall. Figure 18.3 shows the EE comparison of all the chosen materials. The total EE of the base case wall material was found to be 65.07 GJ. In the natural materials, compressed stabilized earth blocks (CSEB) with interior mud plaster and exterior lime plaster, CSEB blocks with lime plaster, straw bale with lime plaster, and rammed earth were considered. Out of the four options in natural materials, straw bale with lime plaster had the highest embodied

Table 18.2 Alternative building materials for retrofit of various building components

Options	*Wall*	*Roof*	*Flooring*
Base case	*Brick*	*RCC*	*PCC, ceramic tiles*
Natural materials			
1	CSEB, mud plaster (int.), lime plaster (ext.)	Limecrete, bamboo reinforcement	Kota stone, cement screed
2	CSEB, lime plaster	Thatch, palmyra beam, and bamboo rafter	Bamboo flooring
3	Strawbale, lime plaster	Thach and bamboo rafters	Limecrete, lime screed
4	Rammed earth	Bamboo roof	Mud flooring
Bio materials			
5	Timbercrete	Mangalore tiles	Linoleum flooring
6	Porotherm	Burnt clay tiles filler slab	Red oxide flooring
7	Mycelium bricks	BFRC	Linoleum
8	Hempcrete	Bamboo reinforced concrete slab	Bamboo boards
9	Rice husk ash brick	Mangalore tiles	Clay tiles
10	Sawdust brick	Bamboo mat corrugated sheet	Bamboo flooring
Salvaged materials			
11	Stone masonry	PCC + China mosaic	Reclaimed stone
12	FalG brick	Broken China mosaic tiles	Reclaimed Tandur stone
13	FalG brick	Salvaged brick	Cement-based Terrazzo tile
14	Medium-density fiber board	Reclaimed timber	Reused timber
15	Fly ash brick	Reused cement + reused steel	Reclaimed broken ceramic tiles

energy of 78.14 GJ, which is 1.2 times more energy-intensive than the base case. Rammed earth has the least embodied of 0.94 GJ which is 0.014 times the base case and 0.012 times less energy-intensive than straw bale with lime plaster. Further, from Figure 18.3, it was observed that CSEB blocks with interior mud plaster and exterior lime plaster and CSEB blocks with lime plaster have almost similar embodied energy values of 20.44 GJ and 25.99 GJ, respectively.

Timbercrete blocks, porotherm blocks, mycelium bricks, hempcrete blocks, and sawdust bricks were considered in the biomaterials category. It was observed from Figure 18.3 that the porotherm blocks had the highest embodied energy of 54.26 GJ followed by timbercrete blocks and hempcrete blocks with 52.07 GJ and 46.98 GJ, respectively. Porotherm blocks are 0.83

Table 18.3 Properties of building materials used in the base case building structure

S. No.	*Component*	*Materials*	*Specification*	*Unit*	*Volume*	*Density*	*EE*	*Unit*
1	**Wall**	Brick	230 mm fired brick	m^3	22.32	2,000	1.2	MJ/kg
		Cement plaster	0.01m	m^3	1.09	1,762	4.8	MJ/kg
		Sand	Cement mortar 1:6 (0.1m)	m^3	0.72	1,442	93.5	MJ/m^3
		Cement		m^3	0.12	2,080	8,096	MJ/ton
2	**Flooring**	PCC	1:1.5:3 (0.1m)	m^3	3.15	2,400	1.1	MJ/kg
		Ceramic tiles	(6mm)	m^3	2.01	2,250	12	MJ/kg
3	**RCC roof**	Cement	01:02:04	m^3	1.17	1,440	8,096	MJ/ton
		Sand		m^3	2.34	1,442	66.6	MJ/m^3
		Aggregate		m^3	4.69	2,548	430	MJ/m^3
		Steel	1% of steel in RCC	m^3	0.08	7,700	8.9	MJ/kg

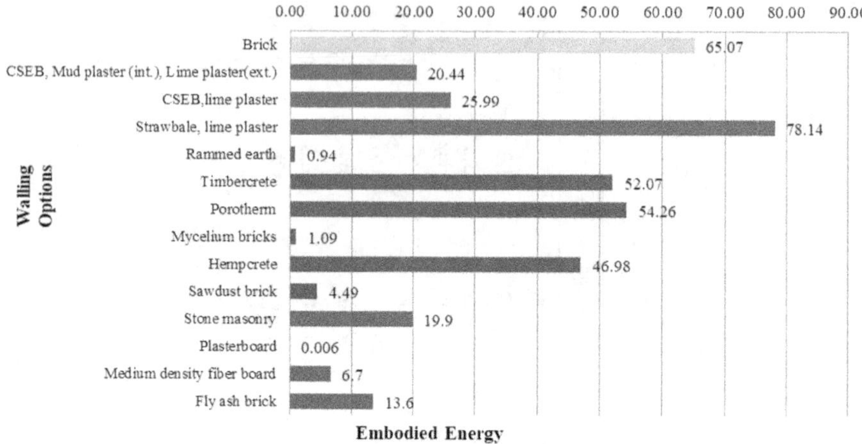

Figure 18.3 EE contribution of retrofit wall materials

times less energy-intensive than the base case. Mycelium brick was found to have the lowest embodied energy of 1.09 GJ among the chosen biomaterial retrofit options which was 0.016 times less energy-intensive than the base case and 0.02 times less energy-intensive than porotherm blocks. Sawdust brick was also found to have a low embodied energy of 4.49 GJ when compared to timbercrete, porotherm, and hempcrete blocks.

In the salvaged materials category, stone masonry, plasterboard, medium-density fiberboard, and fly ash bricks were considered. Figure 18.3 shows that the plaster board had the lowest embodied energy of 0.006 GJ which can be one of the most efficient retrofit options in terms of embodied energy. Medium-density fiberboards had an embodied energy value of 6.7 GJ and were 0.10 times less energy-intensive than fired bricks. Stone masonry and fly ash bricks had embodied energy values of 19.9 GJ and 13.6 GJ, respectively.

The comparison of all the wall retrofit options revealed that plasterboard under salvaged material had almost zero embodied energy while rammed earth, a natural material, had less embodied energy of 0.94 GJ followed by mycelium bricks and sawdust bricks of the biomaterial category. The EE of the natural wall retrofit ranged between 0.94 GJ and 78.14 GJ. Similarly, in the biomaterial retrofit, the EE ranged between 1.09 GJ and 54.26 GJ. The salvaged retrofits were found to have the least EE ranging between 0.006 GJ and 19.9 GJ. The analysis revealed that salvaged retrofit options are the most preferred wall options while considering the EE contribution.

18.4.4 Analysis of EE of roofing retrofits

EE of 13 roofing retrofits with natural, bio, and salvaged materials was assessed and compared with the conventional RCC roof. Figure 18.4 shows the EE

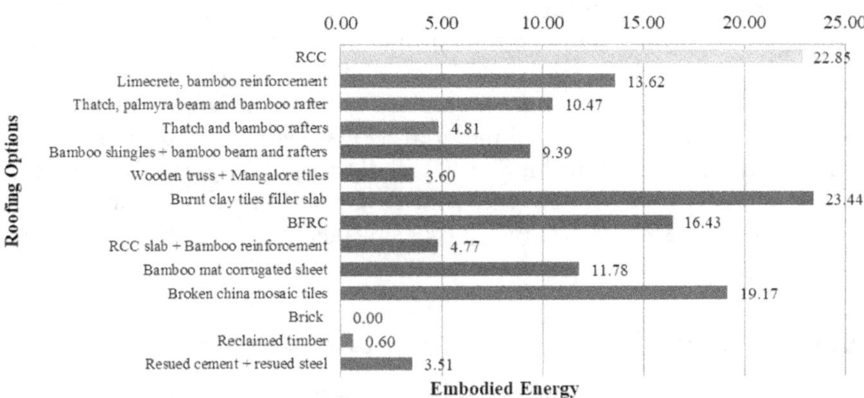

Figure 18.4 EE of roof retrofits

comparison of the chosen retrofits. The total EE of the conventional RCC roof was found to be 22.85 GJ. In the natural materials, limecrete with bamboo reinforcement, thatch with palmyra beams and bamboo rafters, thatch with bamboo rafters, and bamboo shingles with bamboo beams and rafters were considered for the analysis. Limecrete with bamboo reinforcement had the highest EE of 13.62 GJ which is 0.6 times of the base case. Thatch and bamboo rafters among the natural retrofits had the least EE of 4.81 GJ which is 0.2 times of the base case and 0.35 times less energy-intensive than limecrete with bamboo reinforcement. It was also observed that thatch with palmyra beams and bamboo rafters and bamboo shingles with bamboo beams and rafters had almost similar EE values of 10.47 GJ and 9.39 GJ, respectively.

Wooden truss with Mangalore tiles, filler slab with burnt clay tiles, bamboo fiber reinforced concrete (BFRC), RCC slab with bamboo reinforcement, and bamboo mat corrugated sheet were considered as biomaterial retrofits. Figure 18.4 shows that filler slab with burnt clay tiles had the highest EE of 24.44 GJ which is greater than the base case, whereas all other biomaterial retrofits were lesser than the base case. The EE of BFRC was 16.43 GJ and the wooden truss with Mangalore tiles was 3.6 GJ which was the least among the chosen biomaterial roof retrofit and 0.15 times less energy-intensive than the base case. RCC slab with bamboo reinforcement was also found to have a low EE of 4.77 GJ.

In the salvaged roof retrofits, broken China mosaic tiles, salvaged bricks, reclaimed timber, and reused cement and steel were considered. Figure 18.4 shows that salvaged bricks and reclaimed timber had the lowest EE of 0.00 GJ and 0.60 GJ, respectively, and can be the most efficient retrofit options in terms of EE. Reused cement and steel had an EE of 3.51 GJ and is 0.15 times less energy-intensive than the base case. Broken China mosaic tiles had the highest EE value of 19.17 GJ which is 0.83 times less energy-intensive than the base case.

The comparison of all the roofing retrofits revealed that salvaged bricks and reclaimed timber under salvaged material had almost zero EE, while wooden truss with Mangalore tiles with 3.60 GJ had lesser EE among the biomaterial retrofits. Thatch with bamboo rafters can be considered as the best retrofit among the natural materials. Filler slab with burnt clay in the biomaterial retrofit had the highest EE of all the selected options. The EE of natural retrofits ranged between 8 and 14 GJ, biomaterial retrofits ranged between 3.60 and 23.44 GJ, and that of salvaged materials ranged between 0 and 19.17 GJ.

18.4.5 Analysis of EE of flooring materials

Twelve flooring retrofits of natural, bio, and salvaged materials were assessed and compared with the conventional PCC with ceramic tile flooring. Figure 18.5 shows the EE of flooring retrofits. The total EE of the base case flooring was found to be 13.97 GJ.

Kota stone with cement screed, bamboo flooring, limecrete with lime screed, and mud flooring were considered as natural flooring retrofits. From Figure 18.5, it was observed that the bamboo flooring had the highest EE of 15.00 GJ and 1.07 times more energy-intensive than the base case. Kota stone with cement screed was found to have the lowest EE of 1.09 GJ among the chosen natural flooring retrofit which is 0.12 times less energy-intensive than the base case and 0.11 times less energy-intensive than bamboo flooring. Limecrete with lime screed and mud flooring were found to have almost similar EE of 8.51 GJ and 7.02 GJ, respectively.

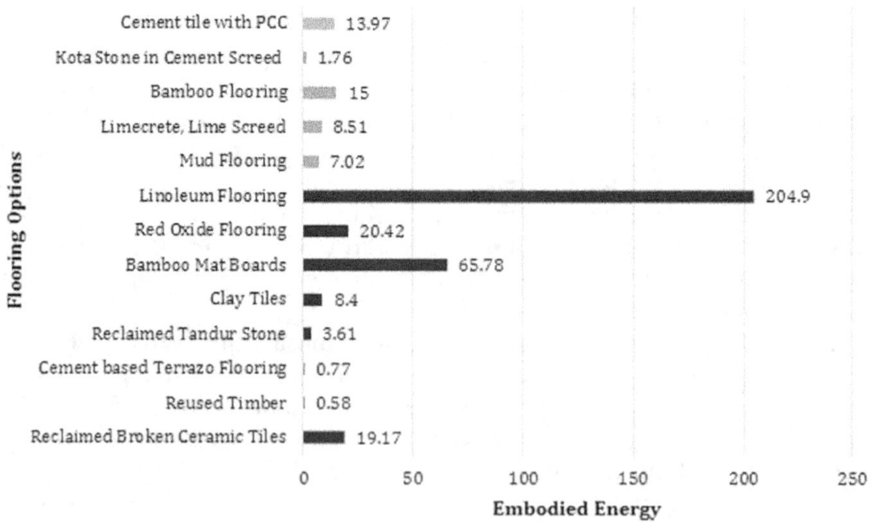

Figure 18.5 EE of floor retrofits

In the biomaterial retrofits, linoleum flooring, red oxide flooring, bamboo mat boards, and clay tiles were considered in the study. Linoleum flooring had the highest EE of 204.9 GJ, which is 14.6 times more energy-intensive than the base case. Clay tiles had the least EE among the biomaterial retrofits. The EE of clay tiles is found to be 8.40 GJ which is 0.6 times that of the base case and 0.04 times less energy-intensive than linoleum flooring. Red oxide flooring had an EE of 20.42 GJ. Bamboo mat boards also had a higher EE (65.78 GJ) than the base which is 0.32 times less energy-intensive than linoleum flooring.

Reclaimed Tandur stone, cement-based terrazzo tile, reused timber, and reclaimed broken ceramic tiles were considered as salvaged retrofit options. Figure 18.5 shows that reused timber and cement-based terrazzo tile are the most efficient retrofit options with 0.58 GJ and 0.77 GJ, respectively. Reclaimed broken ceramic tiles had the highest EE of 19.17 GJ and is 1.37 times more energy intensive than the base case. Reclaimed Tandur stone had an EE of 3.61 GJ which is also considerably lesser than the base case.

The comparison of all the flooring options revealed that reused timber and cement-based terrazzo tiles are the best salvaged retrofits, and Kota stone with cement screed is the best natural retrofit. Linoleum retrofit options in the biomaterials had the highest EE of all the selected options. Further, natural retrofits like limecrete with lime screed and mud flooring, and biomaterial retrofit of clay tiles had similar EE in the range of 7–9 GJ.

18.4.6 Overall EE contribution of retrofits in a single-story residential building

The EE of wall, roof, and floor retrofits was added together to obtain the overall EE of the single-story residential building. Further, the overall EE under each category, namely, natural, bio, and salvaged materials, were analyzed to identify the best retrofit options.

18.4.7 Overall EE of natural materials

Figure 18.6 shows the overall EE of the natural retrofits chosen under wall, roof, and floor. Natural wall retrofit contributed to around 51 to 57 % of the total EE of the building except the straw bale retrofit option which contributed to around 86%. The roof retrofits contributed to around 20 to 54 % of the total EE of the building except the thatch roof with bamboo rafters which contributed to only 5%. The percentage contribution of floor retrofits to the total EE of the building was much less when compared to walls and roofs. Although mud flooring in option 4 contributed to around 41% of the total EE of the building, its EE was only 7.02 GJ. Overall option 4 had the least embodied energy (17.35 GJ) among all the four options considered in the natural materials. The bamboo roofing in option 4 contributed to

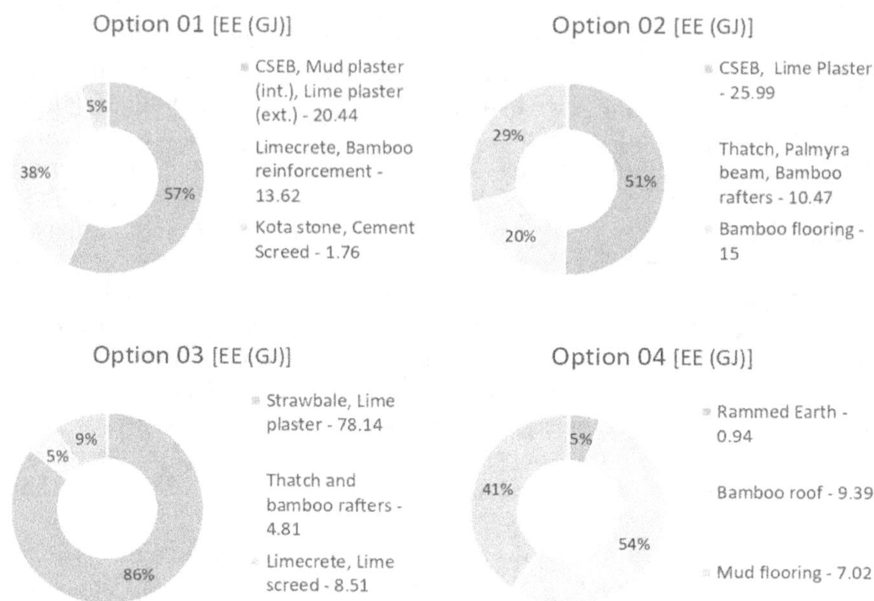

Figure 18.6 Overall EE of natural retrofit options

more than half of the total EE, whereas rammed earth contributed to only 5% of the total EE of the structure. Option 3 with straw bale lime plaster wall had the highest total EE of 91.46 GJ, contributing to around 86% of the total EE of the structure. All four retrofit options chosen contributed to lesser EE when compared with the base case.

18.4.8 Overall EE of biomaterials

Figure 18.7 shows the overall EE of biomaterial under the chosen wall, roof, and floor retrofits. The biomaterial wall retrofit in options 5 to 9 contributed to around 18% to 78% of the total EE of the building except mycelium bricks which contributed to only 1%. The roof retrofit contributed to around 20% to 42% of the total EE except for the wooden truss with Mangalore tiles and RCC slab with bamboo reinforcement which contributed to 1% and 8%, respectively. Linoleum flooring and bamboo mat boards contributed to around 79% of the total EE whereas the other flooring retrofits contributed to around 14% to 40% of the total EE. Option 5 with timbercrete walls, wooden truss with Mangalore tile roof, and linoleum flooring had the maximum EE of 260.59 GJ, wherein linoleum flooring contributed to more than three-quarters of the total EE. Option 9 with sawdust brick wall, bamboo mat corrugated sheet, and bamboo flooring had the least EE with a total EE of 27.78 GJ, wherein the roof retrofit contributed to around 42% and bamboo flooring to around 40%.

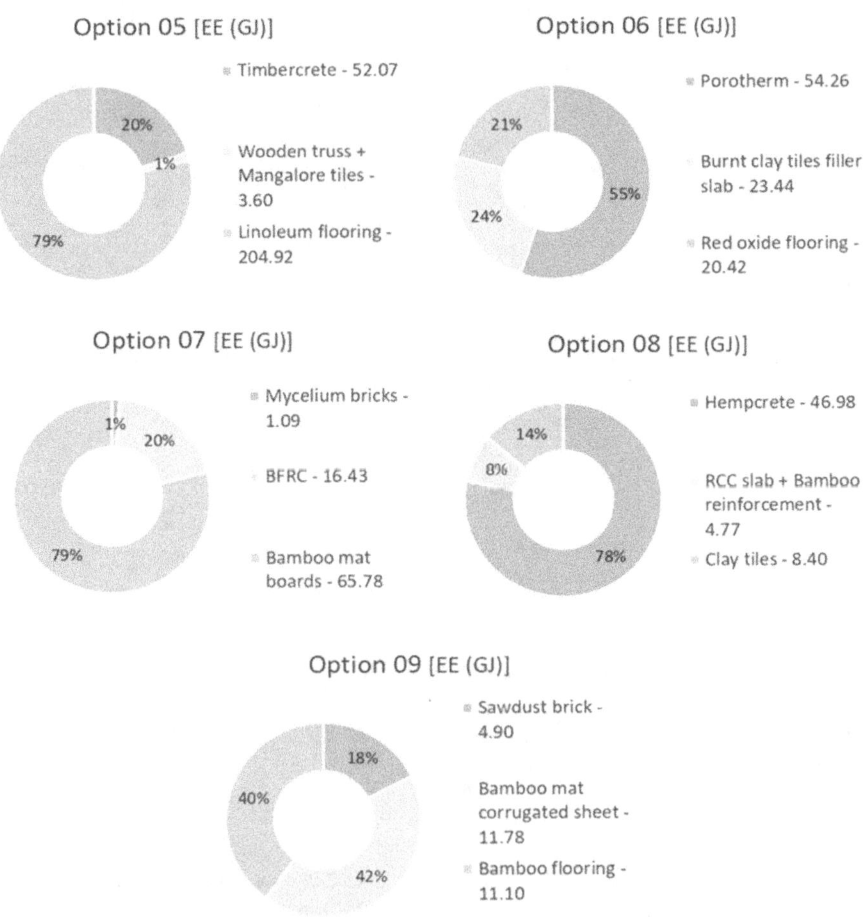

Figure 18.7 Overall EE of biomaterial retrofit options

18.4.9 Overall EE of salvaged materials

Figure 18.8 shows the overall EE of salvaged materials under the chosen wall, roof, and floor retrofits. The EE of the salvaged wall retrofit options 10 to 13 contributed to around 37% to 85% of the total EE of the building, except plasterboard in option 11 which contributed to around 1%. The salvaged roof retrofit contributed to around 10% of the total EE of the building except for broken China mosaic tiles which contributed to 45%. The EE of the salvaged flooring retrofits in options 10 to 13 ranged from 7% to 8% of the total EE of the building except for reclaimed broken ceramic tiles and cement-based terrazzo tile which contributed to 53% and 99%, respectively. Option 11 had the least EE (0.776 GJ) among the salvaged retrofit with plasterboard wall, salvaged brick roof, and cement-based terrazzo tile

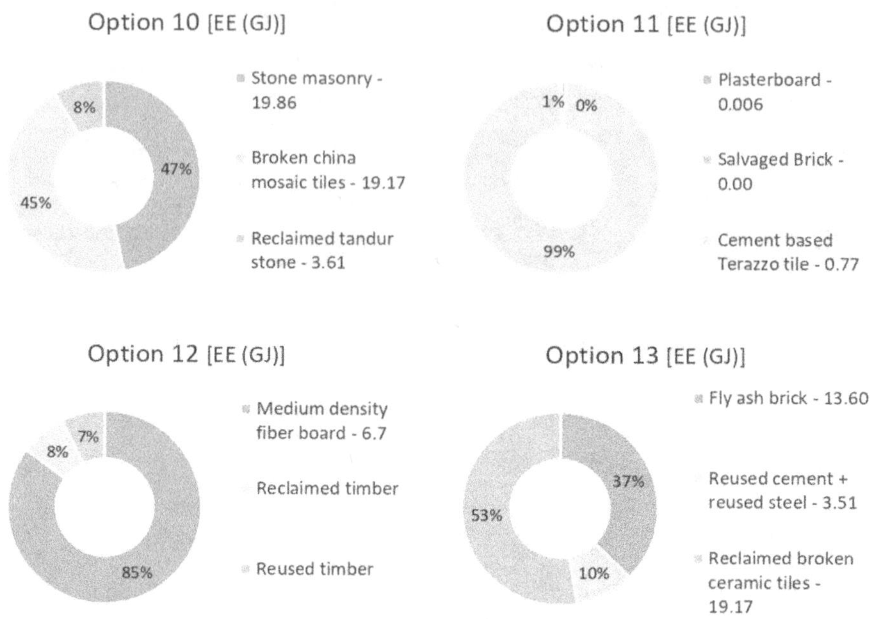

Figure 18.8 Overall EE of salvaged retrofit options

flooring wherein the flooring contributed to 99% of the total EE. Option 10 had the highest EE of 42.638 GJ, with equal contribution from the wall and roof retrofits.

18.4.10 Overall EE contribution of selected options

The total EE of all the retrofit options chosen from natural, bio, and salvaged materials contributed to a lesser EE than the base case except option 5 as shown in Table 18.4 and Figure 18.9. The EE of natural materials ranged from 17.35 GJ to 91.46 GJ. Option 1 (CSEB with mud and lime plaster wall, limecrete with bamboo reinforcement roof, Kota stone with cement screed floor) of natural materials and option 13 (fly ash brick wall, reused cement and reused steel roofing, reclaimed broken ceramic tile floor) of salvaged materials had similar embodied energy values and are 0.35 times less energy-intensive than the base case. Option 5 (timbercrete wall, wooden truss with Mangalore tile roof, and linoleum flooring) in the biomaterials had the highest total EE which is due to the Linoleum flooring despite being environmentally sound. Though the EE value of option 6 (porotherm block wall, burnt clay tile roof, and red oxide flooring) was very close to that of the base case (fired brick wall, RCC roof, and stone flooring), its impact on the environment is lesser than that of the base case. Option 11 (plasterboard, salvaged brick, and cement-based terrazzo tile), in salvaged retrofits,

Table 18.4 Total EE contribution for various options

Option	*Wall*	*Roof*	*Flooring*	*Total EE*
Base case	65.07	22.85	13.97	**101.89**
Natural Materials				
Option 1	20.44	13.62	1.79	**35.82**
Option 2	25.99	10.47	15	**51.46**
Option 3	78.14	4.81	8.51	**91.46**
Option 4	0.94	9.39	7.02	**17.35**
Bio Materials				
Option 5	52.07	3.6	204.92	**260.59**
Option 6	54.26	23.44	20.42	**98.12**
Option 7	1.09	16.43	65.78	**83.3**
Option 8	46.98	4.77	8.4	**60.15**
Option 9	4.49	11.78	11.12	**27.39**
Salvaged Materials				
Option 10	19.9	19.17	3.61	**42.6**
Option 11	0.006	0	0.77	**0.8**
Option 12	6.7	0.6	0.58	**7.9**
Option 13	13.6	3.51	19.17	**36.3**

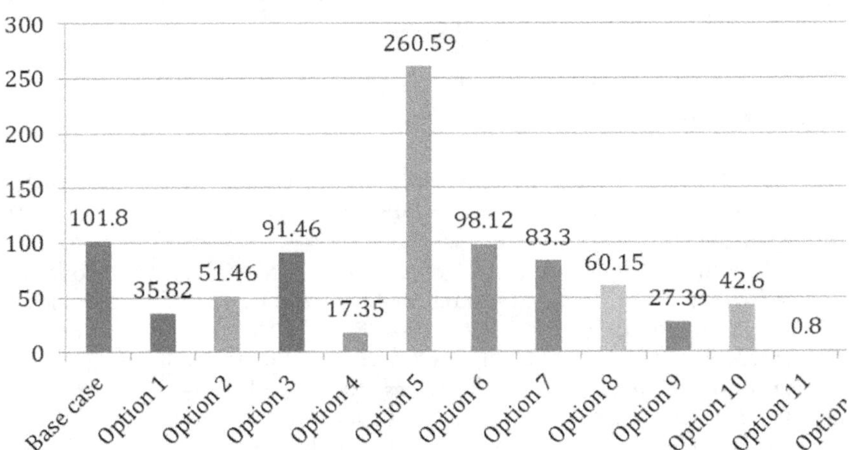

Figure 18.9 Total EE contribution for various options

had the lowest EE values for wall, roof, and floor components among all the 13 options considered. The overall EE values of options 11 and 12 of salvaged retrofits were lesser than that of all natural and biomaterials. However, options 10 and 13 of salvaged retrofits were similar to the EE of natural retrofits. The study found that biomaterial retrofits had higher EE when compared to natural and salvaged retrofits except for option 9 due to the processing of materials.

18.5 CONCLUSION

The study infers that with the use of sustainable retrofits from natural, bio, and salvaged materials, the percentage reduction in the total embodied energy of the building ranged from 10 to 83%, 18 to 73%, and 58 to 99%, respectively. Salvaged materials are one of the least energy-intensive retrofit options for reducing the overall embodied energy of the building. Nevertheless, they also contribute to lesser waste disposal at construction sites. Some of the materials that can be salvaged include steel in the form of grills, woodwork in the form of fenestrations, and other reusable materials from existing buildings. Natural materials were also found to have lesser embodied energy in most retrofit options compared to biomaterials. This is due to the reason that natural materials are raw and don't undergo significant processing in factories/industries. Hence, they contribute to lesser embodied energy. Some of the locally available natural materials in the study area are Cuddapah stone, timber, etc. Biomaterials, however, contribute to lesser EE compared to the base case and are much more environmentally friendly than conventional building materials. Examples of biomaterials include hemp, flax, mycelium, wood wastes, lignin-based fibers, etc. The study concludes that the use of energy-efficient retrofits leads to a considerable reduction in the embodied energy of the building as a whole. An important factor to be considered while selecting energy-efficient retrofits is the embodied energy with regional availability.

REFERENCES

Bansal, D., Minocha, V. K., & Kaur, A. Component wise-embodied energy analysis of affordable houses in India. *Asian Journal of Civil Engineering*, 21(1), 2019, 137–145.

Barrios, G., Huelsz, G., Rojas, J. R., Ochoa, J. M., & Marincic, I. Envelope wall/roof thermal performance parameters for non air-conditioned buildings. *Energy and Buildings*, 50, 2012, 120–127.

Buchanan, A. H., & Honey, B. G. Energy and carbon dioxide implications of building construction. *Energy and Buildings*, 20(3), 1994, 205–217.

Chel, A., & Tiwari, G. N. Performance evaluation and life cycle cost analysis of earth to air heat exchanger integrated with adobe building for New Delhi composite climate. *Energy and Buildings*, 41(1), 2009, 56–66, ISSN 0378-7788.

Debnath, A., Singh, S., & Singh, Y. Comparative assessment of energy requirements for different types of residential buildings in India. *Energy and Buildings*, 23(2), 1995, 141–146.

Demir, I. Effect of organic residues addition on the technological properties of clay bricks. *Waste Management*, 28(3), 2008, 622–627.

Djamil, B. Sun dried clay for sustainable constructions. *International Journal of Applied Engineering Research*, 11(6), 2016, 4628–4633.

Givoni, B. Building design principles for hot humid regions. *Renewable Energy*, 5(5–8), 1994, 908–916. https://doi.org/10.1016/0960-1481(94)90111-2

Haase, M., & Amato, A. An investigation of the potential for natural ventilation and building orientation to achieve thermal comfort in warm and humid climates. *Solar Energy*, 83(3), 2009, 389–399. https://doi.org/10.1016/j.solener.2008.08.015

https://www.cecp_eu.in/uploads/documents/events/63/EU_India_CECP_RI_report_on_low_embodied_energy_building_materials.pdf

IFC. India construction materials database of embodied energy and global warming potential METHODOLOGY REPORT, 2017.

ISO. ISO-6946 – Building components and building elements – Thermal resistance and thermal transmittance – Calculation method, 2007.

Kariyawasam, K. K. G. K. D., & Jayasinghe, C. Cement stabilized rammed earth as a sustainable construction material. *Construction and Building Materials*, 105, 2016, 519–527.

Khitab, A., & Anwar, W. Classical building materials. *Sustainable Infrastructure*, 2020, 304–326.

Kishore, N. K., & Chouhan, J. S. Embodied energy assessment and comparisons for a residential building using conventional and alternative materials in Indian context. *Journal of The Institution of Engineers (India): Series A*, 95, 2014, 117–127.

Kofoworola, O. F., & Gheewala, S. H. Life cycle energy assessment of a typical office building in Thailand. *Energy and Buildings*, 41(10), 2009, 1076–1083.

Kumar, D., Alam, M., Zou, P. X., Sanjayan, J. G., & Memon, R. A. Comparative analysis of building insulation material properties and performance. *Renewable and Sustainable Energy Reviews*, 131, 2020, 110038. https://doi.org/10.1016/j.rser.2020.110038

Mastrucci, A., & Rao, N. D. Bridging India's Housing gap: Lowering costs and CO_2 emissions. *Building Research & Information*, 47(1), 2018, 8–23.

Marzouk, M., & Elshaboury, N. Science mapping analysis of embodied energy in the construction industry. *Energy Reports*, 8, 2022, 1362–1376.

Monahan, J., & Powell, J. An embodied carbon and energy analysis of modern methods of construction in housing: A case study using a lifecycle assessment framework. *Energy and Buildings*, 43(1), 2011, 179–188.

Morel, J., Mesbah, A., Oggero, M., & Walker, P. Building houses with local materials: Means to drastically reduce the environmental impact of construction. *Building and Environment*, 36(10), 2001, 1119–1126.

Oka, T., Suzuki, M., & Konnya, T. The estimation of energy consumption and amount of pollutants due to the construction of buildings. *Energy and Buildings*, 19(4), 1993, 303–311.

Praseeda, K., Mani, M., & Reddy, B. V. Assessing impact of Material transition and thermal Comfort models on embodied and operational energy in vernacular Dwellings (India). *Energy Procedia*, 54, 2014, 342–351.

Praseeda, K., Reddy, B. V., & Mani, M. Embodied energy assessment of building materials in India using process AND input–output analysis. *Energy and Buildings*, 86, 2015, 677–686.

Reddy Venkatarama, B. Sustainable materials for low carbon buildings. *International Journal of Low-Carbon Technologies*, 4(3), 2009, 175–181.

Reddy Venkatarama, B., & Jagadish, K. Embodied energy of common and alternative building materials and technologies. *Energy and Buildings*, 35(2), 2003, 129–137.

Suzki, M., Oka, T., & Okada, K. The estimation of energy consumption and CO_2 emission due to housing construction in Japan. *Energy and Buildings*, 22(2), 1995, 165–169.

Thormark, C. A low energy building in a life cycle—Its embodied energy, energy need for operation and recycling potential. *Building and Environment*, 37(4), 2002, 429–435.

Thormark, C. Conservation of energy and natural resources by recycling building waste. *Resources, Conservation and Recycling*, 33(2), 2001, 113–130.

Thormark, C. Including recycling potential in energy use into the life-cycle of buildings. *Building Research & Information*, 28(3), 2000, 176–183.

Tiwari, P., & Parikh, J. Cost of CO_2 reduction in building construction. *Energy*, 20(6), 1995, 531–547.

UNCHS (*Habitat*). Energy for buildings. Nairobi (HS/250/91/E), 1991, p. 88.

Yu, J., Yang, C., Tian, L., & Liao, D. A study on optimum insulation thicknesses of external walls in hot summer and cold winter zone of China. *Applied Energy*, 86, 2009, 2520–2529.

Zhou, J., Zhang, G., Lin, Y., Wang, H. A new virtual sphere method for estimating the role of thermal mass in natural ventilated buildings. *Energy and Buildings*, 43, 2011, 75–81.

Index

C

D

E

F

J

K

L

M

N

O

P

Z

For Product Safety Concerns and Information please contact our EU representative GPSR@taylorandfrancis.com
Taylor & Francis Verlag GmbH, Kaufingerstraße 24, 80331 München, Germany

www.ingramcontent.com/pod-product-compliance
Lightning Source LLC
LaVergne TN
LVHW020558110826
845149LV00002B/300

* 9 7 8 1 0 3 2 8 0 4 0 6 4 *